Lecture Notes in Computer Science 13179

More information about this subseries at https://link.springer.com/bookseries/7407

Niranjan Balachandran · R. Inkulu (Eds.)

Algorithms and Discrete Applied Mathematics

8th International Conference, CALDAM 2022
Puducherry, India, February 10–12, 2022
Proceedings

 Springer

Editors
Niranjan Balachandran
Indian Institute of Technology Bombay
Mumbai, India

R. Inkulu
Indian Institute of Technology Guwahati
Guwahati, India

ISSN 0302-9743   ISSN 1611-3349 (electronic)
Lecture Notes in Computer Science
ISBN 978-3-030-95017-0   ISBN 978-3-030-95018-7 (eBook)
https://doi.org/10.1007/978-3-030-95018-7

LNCS Sublibrary: SL1 – Theoretical Computer Science and General Issues

This Springer imprint is published by the registered company Springer Nature Switzerland AG
The registered company address is: Gewerbestrasse 11, 6330 Cham, Switzerland

Preface

This volume contains the papers presented at CALDAM 2022 (the 8th International Conference on Algorithms and Discrete Applied Mathematics) held during February 10–12, 2022, at Pondicherry University, Puducherry, India. CALDAM 2022 was organized by the Department of Mathematics, Pondicherry University, and the Association for Computer Science and Discrete Mathematics (ACSDM), India. The program committee consisted of 34 highly experienced and active researchers from various countries.

The conference topics included algorithms, graph theory, computational geometry, and optimization. We received 80 submissions from authors from all over the world. Each paper was extensively reviewed by program committee members and other expert reviewers. The committee decided to accept 24 papers for presentation. The program included three Google invited talks by Timothy M. Chan (University of Illinois at Urbana-Champaign), Daya R. Gaur (University of Lethbridge), and Joseph S. B. Mitchell (Stony Brook University).

As volume editors, we would like to thank the authors of all submissions for considering CALDAM 2022 for the potential presentation of their works. We are very much indebted to the program committee members and the external reviewers for providing serious reviews within a very short period. We thank Springer for publishing the proceedings in the Lecture Notes in Computer Science series. Our sincerest thanks to the invited speakers, Timothy M. Chan, Daya R. Gaur, and Joseph S. B. Mitchell, for accepting our invitation to give a talk. We thank the organizing committee, chaired by S. Francis Raj of Pondicherry University, for conducting CALDAM 2022 smoothly, and Pondicherry University for providing the necessary facilities. We are very grateful to the chair of the steering committee, Subir Ghosh, for his active help, support, and guidance. And, we thank the Program Committee co-chairs of CALDAM 2021, Apurva Mudgal and C R Subrahmanyam, for their timely input throughout. We thank our sponsors, Google Inc. for their financial support and Springer for the best paper presentation awards. We also thank Springer OCS staff for their support.

February 2022

Niranjan Balachandran
R. Inkulu

Organization

Steering Committee

Subir Kumar Ghosh (Chair)	Ramakrishna Mission Vivekananda Educational and Research Institute, India
Gyula O. H. Katona	Alfréd Rényi Institute of Mathematics, Hungarian Academy of Sciences, Hungary
János Pach	École Polytechnique Fédérale De Lausanne (EPFL), Switzerland
Nicola Santoro	Carleton University, Canada
Swami Sarvattomananda	Ramakrishna Mission Vivekananda Educational and Research Institute, India
Chee Yap	Courant Institute of Mathematical Sciences, New York University, USA

Program Committee

Amitabha Bagchi	IIT Delhi, India
Niranjan Balachandran (Co-chair)	IIT Bombay, India
Boštjan Brešar	University of Maribor, Slovenia
Sergio Cabello	University of Ljubljana, Slovenia
Paz Carmi	Ben-Gurion University of the Negev, Israel
Manoj Changat	University of Kerala, India
Sandip Das	ISI Kolkata, India
Josep Diaz	Polytechnic University of Catalonia, Spain
Martin Furer	Pennsylvania State University, USA
Daya Gaur	University of Lethbridge, Canada
Sathish Govindarajan	IISc Bangalore, India
Pavol Hell	Simon Fraser University, Canada
R. Inkulu (Co-chair)	IIT Guwahati, India
Subrahmanyam Kalyanasundaram	IIT Hyderabad, India
Van Bang Le	University of Rostock, Germany
Sanjiv Kapoor	Illinois Institute of Technology, USA
Andrzej Lingas	Lund University, Sweden
Anil Maheshwari	Carleton University, Canada
Bodo Manthey	University of Twente, The Netherlands
Rogers Mathew	IIT Hyderabad, India
Bojan Mohar	Simon Fraser University, Canada
Apurva Mudgal	IIT Ropar, India

Rahul Muthu   DA-IICT, India
Kamal Lochan Patra   NISER Bhubaneswar, India
Iztok Peterin   University of Maribor, Slovenia
Valentin Polishchuk   Linköping University, Sweden
Deepak Rajendraprasad   IIT Palakkad, India
Abhiram Ranade   IIT Bombay, India
Sagnik Sen   IIT Dharwad, India
Rishi Ranjan Singh   IIT Bhilai, India
Michiel Smid   Carleton University, Canada
Joachim Spoerhase   University of Wurzburg, German
C. R. Subramanian   IMSc Chennai, India
Antoine Vigneron   UNIST, South Korea

Organizing Committee

Rajeswari Seshadri   Pondicherry University, India
Malai Subbiah   Pondicherry University, India
S. R. Kannan   Pondicherry University, India
T. Duraivel   Pondicherry University, India
A. Joseph Kennedy   Pondicherry University, India
S. Francis Raj (Chair)   Pondicherry University, India
Syeda Noor Fathima   Pondicherry University, India
I. Subramania Pillai   Pondicherry University, India
Swami Dhyanagamyananda   Ramakrishna Mission Vivekananda Educational
                                  and Research Institute, India
Pritee Khanna   IIITDM Jabalpur, India
Arti Pandey   IIT Ropar, India
Tarkeshwar Singh   BITS Pilani K K Birla Goa Campus, India

Additional Reviewers

Ankush Acharyya   Marko Jakovac
N. R. Arvind   Jesper Jansson
Devsi Bantva   Ce Jin
Manu Basavaraju   Sreejith K. P.
Srimanta Bhattacharya   Anjeneya Swami Kare
Sriram Bhyravarapu   Niraj Khare
Arun Das   Mirosław Kowaluk
Hiranya Kishore Dey   Christos Levcopoulos
Amit Kumar Dhar   Vincenzo Liberatore
Tanja Dravec   Tian Liu
Barun Gorain   Raghunath Reddy M.
Vinod Reddy I.   Atrayee Majumder

Tapas Kumar Mishra
Shuichi Miyazawaki
Fahad Panolan
Pablo Perez-Lantero
Veena Prabhakaran
Sadagopan N.
Francis P.
Venkata Subba Reddy P.
Sajith Padinhatteeri
Narad Rampersad
M. V. Panduranga Rao

Prakash Saivasan
Brahadeesh Sankarnarayanan
Ildikó Schlotter
Elżbieta Tumidajewicz
Karol Wegrzycki
Mariusz Wozniak
Hyeyun Yang
Ismael Gonzalez Yero
Jingru Zhang
Paweł Żyliński

Abstracts of Invited Talks

All-Pairs Shortest Paths and Fine-Grained Complexity

Timothy M. Chan

University of Illinois at Urbana-Champaign, Urbana, USA
tmc@illinois.edu

Abstract. The all-pairs shortest paths (APSP) problem is one of the most fundamental problems in algorithm design and fine-grained complexity. The problem for general weighted dense graphs is conjectured to require close to n^3 time. On the other hand, substantially subcubic algorithms are known in some important special cases via fast matrix multiplication; for example, for directed graphs that are unweighted (or have small integer weights), the current best algorithm due to Zwick (FOCS 1998) had running time near $n^{2.5}$ if the matrix multiplication exponent ω is equal to 2.

In this talk, I will survey the current landscape surrounding the complexity of APSP and its variants, and how the conjectured hardness of APSP in the general and unweighted cases have been used as the basis for establishing conditional lower bounds for other problems. In particular, I will describe recent joint work with Virginia Vassilevska Williams and Yinzhan Xu (ICALP 2021), showing that Zwick's algorithm is in some sense optimal for directed unweighted graphs.

Linear Programming and its Uses in Algorithm Design

Daya R. Gaur

University of Lethbridge, Lethbridge, Canada
gaur@cs.uleth.ca

Abstract. Linear programming has a rich history. In this talk, we focus on its use in algorithm design. We will look at its use in three areas. The first is the design of exact algorithms, and the second is the design of approximate algorithms. Thirdly, its use in the creation of practical algorithms for computationally challenging problems. I will give several examples of how researchers in my group use linear programming to develop exact and approximate algorithms. These illustrative examples will also highlight the computational challenges still remaining. Most of the theory that will be covered is explained nicely in these books [Dantzig and Thapa, 2006; Vazirani, 2003; Lau et al., 2011; Cook et al., 1998]. This talk will be a little tour of the strengths of linear programming and how to use them. This introductory talk will be a mix of theory and practice and no background is assumed.

Approximation Algorithms for Some Geometric Optimization Problems

Joseph S. B. Mitchell

Stony Brook University, Stony Brook, USA
joseph.mitchell@stonybrook.edu

Abstract. We discuss approximation algorithms for some instances of geometric optimization problems, including maximum independent set, dominating set, vehicle routing, and set cover. In all cases the problems are specified by geometric data, such as points, rectangles, polygons, and disks, and the results strongly exploit geometry to yield better results than can be achieved (or at least better than results known so far) in non-geometric settings. We are motivated by applications of computational geometry in sensor networks and mobile robotics, including classic problems on "art galleries" that need to be guarded by static or mobile guards within a polygonal domain. Almost all of these optimization problems are NP-hard even in simple two-dimensional settings. The problems get even harder when we take into account uncertain data, time constraints for scheduled coverage, and routing/connectivity problems in combination with coverage constraints. We discuss selected versions of these geometric optimization problems from the perspective of approximation algorithms and we describe some techniques that have led to new or improved approximation bounds for certain maximum independent set and routing/coverage problems.

Contents

Computational Geometry

Algorithms and Optimization

Graph Theory

A Proof of the Multiplicative 1-2-3 Conjecture

Julien Bensmail[1(✉)], Hervé Hocquard[2], Dimitri Lajou[2], and Éric Sopena[2]

[1] Université Côte d'Azur, CNRS, Inria, I3S, Biot, France
[2] Univ. Bordeaux, CNRS, Bordeaux INP, LaBRI,
UMR 5800, 33400 Talence, France

Abstract. We prove that the product version of the 1-2-3 Conjecture, raised by Skowronek-Kaziów in 2012, is true. Namely, for every connected graph with order at least 3, we can assign labels $1, 2, 3$ to the edges so that no two adjacent vertices are incident to the same product of labels.

Keywords: 1-2-3 Conjecture · Product version · Labels 1, 2, 3

1 Introduction

Let G be a graph. A k-*labelling* $\ell : E(G) \to \{1, \ldots, k\}$ is an assignment of labels $1, \ldots, k$ to the edges of G. From ℓ, we can compute different parameters of interest for all vertices v, such as the *sum* $\sigma_\ell(v)$ of incident labels (being formally $\sigma_\ell(v) = \Sigma_{u \in N(v)} \ell(uv)$), or similarly the *multiset* $\mu_\ell(v)$ of labels incident to v or the *product* $\rho_\ell(v)$ of labels incident to v. We say that ℓ is *s-proper* if σ_ℓ is a proper vertex-colouring of G, *i.e.*, we have $\sigma_\ell(u) \neq \sigma_\ell(v)$ for every edge $uv \in E(G)$. Similarly, we say that ℓ is *m-proper* and *p-proper*, if μ_ℓ and ρ_ℓ, respectively, form proper vertex-colourings of G.

In the context of so-called *distinguishing labellings*, the goal is generally to not only distinguish vertices within some distance according to some parameter computed from labellings (such as the parameters σ_ℓ, μ_ℓ and ρ_ℓ above, to name a few), but also to construct such k-labellings with k as small as possible. We refer the interested reader to [4], which lists hundreds of labelling techniques.

Regarding s-proper, m-proper and p-proper labellings, which are the main focus in this work, we are thus interested, as mentioned above, in finding such k-labellings with k as small as possible, for a given graph G. In other words, we are interested in the parameters $\chi_S(G)$, $\chi_M(G)$ and $\chi_P(G)$ which denote the smallest $k \geq 1$ such that s-proper, m-proper and p-proper, respectively, k-labellings exist (if any). Actually, through greedy labelling arguments, it can be observed that the only connected graph G for which $\chi_S(G)$, $\chi_M(G)$ or $\chi_P(G)$ is not defined, is K_2, the complete graph on 2 vertices. Consequently, these three parameters are

Some proofs in this paper are voluntarily omitted due to space limitation; the interested reader will find them in [3], the full version of the current paper. This work is partially supported by the ANR project HOSIGRA (ANR-17-CE40-0022).

N. Balachandran and R. Inkulu (Eds.): CALDAM 2022, LNCS 13179, pp. 3–14, 2022.
https://doi.org/10.1007/978-3-030-95018-7_1

generally investigated for so-called *nice graphs*, which are those graphs with no connected component isomorphic to K_2.

S-proper, m-proper and p-proper labellings form a subfield of distinguishing labellings, which has been attracting attention due to the so-called 1-2-3 Conjecture, raised, in [6], by Karoński, Łuczak and Thomason in 2004:

1-2-3 Conjecture (sum version). *If G is a nice graph, then $\chi_S(G) \leq 3$.*

Later on, counterparts of the 1-2-3 Conjecture were raised for m-proper and p-proper labellings. Addario-Berry *et al.* first raised, in 2005, the following in [1]:

1-2-3 Conjecture (multiset version). *If G is a nice graph, then $\chi_M(G) \leq 3$.*

while Skowronek-Kaziów then raised, in 2012, the following in [8]:

1-2-3 Conjecture (product version). *If G is a nice graph, then $\chi_P(G) \leq 3$.*

It is worth mentioning that all three conjectures above, if true, would be tight, as attested for instance by complete graphs. Note also that the multiset version of the 1-2-3 Conjecture is, out of the three variants, the easiest one in a sense, as every s-proper or p-proper labelling is also m-proper (thus, proving the sum or product variant of the 1-2-3 Conjecture would prove the multiset one).

To date, the best result towards the sum version of the 1-2-3 Conjecture, proved by Kalkowski, Karoński and Pfender in [5], is that $\chi_S(G) \leq 5$ holds for every nice graph G. Another significant result is due to Przybyło, who recently proved in [7] that even $\chi_S(G) \leq 4$ holds for every nice regular graph G. Karoński, Łuczak and Thomason themselves also proved in [6] that $\chi_S(G) \leq 3$ holds for nice 3-colourable graphs. Regarding the multiset version, for long the best result was the one proved by Addario-Berry, Aldred, Dalal and Reed in [1], stating that $\chi_M(G) \leq 4$ holds for every nice graph G. Building on that result, Skowronek-Kaziów later proved in [8] that $\chi_P(G) \leq 4$ holds for every nice graph G. She also proved that $\chi_P(G) \leq 3$ holds for every nice 3-colourable graph G.

A breakthrough result was recently obtained by Vučković, as he totally proved the multiset version of the 1-2-3 Conjecture in [9]. Due to connections between m-proper and p-proper 3-labellings, we observed in [2] that this result directly implies that $\chi_P(G) \leq 3$ holds for every nice regular graph G. Inspired by Vučković's proof scheme, we were also able to prove that $\chi_P(G) \leq 3$ holds for nice 4-colourable graphs G, and to prove related results that are very close to what is stated in the product version of the 1-2-3 Conjecture.

Building on these results, we prove the following throughout this paper.

Theorem 1. *The product version of the 1-2-3 Conjecture is true. That is, every nice graph admits p-proper 3-labellings.*

2   Proof of Theorem 1

Let us start by introducing some terminology and recalling some properties of p-proper labellings, which will be used throughout the proof. Let G be a graph, and ℓ be a 3-labelling of G. For a vertex $v \in V(G)$ and a label $i \in \{1, 2, 3\}$, we denote by $d_i(v)$ the i-degree of v by ℓ, being the number of edges incident to v that are assigned label i by ℓ. Note then that $\rho_\ell(v) = 2^{d_2(v)} 3^{d_3(v)}$. We say that v is 1-monochromatic if $d_2(v) = d_3(v) = 0$, while we say that v is 2-monochromatic (3-monochromatic, resp.) if $d_2(v) > 0$ and $d_3(v) = 0$ ($d_3(v) > 0$ and $d_2(v) = 0$, resp.). In case v has both 2-degree and 3-degree at least 1, we say that v is bichromatic. We also define the $\{2,3\}$-degree of v as the sum $d_2(v) + d_3(v)$ of its 2-degree and 3-degree. If v is bichromatic, then its $\{2,3\}$-degree is at least 2.

Because ℓ assigns labels $1, 2, 3$, and, in particular, because 2 and 3 are coprime, note that, for every edge uv of G, we have $\rho_\ell(u) \neq \rho_\ell(v)$ when u and v have different 2-degrees, 3-degrees, or $\{2,3\}$-degrees. In particular, u and v cannot be in conflict, i.e., satisfy $\rho_\ell(u) = \rho_\ell(v)$, if u and v are i-monochromatic and j-monochromatic for $i \neq j$, or if u is monochromatic while v is bichromatic.

Before going into the proof of Theorem 1, let us start by giving an overview of it. Let G be a nice graph. Our goal is to build a p-proper 3-labelling ℓ of G. We can clearly assume that G is connected. We also set $t = \chi(G)$, where, recall, $\chi(G)$ refers to the chromatic number[1] of G. In particular, $t \geq 2$.

In what follows, we construct ℓ through three main steps. First, we need to partition the vertices of G in a way satisfying specific cut properties, forming what we call a valid partition of $V(G)$ (see later Definition 1 for a more formal definition). In short, a valid partition $\mathcal{V} = (V_1, \ldots, V_t)$ is a partition of $V(G)$ into t independent sets $V_1, \ldots, V_t$ fulfilling two main properties, being, roughly put, that 1) every vertex v in some part V_i with $i > 1$ has an incident upward edge to every part V_j with $j < i$, and 2) for every connected component of $G[V_1 \cup V_2]$ having only one edge, we can freely swap its two vertices in V_1 and V_2 while preserving the main properties of a valid partition.

Once we have this valid partition $\mathcal{V}$ in hand, we can then start constructing ℓ. The main part of the labelling process, Step 2 below, consists in starting from all edges of G being assigned label 1 by ℓ, and then processing the vertices of $V_3, \ldots, V_t$ one after another, possibly changing the labels by ℓ assigned to some of their incident edges, so that certain product types are achieved by ρ_ℓ. These desired product types can be achieved due to the many upward edges that some vertices are incident to (in particular, the deeper a vertex lies in $\mathcal{V}$, the more upward edges it is incident to). The product types we achieve for the vertices depend on the part V_i of $\mathcal{V}$ they belong to. In particular, the modifications we make on ℓ guarantee that all vertices in $V_3, \ldots, V_t$ are bichromatic, every two vertices in V_i and V_j with $i, j \in \{3, \ldots, t\}$ and $i \neq j$ have different 2-degrees or 3-degrees, all vertices in V_2 are 1-monochromatic or 2-monochromatic, and all vertices in V_1 are 1-monochromatic or 3-monochromatic. By itself, achieving these product types makes ℓ almost p-proper, in the sense that the only possible conflicts are between 1-monochromatic vertices in V_1 and V_2. An important point also, is that, through these label modifications, we will make sure that all edges

---

[1] Recall that a proper k-vertex-colouring of a graph G is a partition $(V_1, \ldots, V_k)$ of $V(G)$ where all V_i's are independent. The chromatic number $\chi(G)$ of G is the smallest $k \geq 1$ such that proper k-vertex-colourings of G exist. G is k-colourable if $\chi(G) \leq k$.

of $G[V_1 \cup V_2]$ remain assigned label 1, and no vertex in $V_3 \cup \cdots \cup V_t$ has 3-degree 1, 2-degree at least 2, and odd $\{2,3\}$-degree; in last Step 3 below, we will use that last fact to remove remaining conflicts by allowing some vertices of $V_1 \cup V_2$ to become *special*, i.e., make such vertices v satisfy $d_3(v) = 1$, $d_2(v) \geq 2$, and $d_2(v) + d_3(v) \equiv 1 \bmod 2$, while making sure that the products of the vertices in $V_3 \cup \cdots \cup V_t$ are not altered.

Step 3 is designed to get rid of the last conflicts between the adjacent 1-monochromatic vertices of V_1 and V_2 without introducing new ones in G. To that end, we will consider the set $\mathcal{H}$ of the connected components of $G[V_1 \cup V_2]$ having conflicting vertices, and, if needed, modify the labels assigned by ℓ to some of their incident edges so that no conflicts remain, and no new conflicts are created in G. To make sure that no new conflicts are created between vertices in $V_1 \cup V_2$ and vertices in $V_3 \cup \cdots \cup V_t$, we will modify labels while making sure that all vertices in $V_1 \cup V_2$ are monochromatic or special. An important point also, is that the fixing procedures we introduce require the number of edges in a connected component of $\mathcal{H}$ to be at least 2. Because of that, once Step 2 ends, we must ensure that $\mathcal{H}$ does not contain a connected component with only one edge incident to two 1-monochromatic vertices. To guarantee this, we will also make sure, during Step 2, to modify labels and the partition $\mathcal{V}$ slightly so that $\mathcal{H}$ has no such configuration.

Step 1: Constructing a valid partition

Let $\mathcal{V} = (V_1, \ldots, V_t)$ be a partition of $V(G)$ where each V_i is an independent set. Note that such a partition exists, as, for instance, any proper t-vertex-colouring of G forms such a partition of $V(G)$. For every vertex $u \in V_i$, an incident *upward edge* (*downward edge*, resp.) is an edge uv for which v belongs to some V_j with $j < i$ ($j > i$, resp.). Note that all vertices in V_1 have no incident upward edges, while all vertices in V_t have no incident downward edges.

We denote by $M_0(\mathcal{V})$ (also denoted M_0 when the context is clear) the set of isolated edges in the subgraph $G[V_1 \cup V_2]$ of G induced by the vertices of $V_1 \cup V_2$. That is, M_0 contains the edges of the connected components of $G[V_1 \cup V_2]$ that consist in one edge only. To lighten the exposition, whenever referring to the vertices of M_0, we mean the vertices of G incident to the edges in M_0.

For an edge $uv \in M_0$ with $u \in V_1$ and $v \in V_2$, *swapping* uv consists in modifying the partition $\mathcal{V}$ by removing u from V_1 (v from V_2, resp.) and adding it to V_2 (V_1, resp.). In other words, we exchange the parts to which u and v belong. Note that if V_1 and V_2 are independent sets before the swap, then, because $uv \in M_0$, by definition the resulting new V_1 and V_2 remain independent. Also, the set M_0 is unchanged by the swap operation.

We can now give a formal definition for the notion of valid partition.

Definition 1 (Valid partition). *For a t-colourable graph G, a partition $\mathcal{V} = (V_1, \ldots, V_t)$ of $V(G)$ is* valid *(for G) if $\mathcal{V}$ satisfies the following properties.*

($\mathcal{I}$) *Every V_i is an independent set.*
($\mathcal{P}_1$) *Every vertex in some V_i with $i \geq 2$ has a neighbour in V_j for every $j < i$.*

(S) *For every sequence $(e_i)_i$ of edges of $M_0(\mathcal{V})$, successively swapping every e_i (in any order) results in a partition $\mathcal{V}'$ satisfying Properties ($\mathcal{I}$) and ($\mathcal{P}_1$).*

Note that Property (S) implies the following property:

($\mathcal{P}_2$) *Swapping any number of edges of $M_0(\mathcal{V})$ results in a valid partition $\mathcal{V}'$.*

To prove Theorem 1, as mentioned earlier, to start constructing ℓ we need to have a valid partition of G in hand. The following result guarantees its existence.

Lemma 1. *Every nice t-colourable graph G admits a valid partition.*

Proof. For a partition $\mathcal{V} = (V_1, \ldots, V_t)$ of $V(G)$ where each V_i is independent (such a partition exists, as attested by any proper t-vertex-colouring of G), set $f(\mathcal{V}) = \sum_{k=1}^{t} k \cdot |V_i|$. Among all possible $\mathcal{V}$'s, consider a $\mathcal{V}$ that minimises $f(\mathcal{V})$.

Suppose that there is a vertex $u \in V_i$ with $i \geq 2$ for which Property ($\mathcal{P}_1$) does not hold, *i.e.*, there is a $j < i$ such that u has no incident upward edge to V_j. By moving u to V_j, we obtain another partition $\mathcal{V}'$ of $V(G)$ where every part is an independent set. However, note that $f(\mathcal{V}') = f(\mathcal{V}) + j - i < f(\mathcal{V})$, a contradiction to the minimality of $\mathcal{V}$. From this, we deduce that every partition $\mathcal{V}$ minimising f must satisfy Property ($\mathcal{P}_1$).

Let now $\mathcal{V}'$ be the partition of $V(G)$ obtained by successively swapping edges of $M_0(\mathcal{V})$. Recall that the swapping operation preserves Property ($\mathcal{I}$) and observe that $f(\mathcal{V}) = f(\mathcal{V}')$. Hence, $\mathcal{V}'$ minimises f and thus satisfies Properties ($\mathcal{I}$) and ($\mathcal{P}_1$). Thus Property (S) also holds, and $\mathcal{V}$ is a valid partition of G. □

From here, we assume that we have a valid partition $\mathcal{V} = (V_1, \ldots, V_t)$ of G.

Step 2: Labelling the upward edges of $V_3, \ldots, V_t$

From G and $\mathcal{V}$, our goal now is to construct a 3-labelling ℓ of G achieving certain properties, the most important of which being that the only possible conflicts are between pairs of vertices of V_1 and V_2 that do not form an edge of M_0. The following result sums up the exact conditions we want ℓ to fulfil. Recall that a vertex v is special by ℓ, if $d_3(v) = 1$, $d_2(v) \geq 2$ and $d_2(v) + d_3(v)$ is odd. Note that special vertices are bichromatic.

Lemma 2. *For every nice graph G and every valid partition $(V_1, \ldots, V_t)$ of G, there exists a 3-labelling ℓ of G such that:*

1. *all vertices of V_1 are either 1-monochromatic or 3-monochromatic,*
2. *all vertices of V_2 are either 1-monochromatic or 2-monochromatic,*
3. *all vertices of $V_3 \cup \cdots \cup V_t$ are bichromatic,*
4. *no vertex is special,*
5. *if $u \in V_1$ and $v \in V_2$ are adjacent, then $\ell(uv) = 1$,*
6. *if two vertices u and v are in conflict, then $u \in V_1$ and $v \in V_2$ (or vice versa), and at least one of u or v has a neighbour w in $V_1 \cup V_2$.*

Proof. From now on, we fix the valid partition $\mathcal{V} = (V_1, \ldots, V_t)$ of G. During the construction of ℓ, we may have, however, to swap some edges of M_0, resulting in a different valid partition of G. Abusing the notations, for simplicity we will still denote by $\mathcal{V}$ any valid partition of G obtained this way, through swapping edges. Recall that valid partitions are closed under swapping edges of M_0 (Property $(\mathcal{P}_2)$ of Definition 1).

Our goal is to design ℓ so that it not only satisfies the four colour properties of Items 1 to 4 of the statement, but also achieves the following refined product types, for every vertex v in a part V_i of $\mathcal{V}$:

- $v \in V_1$: v is 1-monochromatic or 3-monochromatic;
- $v \in V_2$: v is 1-monochromatic or 2-monochromatic;
- $v \in V_3$: v is bichromatic with 2-degree 1 and even $\{2,3\}$-degree;
- $v \in V_4$: v is bichromatic with 3-degree 2 and odd $\{2,3\}$-degree;
- $v \in V_5$: v is bichromatic with 2-degree 2 and even $\{2,3\}$-degree;
- ...
- $v \in V_{2n}$, $n \geq 3$: v is bichromatic with 3-degree n and odd $\{2,3\}$-degree;
- $v \in V_{2n+1}$, $n \geq 3$: v is bichromatic with 2-degree n and even $\{2,3\}$-degree;
- ...

We start from ℓ assigning label 1 to all edges of G. Let us now describe how to modify ℓ so that the conditions above are met for all vertices. We consider the vertices of $V_t, \ldots, V_3$ following that order, from "bottom to top", and modify labels assigned to upward edges. An important condition we will maintain, is that every vertex in an odd part V_{2n+1} ($n \geq 0$) has all its incident downward edges (if any) labelled 3 or 1, while every vertex in an even part V_{2n} ($n \geq 1$) has all its incident downward edges (if any) labelled 2 or 1. Note that this is trivially satisfied for the vertices in V_t, since they have no incident downward edges.

At any point in the process, let M be the set of edges of M_0 for which both ends are 1-monochromatic (initially, $M = M_0$). When treating a vertex $u \in V_3 \cup \cdots \cup V_t$, we define M_u as the subset of edges of M having an end that is a neighbour of u. For every edge $e \in M_u$, we choose one end of e that is a neighbour of u and we add it to a set S_u. Note that $|S_u| = |M_u|$. Another goal during the labelling process, to fulfil Item 6, is to label the edges incident to u so that at least one end of every edge in M_u is no longer 1-monochromatic. Note that the set M_u considered when labelling the edges incident to u is not necessarily the set of edges of M_0 incident to a neighbour of u, as, during the whole process, some of these edges might be removed from M when dealing with previous vertices in $V_3 \cup \cdots \cup V_t$.

Let us now consider the vertices in $V_t, \ldots, V_3$ one by one, following that order. Let thus $u \in V_i$ be a vertex that has not been treated yet, with $i \geq 3$. Recall that every vertex belonging to some V_j with $j > i$ was treated earlier on, and thus has its desired product. Suppose that $i = 2n$ with $n \geq 2$ ($i = 2n + 1$ with $n \geq 1$, resp.). Recall also that u is assumed to have all its incident downward edges labelled 1 or 2 (3, resp.), due to how vertices in V_j's with $j > i$ have been treated earlier on, and to have all its incident upward edges labelled 1.

If $M_u \neq \emptyset$, then we swap edges of M_u, if necessary, so that every vertex in S_u belongs to V_2 (V_1, resp.). This does not invalidate any of our invariants since both ends of an edge in S_u are 1-monochromatic.

In any case, by Property ($\mathcal{P}_1$), we know that, for every $j < i$, there is a vertex $x_j \in V_j$ which is a neighbour of u. In particular, the vertex x_1 (x_2, resp.) does not belong to S_u (but may be the other end of an edge in M_u). We label the edges $ux_3, ux_5, \ldots, ux_{2n-1}$ with 3 ($ux_4, ux_6, \ldots, ux_{2n}$ with 2, resp.). Note that, at this point, $d_3(u) = n - 1$ ($d_2(u) = n - 1$, resp.). To finish dealing with u, we need to distinguish two cases depending on whether M_u is empty or not.

- Suppose first that $M_u = \emptyset$. Label ux_1 with 3 (ux_2 with 2, resp.). Now u has the desired 3-degree (2-degree, resp.). If $i > 3$, then label ux_{i-2} with 2 (3, resp.) so that u is sure to be bichromatic. If $i > 3$ and the $\{2, 3\}$-degree of u does not have the desired parity, then label ux_2 with 2 (ux_1 with 3, resp.). If $u \in V_3$ and the $\{2, 3\}$-degree of u is even, then u is already bichromatic since $d_2(u) = 1$. If $u \in V_3$ and the $\{2, 3\}$-degree of u is odd, then label ux_1 with 3 to adjust the parity of the $\{2, 3\}$-degree of u and make u bichromatic. In all cases, u gets bichromatic with 3-degree n (2-degree n, resp.) and odd $\{2, 3\}$-degree (even $\{2, 3\}$-degree, resp.), which is what is desired for u.
- Suppose now that $M_u \neq \emptyset$. Let $z \in S_u$ and let e be the edge of M_u containing z. For every $w \in S_u \setminus \{z\}$, we label the edge uw with 2 (3, resp.). Then:
  - If $d_2(u) + d_3(u)$ is odd (even, resp.), then label uz with 2 (3, resp.) and ux_1 with 3 (ux_2 with 2, resp.). In this case, every edge in M_u is incident to at least one vertex which is not 1-monochromatic, while u is bichromatic with 3-degree n (2-degree n, resp.) and odd $\{2, 3\}$-degree (even $\{2, 3\}$-degree, resp.).
  - If $d_2(u) + d_3(u)$ is even (odd, resp.) and $d_2(u) > 0$ ($d_3(u) > 0$, resp.), then swap e and label uz with 3 (2, resp.). Note that, after the swap of e, we have $z \in V_1$ ($z \in V_2$, resp.). In this case, every edge in M_u is incident to at least one vertex which is not 1-monochromatic, while u is bichromatic with 3-degree n (2-degree n, resp.) and odd $\{2, 3\}$-degree (even $\{2, 3\}$-degree, resp.).
  - The last case is when $d_2(u) + d_3(u)$ is even (odd, resp.) and $d_2(u) = 0$ ($d_3(u) = 0$, resp.). If $i > 4$, then we can label ux_{i-2} with 2 (3, resp.) and fall back into one of the previous cases. If $i = 4$, then the only edge labelled 3 is the edge ux_3 which implies that $d_3(u) = 1$, which is impossible since $d_2(u) = 0$ and $d_2(u) + d_3(u)$ is odd. If $i = 3$, then the conditions of this case imply that $d_2(u) = 1$ while every upward edge incident to u is labelled 1 or 3 and similarly for every incident downward edge; this case thus cannot occur.

To finish, we remove the edges of M_u from M since their two ends are not both 1-monochromatic any more.

At the end of this process, all vertices in V_1 are 1-monochromatic or 3-monochromatic, while all vertices in V_2 are 1-monochromatic or 2-monochromatic. Every vertex in $V_3 \cup \cdots \cup V_t$ is bichromatic and there are no conflicts involving any pair of these vertices. Indeed if $a \in V_i$ and $b \in V_j$ are adjacent with $i > j \geq 3$, then either i and j do not have the same parity, in which case a and b do not have the same $\{2,3\}$-degree; or both i and j are even (odd, resp.) and $d_3(a) = \frac{i}{2} \neq \frac{j}{2} = d_3(b)$ $(d_2(a) = \frac{i-1}{2} \neq \frac{j-1}{2} = d_2(b)$, resp.). Note also that no vertex in G is special, as special vertices have 3-degree 1, 2-degree at least 2, and odd $\{2,3\}$-degree. Also, we did not relabel any edge in the cut (V_1, V_2).

Finally, suppose that there is a conflict between two vertices u and v. Previous remarks imply that $u \in V_1$ and $v \in V_2$ (or *vice versa*) and that both u and v are 1-monochromatic. If none of u and v has another neighbour w in $V_1 \cup V_2$, then the edge uv belongs to the set M_0. Since G is nice, one of u or v must have a neighbour z in $V_3 \cup \cdots \cup V_t$. Hence $uv \in M_z$. Recall also that we relabelled the edges incident to z in such a way that, for every edge of M_z, at least one incident vertex became 2-monochromatic or 3-monochromatic, a contradiction to the existence of u and v. Hence, all properties of the lemma hold.   □

Step 3: Labelling the edges between V_1 and V_2

From now on, we will modify a 3-labelling ℓ of G obtained by applying Lemma 2. We denote by $\mathcal{H}$ the set of the connected components of $G[V_1 \cup V_2]$ that contain two adjacent vertices $u \in V_1$ and $v \in V_2$ having the same product by ℓ. By Items 1 and 2 of Lemma 2, such u and v are 1-monochromatic. Also, by Item 6 of Lemma 2, recall that every connected component of $\mathcal{H}$ has at least two edges. In what follows, we only relabel edges of some connected components $H \in \mathcal{H}$ with making sure that their vertices (in $V_1 \cup V_2$) are monochromatic or special. This ensures that only vertices of H have their product affected, thus that no new conflicts involving vertices in $V_3 \cup \cdots \cup V_t$ are created.

For a subgraph X of $H \in \mathcal{H}$ (possibly $X = H$), if, after having relabelled edges of X, no conflict remains between vertices of X and all vertices of X are either monochromatic or special, then we say that X satisfies Property $(\mathcal{P}_3)$.

Lemma 3. *If we can relabel the edges of every $H \in \mathcal{H}$ so that every H satisfies Property $(\mathcal{P}_3)$, then the resulting 3-labelling is p-proper.*

Proof. This is because if we get rid of all conflicts in $\mathcal{H}$, then the only possible remaining conflicts are between vertices in $V_1 \cup V_2$ and in $V_3 \cup \cdots \cup V_t$. In particular, recall that any two vertices of two distinct connected components $H_1, H_2 \in G[V_1 \cup V_2]$ cannot be adjacent. Note also that, because we only relabelled edges in $\mathcal{H}$, the vertices in $V_3 \cup \cdots \cup V_t$ retain the product types described in Lemma 2. In particular, they remain bichromatic and none of them is special. Thus, they cannot be in conflict with the vertices in $V_1 \cup V_2$.   □

In order to show that we can relabel the edges of every $H \in \mathcal{H}$ so that it fulfils Property $(\mathcal{P}_3)$, the following result will be particularly handy.

Lemma 4. *For every integer $s \in \{2,3\}$, every connected bipartite graph H whose edges are labelled 1 or s, and any vertex v in any part $V_i \in \{V_1, V_2\}$ of H, we can relabel the edges of H with 1 and s so that $d_s(u)$ is odd (even, resp.) for every $u \in V_i \setminus \{v\}$, and $d_s(u)$ is even (odd, resp.) for every $u \in V_{3-i}$.*

Proof. As long as H has a vertex u different from v that does not satisfy the desired condition, apply the following. Choose P any path from u to v, which exists by the connectedness of H. Now follow P from u to v, and change the labels of the traversed edges from 1 to s and *vice versa*. It can be noted that this alters the parity of the s-degrees of u and v, while this does not alter that parity for any of the other vertices of H. Thus, this makes u satisfy the desired condition, while the situation did not change for the other vertices different from u and v. Thus, once this process ends, all vertices of H different from v have their s-degree being as desired by the resulting labelling. □

We are now ready to treat the connected components $H \in \mathcal{H}$ independently, so that they all meet Property ($\mathcal{P}_3$). To ease the reading, we distinguish several cases depending on the types and on the degrees of the vertices that H includes. In each of the successive cases we consider, it is implicitly assumed that H does not meet the conditions of any previous case.

Claim 1. *If $H \in \mathcal{H}$ has a 3-monochromatic vertex $v \in V_1$, or a 1-monochromatic vertex $v_1 \in V_1$ having two 1-monochromatic neighbours $u_1, u_2 \in V_2$ with degree 1 (in H), then we can relabel edges of H so that H satisfies Property ($\mathcal{P}_3$).*

Proof. Recall that all edges of H are assigned label 1; thus, if a vertex of H is 3-monochromatic, then it must be due to incident downward edges to $V_3, \ldots, V_t$.

If H has a 1-monochromatic vertex $v_1 \in V_1$ that is adjacent to two degree-1 1-monochromatic vertices $u_1, u_2 \in V_2$, then we set $\ell(v_1u_1) = \ell(v_1u_2) = 3$. Note that u_1 and u_2 become 3-monochromatic with 3-degree 1, and are thus no longer in conflict with v_1, as it becomes 3-monochromatic with 3-degree 2. Note that either we got rid of all conflicts in H and H now satisfies Property ($\mathcal{P}_3$) as desired, or conflicts between other 1-monochromatic vertices of H remain. In the latter case, we continue with the following arguments.

Assume H has remaining conflicts, and that H has a 3-monochromatic vertex $v \in V_1$ (and, due to the previous process, perhaps 3-monochromatic vertices u_1 and u_2 in V_2, in which case their 3-degree (and degree in H) is precisely 1, while their unique neighbour v in $V_1 \cap V(H)$ is 3-monochromatic with 3-degree 2). Let X be the set of all 3-monochromatic vertices of H belonging to V_1. Let $C_1, \ldots, C_q$ denote the $q \geq 1$ connected components of $H - X$ that do not consist in a 3-monochromatic vertex of V_2 (the vertices u_1 and u_2 we dealt with earlier on). For every C_i, we choose arbitrarily a vertex $x_i \in X$ and a vertex $y_i \in C_i$ such that x_i and y_i are adjacent in H. Note that the vertices of C_i are either 1-monochromatic or 2-monochromatic (in which case they belong to V_2), since all 3-monochromatic vertices of H are part of X (or are the vertices u_1 and u_2 dealt with earlier on, which we have omitted and are not part of the C_i's).

By Lemma 4, in every C_i we can relabel the edges with 1 and 2 so that all vertices in $(V_2 \cap V(C_i)) \setminus \{y_i\}$ are 2-monochromatic with odd 2-degree, while all vertices in $V_1 \cap V(C_i)$ are 2-monochromatic with even 2-degree or possibly 1-monochromatic if their even 2-degree is 0. In particular, recall that y_i must be 1-monochromatic or 2-monochromatic. If y_i has odd 2-degree, then there are no conflicts between vertices of C_i. If y_i has even non-zero 2-degree, then we set $\ell(x_i y_i) = 3$, thereby making y_i special.

Let Y be the set of all 1-monochromatic y_i's having a 1-monochromatic neighbour w_i in C_i. Let H' be the subgraph of H induced by $Y \cup X$. Note that every edge of H' is labelled 1. Let now $Q_1, \ldots, Q_p$ denote the connected components of H' and choose $x_k \in X \cap V(Q_k)$ for every $k \in \{1, \ldots, p\}$. For every k, we apply Lemma 4 with labels 1 and 3 so that all vertices in $V_2 \cap V(Q_k)$ get 3-monochromatic with odd 3-degree, while all vertices in $V_1 \cap V(Q_k) \setminus \{x_k\}$ get 3-monochromatic with even 3-degree or possibly 1-monochromatic (3-degree 0).

If x_k is involved in a conflict with a vertex $y_i \in V_2 \cap V(Q_k)$, then this is because x_k has odd 3-degree. Then:

- If $\ell(x_k y_i) = 3$, then $d_3(y_i) = d_3(x_k) \geq 3$ since $x_k \in X$ (x_k must thus be incident to at least one other edge labelled 3, either a downward edge to $V_3, \ldots, V_t$ or an edge incident to u_1 (and similarly an edge incident to u_2)). We here assign label 1 to the edge $x_k y_i$ and label 3 to the edge $y_i w_i$. This way, x_k gets even 3-degree while the 3-degree of y_i does not change. Note that y_i and w_i are not in conflict since $d_3(w_i) = 1$ and $d_3(y_i) \geq 3$.
- Otherwise, if $\ell(x_k y_i) = 1$, then we assign label 3 to the edge $x_k y_i$ and label 3 to the edge $y_i w_i$. This way, x_k gets even 3-degree while the 3-degree of y_i remains odd and must be at least 3. Again y_i and w_i are not in conflict since $d_3(w_i) = 1$ and $d_3(y_i) \geq 3$.

We claim that we got rid of all conflicts in H. Indeed, consider two adjacent vertices $a \in V_1 \cap V(H)$ and $b \in V_2 \cap V(H)$. Suppose first that a and b belong to some C_i. Note that, with the exception of y_i and maybe of the vertex w_i (if it exists and $y_i \in Y$), every vertex of C_i is 1-monochromatic or 2-monochromatic, the vertices of $V_1 \cap V(C_i)$ having even 2-degree and the vertices of $V_2 \cap V(C_i)$ having odd 2-degree. Thus, no conflict involves two of these vertices. Suppose now that $b = y_i$. If y_i is 2-monochromatic with odd 2-degree, then there is no conflict involving y_i in C_i since all of its neighbours in C_i have even 2-degree. If y_i is special, then it is the only special vertex of C_i, so, here again, it cannot be involved in a conflict. If $y_i \notin Y$ and y_i is 1-monochromatic, then y_i has no other 1-monochromatic neighbour in C_i by definition of Y. If $y_i \in Y$, then y_i is 3-monochromatic with odd 3-degree, the only other possible 3-monochromatic neighbour of y_i in C_i being w_i, but we showed previously that their 3-degrees differ. Thus, in all cases, there cannot be conflicts between vertices of C_i.

We are left with the case where a and b do not belong to the same C_i. In particular, this implies that $a \in X$ and that a is 3-monochromatic. The only possible 3-monochromatic vertices in V_2 are the vertices of Y, which have odd 3-degree, and the 3-monochromatic vertices u_1 and u_2 with 3-degree 1 and degree 1 in H which might have been created at the very beginning of the proof. If $b \in Y$, then, due to the application of Lemma 4 above, the only vertex of X which can have odd 3-degree is some x_k, but for this vertex we either ensured that it was involved in no conflict, or we tweaked the labelling so that it got even 3-degree without modifying the labelling properties obtained through Lemma 4. If b is u_1

or u_2, then b has only one neighbour v. Note that the edges vu_1 and vu_2 are still labelled 3 as they are not part of the Q_i's, and, thus, $d_3(b) = 1$ and $d_3(v) \geq 2$. Hence, there is no conflict between vertices of X and other vertices of H. This implies that H satisfies Property $(\mathcal{P}_3)$. □

We can thus assume that H does not meet any of the conditions in Claim 1. The next step is showing that we can treat H in a similar way, in case H contains a 1-monochromatic vertex $u \in V_2$ with at least two neighbours in H. This can be proved similarly as Claim 1, by investigating the structure of H and making use of Lemma 4 to relabel edges of H in such a way that all remaining conflicts are located in very precise places of H (so that we can then handle them one by one). The formal proof being long, tedious, and in the same vein as that of Claim 1, due to space limitation we omit it from this paper. The interested reader will find the whole proof in [3], the full version of the current paper.

Claim 2. *If H has a 1-monochromatic vertex $u \in V_2$ with at least two neighbours in H, then we can relabel edges of H so that H satisfies Property $(\mathcal{P}_3)$.*

Assuming H does not meet any of the conditions in Claims 1 and 2, final arguments allow to relabel edges of H to get rid of all its conflicts.

Claim 3. *We can relabel edges of H so that it satisfies Property $(\mathcal{P}_3)$.*

Proof. Let $v \in V_1$ and $u \in V_2$ be two adjacent 1-monochromatic vertices of H (which must exist as otherwise H would satisfy Property $(\mathcal{P}_3)$). Because H has at least two edges (as otherwise it would belong to M, not to $\mathcal{H}$), at least one of v and u must have another neighbour in H. Since Claim 2 does not apply, note that u must have degree 1 in H (since all neighbours of u in H must be 1-monochromatic due to Claim 1 not applying). So v is also adjacent to $k \geq 1$ vertices $x_1, \ldots, x_k \in V_2$ different from u, which must all be 2-monochromatic (because of incident downward edges to $V_3, \ldots, V_t$; recall that all edges of H are labelled 1) as otherwise Claim 2 would apply.

Set $H' = H - u$. According to Lemma 4, we can relabel edges in H' with 1 and 2 so that all vertices in $(V_1 \cap V(H')) \setminus \{v\}$ have odd 2-degree, while all vertices in $V_2 \cap V(H')$ have even 2-degree. Recall that u is 1-monochromatic. Thus, if also v is 2-monochromatic with odd 2-degree, then we are done. Assume thus that v is 2-monochromatic with even 2-degree.

- Assume first that the 2-degree of v is even at least 2. In that case, set $\ell(vu) = 3$. This way, u becomes 3-monochromatic, while v becomes special.
- Assume now v is 1-monochromatic. This implies that $\ell(vx_1) = 1$. Change $\ell(vx_1)$ to 3. This way, x_1 becomes special (recall its 2-degree is even and at least 1, due to incident downward edges), while v becomes 3-monochromatic. Note that u remains 1-monochromatic.

In both cases, it can be checked that H now fulfils Property $(\mathcal{P}_3)$. □

At this point, we dealt with all connected components of $\mathcal{H}$, and the resulting labelling ℓ of G is p-proper by Lemma 3. The whole proof is thus complete.

References

1. Addario-Berry, L., Aldred, R.E.L., Dalal, K., Reed, B.A.: Vertex colouring edge partitions. J. Comb. Theory Ser. B **94**(2), 237–244 (2005)
2. Bensmail, J., Hocquard, H., Lajou, D., Sopena, É.: Further evidence towards the Multiplicative 1-2-3 Conjecture. Discrete Appl. Math. **307**, 135–144 (2022). https://doi.org/10.1016/j.dam.2021.10.014
3. Bensmail, J., Hocquard, H., Lajou, D., Sopena, É.: A proof of the Multiplicative 1-2-3 Conjecture. http://arxiv.org/abs/2108.10554 (2021)
4. Gallian, J.A.: A dynamic survey of graph labeling. Electron. J. Comb. **6**, DS6 (1998)
5. Kalkowski, M., Karoński, M., Pfender, F.: Vertex-coloring edge-weightings: towards the 1-2-3 Conjecture. J. Comb. Theory Ser. B **100**, 347–349 (2010)
6. Karoński, M., Łuczak, T., Thomason, A.: Edge weights and vertex colours. J. Comb. Theory Ser. B **91**, 151–157 (2004)
7. Przybyło, J.: The 1-2-3 Conjecture almost holds for regular graphs. J. Comb. Theory Ser. B **147**, 183–200 (2021)
8. Skowronek-Kaziów, J.: Multiplicative vertex-colouring weightings of graphs. Inf. Process. Lett. **112**(5), 191–194 (2012)
9. Vučković, B.: Multi-set neighbor distinguishing 3-edge coloring. Discrete Math. **341**, 820–824 (2018)

Chromatic Bounds for Some Subclasses of $(P_3 \cup P_2)$-free graphs

Athmakoori Prashant[1]($\boxtimes$)(iD), P. Francis[2](iD), and S. Francis Raj[1](iD)

[1] Department of Mathematics, Pondicherry University, Puducherry 605014, India
11994prashant@gmail.com, francisraj_s@pondiuni.ac.in
[2] Department of Computer Science, Indian Institute of Technology,
Palakkad 678557, India
pfrancis@iitpkd.ac.in

Abstract. The class of $2K_2$-free graphs has been well studied in various contexts in the past. It is known that the class of $\{2K_2, 2K_1 + K_p\}$-free graphs and $\{2K_2, (K_1 \cup K_2) + K_p\}$-free graphs admits a linear χ-binding function. In this paper, we study the classes of $(P_3 \cup P_2)$-free graphs which is a superclass of $2K_2$-free graphs. We show that $\{P_3 \cup P_2, 2K_1 + K_p\}$-free graphs and $\{P_3 \cup P_2, (K_1 \cup K_2) + K_p\}$-free graphs also admits a linear χ-binding function. In addition, we give tight chromatic bounds for $\{P_3 \cup P_2, HVN\}$-free graphs and $\{P_3 \cup P_2, diamond\}$-free graphs, and it can be seen that the latter is an improvement of the existing bound given by A. P. Bharathi and S. A. Choudum [1].

Keywords: Chromatic number · χ-binding function · $(P_3 \cup P_2)$-free graphs · Perfect graphs

2000 AMS Subject Classification: 05C15 · 05C75

1 Introduction

All graphs considered in this paper are simple, finite and undirected. Let G be a graph with vertex set $V(G)$ and edge set $E(G)$. For any positive integer k, a *proper k-coloring* of a graph G is a mapping $c : V(G) \rightarrow \{1, 2, \ldots, k\}$ such that adjacent vertices receive distinct colors. If a graph G admits a proper k-coloring, then G is said to be *k-colorable*. The *chromatic number*, $\chi(G)$, of a graph G is the smallest k such that G is k-colorable. Let P_n, C_n and K_n respectively denote the path, the cycle and the complete graph on n vertices. For $S, T \subseteq V(G)$, let $N_T(S) = N(S) \cap T$ (where $N(S)$ denotes the set of all neighbors of S in G), let $\langle S \rangle$ denote the subgraph induced by S in G and let $[S, T]$ denote the set of all edges with one end in S and the other end in T. If every vertex in S is adjacent with every vertex in T, then $[S, T]$ is said to be complete. For any graph G, let $\overline{G}$ denote the complement of G.

Let $\mathcal{F}$ be a family of graphs. We say that G is $\mathcal{F}$-*free* if it does not contain any induced subgraph which is isomorphic to a graph in $\mathcal{F}$. For a fixed graph H, let us denote the family of H-free graphs by $\mathcal{G}(H)$. For any two disjoint graphs G_1 and G_2 let $G_1 \cup G_2$ and $G_1 + G_2$ denote the *union* and the *join* of G_1 and G_2 respectively. Let $\omega(G)$ and $\alpha(G)$ denote the *clique number* and *independence number* of a graph G respectively. When there is no ambiguity, $\omega(G)$ will be denoted by ω. A graph G is said to be *perfect* if $\chi(H) = \omega(H)$, for every induced subgraph H of G.

In order to determine an upper bound for the chromatic number of a graph in terms of their clique number, the concept of χ-binding functions was introduced by A. Gyárfás in [4]. A class $\mathcal{G}$ of graphs is said to be χ-*bounded* [4] if there is a function f (called a χ-binding function) such that $\chi(G) \leq f(\omega(G))$, for every $G \in \mathcal{G}$. We say that the χ-binding function f is *special linear* if $f(x) = x + c$, where c is a constant.

The family of $2K_2$-free graphs has been well studied. A. Gyárfás in [4] posed a problem which asks for the order of magnitude of the smallest χ-binding function for $\mathcal{G}(2K_2)$. In this direction, T. Karthick et al. in [5] proved that the families of $\{2K_2, H\}$-free graphs, where $H \in \{HVN, diamond, K_1 + P_4, K_1 + C_4, \overline{P_5}, \overline{P_2 \cup P_3}, K_5 - e\}$ admit a special linear χ-binding functions. The bounds for $\{2K_2, K_5-e\}$-free graphs and $\{2K_2, K_1+C_4\}$-free graphs were later improved by Athmakoori Prashant et al., in [7,8]. In [2], C. Brause et al., improved the χ-binding function for $\{2K_2, K_1 + P_4\}$-free graphs to $\max\{3, \omega(G)\}$. Also they proved that for $s \neq 1$ or $\omega(G) \neq 2$, the class of $\{2K_2, (K_1 \cup K_2) + K_s\}$-free graphs with $\omega(G) \geq 2s$ is perfect and for $r \geq 1$, the class of $\{2K_2, 2K_1 + K_r\}$-free graphs with $\omega(G) \geq 2r$ is perfect. Clearly when $s = 2$ and $r = 3$, $(K_1 \cup K_2) + K_s \cong HVN$ and $2K_1 + K_r \cong K_5 - e$ which implies that the class of $\{2K_2, HVN\}$-free graphs and $\{2K_2, K_5-e\}$-free graphs are perfect for $\omega(G) \geq 4$ and $\omega(G) \geq 6$ respectively which improved the bounds given in [5].

Motivated by C. Brause et al., and their work on $2K_2$-free graphs in [2], we started looking at $(P_3 \cup P_2)$-free graphs which is a superclass of $2K_2$-free graphs. In [1], A. P. Bharathi et al., obtained a $O(\omega^3)$ upper bound for the chromatic number of $(P_3 \cup P_2)$-free graphs and obtained sharper bounds for $\{P_3 \cup P_2, diamond\}$)-free graphs. In this paper, we obtain linear χ-binding functions for the class of $\{P_3 \cup P_2, (K_1 \cup K_2) + K_p\}$-free graphs and $\{P_3 \cup P_2, 2K_1 + K_p\}$-free graphs. In addition, for $\omega(G) \geq 3p - 1$, we show that the class of $\{P_3 \cup P_2, (K_1 \cup K_2) + K_p\}$-free graphs admits a special linear χ-binding function $f(x) = \omega(G) + p - 1$ and the class of $\{P_3 \cup P_2, 2K_1 + K_p\}$-free graphs are perfect. In addition, we give a tight χ-binding function for $\{P_3 \cup P_2, HVN\}$-free graphs and $\{P_3 \cup P_2, diamond\}$-free graphs. This bound for $\{P_3 \cup P_2, diamond\}$-free graphs turns out to be an improvement of the existing bound obtained by A. P. Bharathi et al., in [1].

Some graphs that are considered as forbidden induced subgraphs in this paper are given in Fig. 1. Notations and terminologies not mentioned here are as in [10].

paw   diamond   HVN

Fig. 1. Some special graphs

2   Preliminaries

Throughout this paper, we use a particular partition of the vertex set of a graph G as defined initially by S. Wagon in [9] and later improved by A. P. Bharathi et al., in [1] as follows. Let $A = \{v_1, v_2, \ldots, v_\omega\}$ be a maximum clique of G. Let us define the *lexicographic ordering* on the set $L = \{(i,j) : 1 \leq i < j \leq \omega\}$ in the following way. For two distinct elements $(i_1, j_1), (i_2, j_2) \in L$, we say that (i_1, j_1) precedes (i_2, j_2), denoted by $(i_1, j_1) <_L (i_2, j_2)$ if either $i_1 < i_2$ or $i_1 = i_2$ and $j_1 < j_2$. For every $(i,j) \in L$, let

$$C_{i,j} = \{v \in V(G)\backslash A : v \notin N(v_i) \cup N(v_j)\} \backslash \left\{ \bigcup_{(i',j') <_L (i,j)} C_{i',j'} \right\}.$$ Note that, for

any $k \in \{1, 2, \ldots, j\}\backslash\{i,j\}$, $[v_k, C_{i,j}]$ is complete. Hence $\omega(\langle C_{i,j} \rangle) \leq \omega(G) - j + 2$.

For $1 \leq k \leq \omega$, let us define $I_k = \{v \in V(G)\backslash A : v \in N(v_i)$, for every $i \in \{1, 2, \ldots, \omega\}\backslash\{k\}\}$. Since A is a maximum clique, for $1 \leq k \leq \omega$, I_k is an independent set and for any $x \in I_k$, $xv_k \notin E(G)$. Clearly, each vertex in $V(G)\backslash A$ is non-adjacent to at least one vertex in A. Hence those vertices will be contained either in I_k for some $k \in \{1, 2, \ldots, \omega\}$, or in $C_{i,j}$ for some $(i,j) \in L$. Thus $V(G) = A \cup \left(\bigcup_{k=1}^{\omega} I_k \right) \cup \left(\bigcup_{(i,j) \in L} C_{i,j} \right)$. Sometimes, we use the partition $V(G) = V_1 \cup V_2$, where $V_1 = \bigcup_{1 \leq k \leq \omega} (\{v_k\} \cup I_k) = \bigcup_{1 \leq k \leq \omega} U_k$ and $V_2 = \bigcup_{(i,j) \in L} C_{i,j}$.

Let us recall a result on $(P_3 \cup P_2)$-free graphs given by A. P. Bharathi et al., in [1].

Theorem 1 [1]. *If a graph G is $(P_3 \cup P_2)$-free, then $\chi(G) \leq \frac{\omega(G)(\omega(G)+1)(\omega(G)+2)}{6}$.*

Without much difficulty one can make the following observations on $(P_3 \cup P_2)$-free graphs.

Fact 2. *Let G be a $(P_3 \cup P_2)$-free graph. For $(i,j) \in L$, the following holds.*

(i) Each $C_{i,j}$ is a disjoint union of cliques, that is, $\langle C_{i,j} \rangle$ is P_3-free.

(ii) For every integer $s \in \{1, 2, \ldots, j\}\backslash\{i,j\}$, $N(v_s) \supseteq \{C_{i,j} \cup A \cup (\bigcup_{k=1}^{\omega} I_k)\}\backslash\{v_s \cup I_s\}$.

3 $\{P_3 \cup P_2, (K_1 \cup K_2) + K_p\}$-free graphs

Let us start Sect. 3 with some observations on $((K_1 \cup K_2) + K_p)$-free graphs.

Proposition 1. *Let G be a $((K_1 \cup K_2) + K_p)$-free graph with $\omega(G) \geq p + 2$, $p \geq 1$. Then G satisfies the following.*

(i) *For $k, \ell \in \{1, 2, \ldots, \omega(G)\}$, $[I_k, I_\ell]$ is complete. Thus, $\langle V_1 \rangle$ is a complete multipartite graph with $U_k = \{v_k\} \cup I_k$, $1 \leq k \leq \omega(G)$ as its partitions.*

(ii) *For $j \geq p + 2$ and $1 \leq i < j$, $C_{i,j} = \emptyset$.*

(iii) *For $x \in V_2$, x has neighbors in at most $(p - 1)$ U_ℓ's where $\ell \in \{1, 2, \ldots, \omega(G)\}$.*

Proposition 2. *Let G be a $\{P_3 \cup P_2, (K_1 \cup K_2) + K_p\}$-free graph with $\omega(G) \geq p + 2$, $p \geq 1$. Then G satisfies the following.*

(i) *For $(i, j) \in L$ such that $j \leq p + 1$, if $\omega(\langle C_{i,j} \rangle) \geq p - j + 4$, then $\omega(\langle C_{k,j} \rangle) \leq 1$ for $k \neq i$ and $1 \leq k \leq j - 1$.*

(ii) *If $\omega(G) = (p + 2 + k)$, $k \geq 0$, then $\left\langle \left(\bigcup_{j=\max\{2, p+1-\lfloor \frac{k}{2} \rfloor\}}^{p+1} \left(\bigcup_{i=1}^{j-1} C_{i,j} \right) \right) \right\rangle$ is P_3-free.*

As a consequence of Proposition 1, we obtain Corollary 1 which is a result due to S. Olariu in [6].

Corollary 1 [6]. *Let G be a connected graph. Then G is paw-free graph if and only if G is either K_3-free or complete multipartite.*

Now, for $p \geq 1$ and $\omega \geq \max\{3, 3p - 1\}$, let us determine the structural characterization and the chromatic number of $\{P_3 \cup P_2, (K_1 \cup K_2) + K_p\}$-free graphs.

Theorem 3. *Let p be a positive integer and G be a $\{P_3 \cup P_2, (K_1 \cup K_2) + K_p\}$-free graph with $V(G) = V_1 \cup V_2$. If $\omega(G) \geq \max\{3, 3p - 1\}$, then (i) $\langle V_1 \rangle$ is a complete multipartite graph with partition $U_1, U_2, \ldots, U_\omega$, (ii) $\langle V_2 \rangle$ is P_3-free graph and (iii) $\chi(G) \leq \omega(G) + p - 1$.*

Proof. Let $p \geq 1$ and G be a $\{P_3 \cup P_2, (K_1 \cup K_2) + K_p\}$-free graph with $\omega(G) \geq \max\{3, 3p-1\}$. By (i) of Proposition 1, we see that $\langle V_1 \rangle$ is a complete multipartite graph with the partition $U_k = \{v_k\} \cup I_k$, $1 \leq k \leq \omega$. By (i) of Fact 2, each $\langle C_{i,j} \rangle$ is P_3-free, for every $(i, j) \in L$. Also without much difficulty we can show that $\langle V_2 \rangle$ is P_3-free. Now, let us exhibit an $(\omega + p - 1)$-coloring for G using $\{1, 2, \ldots, \omega + p - 1\}$ colors. For $1 \leq k \leq \omega$, give the color k to the vertices of U_k. Let H be a component in $\langle V_2 \rangle$. Clearly, each vertex in H is adjacent to at most $p - 1$ colors given to the vertices of V_1 and is adjacent to $\omega(H) - 1$ vertices of H. Since $\omega(H) \leq \omega(G)$, each vertex in H is adjacent to at most $\omega(G) + p - 2$ colors. Hence, there is a color available for each vertex in H. Similarly, all the components of $\langle V_2 \rangle$ can be colored properly. Hence, $\chi(G) \leq \omega(G) + p - 1$.

Even though we are not able to show that the bound given in Theorem 3 is tight, we can observe that the upper bound cannot be made smaller than $\omega + \lceil \frac{p-1}{2} \rceil$ by providing the following example. For $p \geq 1$, consider the graph G^* with $V(G^*) = X \cup Y \cup Z$, where $X = \overset{\omega}{\underset{i=1}{\cup}} x_i$, $Y = \overset{\omega}{\underset{i=1}{\cup}} y_i$ and $Z = \overset{p-1}{\underset{i=1}{\cup}} z_i$ if $p \geq 2$ else $Z = \emptyset$ and edge set $E(G^*) = \{\{x_i x_j\} \cup \{y_i y_j\} \cup \{z_r z_s\} \cup \{x_m y_n\} \cup \{y_m z_n\} \cup \{x_n z_n\}$, where $1 \leq i, j, m \leq \omega$, $1 \leq r, s, n \leq p-1$, $i \neq j$, $r \neq s$ and $m \neq n\}$. Without much difficulty, one can observe that G^* is a $\{P_3 \cup P_2, (K_1 \cup K_2) + K_p\}$-free graph, $\omega(G^*) = \omega$ and $\alpha(G^*) = 2$. Hence, $\chi(G^*) \geq \left\lceil \frac{|V(G^*)|}{\alpha(G^*)} \right\rceil = \omega(G^*) + \lceil \frac{p-1}{2} \rceil$.

Next, we obtain a linear χ-binding function for $\{P_3 \cup P_2, (K_1 \cup K_2) + K_p\}$-free graphs with $\omega \geq 3$.

Theorem 4. *Let p be an integer greater than 1. If G is a $\{P_3 \cup P_2, (K_1 \cup K_2) + K_p\}$-free graph, then*

$$\chi(G) \leq \begin{cases} \omega(G) + \sum_{j=2}^{p+1} (j-1)(p-j+3) & \text{for } 3 \leq \omega(G) \leq p+1 \\ \omega(G) + 7(p-1) + \sum_{j=4}^{p-\lfloor \frac{k}{2} \rfloor} (j-1)(p-j+3) & \text{for } \omega(G) = (p+2+k), 0 \leq k \leq 2p-5 \\ \omega(G) + 4p - 3 & \text{for } \omega(G) = 3p - 2 \\ \omega(G) + p - 1 & \text{for } \omega(G) \geq 3p - 1. \end{cases}$$

Proof. Let G be a $\{P_3 \cup P_2, (K_1 \cup K_2) + K_p\}$-free graph with $p \geq 2$. For $\omega(G) \geq 3p - 1$, the bound follows from Theorem 3. By (ii) of Proposition 1, we see that $C_{i,j} = \emptyset$ for all $j \geq p+2$. We know that $V(G) = V_1 \cup V_2$, where $V_1 = \underset{1 \leq \ell \leq \omega}{\cup} (\{v_\ell\} \cup I_\ell)$ and $V_2 = \underset{(i,j) \in L}{\cup} C_{i,j}$. Clearly, the vertices of V_1 can be colored with $\omega(G)$ colors. Let us find an upper bound for $\chi(\langle V_2 \rangle)$. First, let us consider the case when $3 \leq \omega(G) \leq p+1$. For $1 \leq i < j \leq p+1$, one can observe that $\omega(\langle C_{i,j} \rangle) \leq \omega(G) - j + 2 \leq p - j + 3$. Thus one can properly color the vertices of $\left(\overset{j-1}{\underset{i=1}{\cup}} C_{i,j} \right)$ with at most $(j-1)(p-j+3)$ colors and hence $\chi(G) \leq \chi(\langle V_1 \rangle) + \chi(\langle V_2 \rangle) \leq \omega(G) + \sum_{j=2}^{p+1} (j-1)(p-j+3)$.

Next, let us consider $\omega(G) = (p+2+k)$, where $0 \leq k \leq 2p-4$. By using (ii) of Proposition 2, $\left\langle \left(\overset{p+1}{\underset{j=p-\lfloor \frac{k}{2} \rfloor+1}{\cup}} \left(\overset{j-1}{\underset{i=1}{\cup}} C_{i,j} \right) \right) \right\rangle$ is a P_3-free graph. As in Theorem 3, one can color the vertices of $V(G) \backslash \left\{ \overset{p-\lfloor \frac{k}{2} \rfloor}{\underset{j=2}{\cup}} \left(\overset{j-1}{\underset{i=1}{\cup}} C_{i,j} \right) \right\}$ with at most $\omega(G) + p - 1$ colors. For $k = 2p - 4$, $\omega(G) = 3p - 2$ and we see that $\langle V_2 \backslash C_{1,2} \rangle$ is P_3-free and $\chi(\langle V(G) \backslash C_{1,2} \rangle) \leq \omega(G) + p - 1$. Therefore, $\chi(G) \leq \chi(\langle V(G) \backslash C_{1,2} \rangle) + \chi(\langle C_{1,2} \rangle) = \omega(G) + 4p - 3$. Finally, for $0 \leq k \leq 2p - 5$, by using (i) of Proposition 2 and by using similar strategies but with a little more involvement, we can show that $\left(\overset{j-1}{\underset{i=1}{\cup}} C_{i,j} \right)$, where $2 \leq j \leq p - \lfloor \frac{k}{2} \rfloor$ can be properly colored using $\omega(G)$ colors when $\omega(\langle C_{i,j} \rangle) \geq p - j + 4$, for some $i \in \{1, 2, \ldots, j - 1\}$ or by using

$(j-1)(p-j+3)$ colors when $\omega(\langle C_{i,j}\rangle) \leq p-j+3$, for every $i \in \{1,2,\ldots,j-1\}$. For $4 \leq j \leq p-\lfloor\frac{k}{2}\rfloor$, one can observe that $(j-1)(p-j+3) \geq \omega(G)$ and hence the vertices of $\left(\bigcup\limits_{j=4}^{p-\lfloor\frac{k}{2}\rfloor}\left(\bigcup\limits_{i=1}^{j-1} C_{i,j}\right)\right)$ can be properly colored with $\sum\limits_{j=4}^{p-\lfloor\frac{k}{2}\rfloor}(j-1)(p-j+3)$ colors. When $j=3$, $(j-1)(p-j+3) = 2p$. Since $2p-5 \geq 0$, $p \geq 3$. Also, since $\omega(G) \geq 4$, the vertices of $C_{1,2}$ and $(C_{1,3} \cup C_{2,3})$ can be properly colored with at most $(3p-3)$ colors each. Hence, $\chi(G) \leq (\omega(G)+p-1) + 2(3p-3) + \sum\limits_{j=4}^{p-\lfloor\frac{k}{2}\rfloor}(j-1)(p-j+3) = \omega(G) + 7(p-1) + \sum\limits_{j=4}^{p-\lfloor\frac{k}{2}\rfloor}(j-1)(p-j+3)$.

The bound obtained in Theorem 4 is not optimal. This can be seen in Theorem 5. Note that when $p=2$, $(K_1 \cup K_2) + K_p \cong HVN$.

Theorem 5. *If G is a $\{P_3 \cup P_2, HVN\}$-free graph with $\omega(G) \geq 4$, then $\chi(G) \leq \omega(G)+1$.*

The graph G^* (defined next to Theorem 3) shows that the bound given in Theorem 5 is tight.

4 $\{P_3 \cup P_2, 2K_1 + K_p\}$-free graphs

Let us start Sect. 4 by observing that any $\{P_3 \cup P_2, 2K_1 + K_p\}$-free graph is also a $\{P_3 \cup P_2, (K_1 \cup K_2) + K_p\}$-free graph. Hence the properties established for $\{P_3 \cup P_2, (K_1 \cup K_2) + K_p\}$-free graphs is also true for $\{P_3 \cup P_2, 2K_1 + K_p\}$-free graphs.

One can observe that by using techniques similar to the one's used in Theorem 3 and by Strong Perfect Graph Theorem [3], any $\{P_3 \cup P_2, 2K_1 + K_p\}$-free graph is perfect, when $\omega \geq 3p-1$.

Theorem 6. *Let p be a positive integer and G be a $\{P_3 \cup P_2, 2K_1 + K_p\}$-free graph with $V(G) = V_1 \cup V_2$. If $\omega(G) \geq 3p-1$, then $\langle V_1 \rangle$ is complete, $\langle V_2 \rangle$ is P_3-free and G is perfect.*

As a consequence of Theorem 4 and Theorem 6, without much difficulty one can observe Proposition 3 and Corollary 2.

Proposition 3. *Let G be a $\{P_3 \cup P_2, 2K_1 + K_p\}$-free graph. If $\omega(\langle C_{1,2}\rangle) \geq 2p$, then $\chi(G) = \omega(G)$.*

Corollary 2. *Let p be an integer greater than 1. If G is a $\{P_3 \cup P_2, 2K_1 + K_p\}$-free graph, then*

$$\chi(G) \leq \begin{cases} \omega(G) + \sum\limits_{j=2}^{p+1}(j-1)(p-j+3) & \text{for } 3 \leq \omega(G) \leq p+1 \\ \omega(G) + 2p - 1 + \sum\limits_{j=3}^{p-\lfloor\frac{k}{2}\rfloor}(j-1)(p-j+3) & \text{for } \omega(G) = (p+2+k), 0 \leq k \leq 2p-5 \\ \omega(G) + 2p - 1 & \text{for } \omega(G) = 3p-2 \\ \omega(G) & \text{for } \omega(G) \geq 3p-1. \end{cases}$$

When $p = 2$, $2K_1 + K_p \cong diamond$ and hence by Theorem 6, $\{P_3 \cup P_2, diamond\}$-free graphs are perfect for $\omega(G) \geq 5$ which was shown by A. P. Bharathi et al., in [1].

Theorem 7 [1].   *If G is a $\{P_3 \cup P_2, diamond\}$-free graph then*

$$\chi(G) \leq \begin{cases} 4 \text{ for } \omega(G) = 2 \\ 6 \text{ for } \omega(G) = 3 \text{ and } G \text{ is perfect if } \omega(G) \geq 5. \\ 5 \text{ for } \omega(G) = 4 \end{cases}$$

We can further improve the bound given in Theorem 7 by obtaining a $\omega(G)$-coloring when $\omega(G) = 4$.

Theorem 8. *If G is a $\{P_3 \cup P_2, diamond\}$-free graph then*

$$\chi(G) \leq \begin{cases} 4 \text{ for } \omega(G) = 2 \\ 6 \text{ for } \omega(G) = 3 \text{ and } G \text{ is perfect if } \omega(G) \geq 5. \\ 4 \text{ for } \omega(G) = 4. \end{cases}$$

Acknowledgment. The first author's research was supported by the Council of Scientific and Industrial Research, Government of India, File No: 09/559(0133)/2019-EMR-I. The second author's research was supported by Post Doctoral Fellowship at Indian Institute of Technology, Palakkad.

References

1. Bharathi, A.P., Choudum, S.A.: Colouring of $(P_3 \cup P_2)$-free graphs. Graphs Comb. **34**(1), 97–107 (2018)
2. Brause, C., Randerath, B., Schiermeyer, I., Vumar, E.: On the chromatic number of $2K_2$-free graphs. Discrete Appl. Math. **253**, 14–24 (2019)
3. Chudnovsky, M., Robertson, N., Seymour, P., Thomas, R.: The strong perfect graph theorem. Ann. Math. **164**, 51–229 (2006)
4. Gyárfás, A.: Problems from the world surrounding perfect graphs. Zastosowania Matematyki Applicationes Mathematicae **19**(3–4), 413–441 (1987)
5. Karthick, T., Mishra, S.: Chromatic bounds for some classes of $2K_2$-free graphs. Discrete Math. **341**(11), 3079–3088 (2018)
6. Olariu, S.: Paw-free graphs. Inf. Process. Lett. **28**(1), 53–54 (1988)
7. Prashant, A., Francis Raj, S., Gokulnath, M.: Chromatic bounds for the subclasses of pK_2-free graphs. arXiv preprint arXiv:2102.13458 (2021)
8. Prashant, A., Gokulnath, M.: Chromatic bounds for the subclasses of pK_2-free graphs. In: Mudgal, A., Subramanian, C.R. (eds.) CALDAM 2021. LNCS, vol. 12601, pp. 288–293. Springer, Cham (2021). https://doi.org/10.1007/978-3-030-67899-9_23
9. Wagon, S.: A bound on the chromatic number of graphs without certain induced subgraphs. J. Comb. Theory Ser. B **29**(3), 345–346 (1980)
10. West, D.B.: Introduction to Graph Theory. Prentice-Hall of India Private Limited, Upper Saddle River (2005)

List Homomorphisms to Separable Signed Graphs

Jan Bok[1]([⊠])[iD], Richard Brewster[2][iD], Tomás Feder[3], Pavol Hell[4][iD],
and Nikola Jedličková[5][iD]

[1] Computer Science Institute, Faculty of Mathematics and Physics,
Charles University, Prague, Czech Republic
bok@iuuk.mff.cuni.cz
[2] Department of Mathematics and Statistics, Thompson Rivers University,
Kamloops, Canada
rbrewster@tru.ca
[3] 268 Waverley Street, Palo Alto, USA
tomas@theory.stanford.edu
[4] School of Computing Science, Simon Fraser University, Burnaby, Canada
pavol@cs.sfu.ca
[5] Department of Applied Mathematics, Faculty of Mathematics and Physics,
Charles University, Prague, Czech Republic
jedlickova@kam.mff.cuni.cz

Abstract. The complexity of the list homomorphism problem for signed
graphs appears difficult to classify. Existing results focus on special
classes of signed graphs, such as trees [1] and reflexive signed graphs [18].
Irreflexive signed graphs are the heart of the problem, and Kim and Siggers have formulated a conjectured classification for these signed graphs.
We focus on a special case of irreflexive signed graphs, namely those in
which the unicoloured edges form a spanning path or cycle, and classify
the complexity of list homomorphisms to these signed graphs. In particular, our results confirm the conjecture of Kim and Siggers for this class
of signed graphs.

1 Motivation and Background

We investigate the complexity of (list) homomorphism problems for signed
graphs. The complexity of homomorphism (and list homomorphism) problems
is a popular topic. For undirected graphs, it was shown in [16] that the problem of deciding the existence of homomorphisms from an input graph to a fixed
graph H is polynomial if H is bipartite or has a loop, and is NP-complete otherwise. For general structures H, the corresponding problem lead to the so-called
Dichotomy Conjecture [12,17], which was only recently established [8,25]. In the
list homomorphism problem for H, the input contains with each input graph
also lists of allowed images for each vertex. (The precise definitions are given
below.) The list homomorphism problems have generally a nicer behaviour than

© Springer Nature Switzerland AG 2022
N. Balachandran and R. Inkulu (Eds.): CALDAM 2022, LNCS 13179, pp. 22–35, 2022.
https://doi.org/10.1007/978-3-030-95018-7_3

the homomorphism problems, because the lists facilitate recursion to subproblems. For undirected graphs, the list homomorphism problem is polynomial if H is a bi-arc graph (see below), and is NP-complete otherwise [9,10]. Even for general structures H, where the list version is equivalent to a special case of the basic version, the classification for the list version was achieved a decade earlier [7].

Signed graphs are related to graphs with two symmetric binary relations; in addition, they are equipped with an operation of *switching* (explained below). The possibility of switching poses challenges when classifying the complexity of homomorphisms, as the problem no longer appears to be a homomorphism problem for relational structures. Nevertheless, it can be shown that it is equivalent to such a problem and hence the results from [8,25] imply that there these problems also enjoy a dichotomy of polynomial versus NP-complete. For homomorphisms of signed graphs (without lists), a concrete dichotomy classification was conjectured in [4], and proved in [6]. Interestingly, for signed graphs, the list version no longer seems easier to classify, and the progress towards a classification has been slow [1,4,18]. In this paper we focus on one particular class of signed graphs and provide a full classification of complexity of the corresponding homomorphism problem. In particular, our results confirm a conjecture from [18] for this class of signed graphs.

A *signed graph* $\widehat{G}$ consists of a set $V(G)$ and two symmetric binary relations $+, -$. We also view $\widehat{G}$ as a graph G with the vertex set $V(G)$, the edge set $+ \cup -$ (the *underlying graph of* $\widehat{G}$), and a mapping $\sigma : E(G) \rightarrow \{+, -\}$, assigning a sign ($+$ or $-$) to each edge of G. (A loop is considered to be an edge.) Two signed graphs are considered *(switching-) equivalent* if one can be obtained from the other by a sequence of *switchings*; switching at a vertex v results in changing the signs of all edges incident to v. We will usually view signs of edges as colours, and view positive edges as *blue*, and negative edges as *red*. It will be convenient to call a red-blue pair of edges with the same endpoint(s) a *bicoloured edge*; however, it is important to keep in mind that formally they are two distinct edges.

The study of signed graphs seems to have originated in [14,15], and was most notably advanced in the papers of Zaslavsky [20–24]. Guenin [13] pioneered the investigation of homomorphisms of signed graphs; see also, e.g., [5] and [19].

A *homomorphism* of the signed graph $\widehat{G}$ to the signed graph $\widehat{H}$ is a mapping $f : V(G) \rightarrow V(H)$ for which there exists a signed graph $\widehat{G'}$ equivalent to $\widehat{G}$ such that f preserves both relations $+$ and $-$. A *list homomorphism* of $\widehat{G}$ to $\widehat{H}$, with respect to the lists $L(v) \subseteq V(H), v \in V(G)$, is a homomorphism f of $\widehat{G}$ to $\widehat{H}$ such that $f(v) \in L(v)$ for all $v \in V(G)$. Let $\widehat{H}$ be a fixed signed graph. The *homomorphism problem* for $\widehat{H}$ takes as input a signed graph $\widehat{G}$ and asks whether there exists a homomorphism of $\widehat{G}$ to $\widehat{H}$. The *list homomorphism problem* for $\widehat{H}$ takes an input a signed graph $\widehat{G}$ with lists $L(v) \subseteq V(H)$ for every $v \in V(G)$, and asks whether there exists a homomorphism f of the signed graph $\widehat{G}$ to $\widehat{H}$ such that $f(v) \in L(v)$ for every $v \in V(G)$. A subgraph $\widehat{G}$ of the signed graph $\widehat{H}$ is the *signed core* of $\widehat{H}$ if there is signed graph homomorphism f of $\widehat{H}$ to $\widehat{G}$, and every homomorphism of the signed graph $\widehat{G}$ to itself is a bijection on $V(G)$. It is easy to see that the signed core of any signed graph is unique up to isomorphism and switching equivalence. The dichotomy classification conjectured in [4] and proved in [6] is as follows. (In counting edges we count each unicoloured edge as one and each bicoloured edge as two.)

Theorem 1 [6]. *The homomorphism problem for the signed graph $\widehat{H}$ is polynomial-time solvable if the signed core of $\widehat{H}$ has at most two edges, and is NP-complete otherwise.*

In an earlier paper [3], cf. [1], we have classified the complexity of the list homomorphism problem for signed graphs with only unicoloured edges. A signed graph is *balanced* if it is equivalent to one without red edges (and bicoloured edges), and is *anti-balanced* if it is equivalent to one without blue edges (and bicoloured edges); here we view a bicoloured edge as both blue and red. We say that a signed graph is *weakly balanced* (*weakly anti-balanced*) if it is equivalent to one in which all edges are bicoloured or blue (respectively red). (Previously [1] we used the slightly awkward terms 'uni-balanced' and 'anti-uni-balanced'.)

Let C be a fixed circle with two specified points n and s. A *bi-arc graph* is a graph H such that each vertex $v \in V(H)$ can be associated with a pair of intervals N_v, S_v where N_v contains n but not s and S_v contains s but not n satisfying the following conditions: (i) N_v intersects S_w if and only if S_v intersects N_w, and (ii) N_v intersects S_w if and only if vw is not an edge of H. This class of graphs includes all interval graphs: a reflexive graph is a bi-arc graph if and only if it is an interval graph. Moreover, an irreflexive graph is a bi-arc graph if and only if it is bipartite and its complement is a circular arc graph [10].

Theorem 2 [3]. *Suppose $\widehat{H}$ is a connected signed graph without bicoloured edges. If the underlying graph H is a bi-arc graph, and $\widehat{H}$ is balanced or anti-balanced, then the list homomorphism problem for $\widehat{H}$ is polynomial-time solvable. Otherwise, the problem is NP-complete.*

Additionally, in [1] we have classified the complexity of the list homomorphism problems for signed trees. The general classification is quite technical, but we will give a simplified description in the special case of irreflexive trees $\widehat{H}$. The following concept plays an important role. Let U, D be two walks in $\widehat{H}$ of equal length. Suppose U has vertices $u = u_0, u_1, \ldots, u_k = v$, and D has vertices $u = d_0, d_1, \ldots, d_k = v$. We say that (U, D) is a *chain*, provided $uu_1, d_{k-1}v$ are unicoloured edges and $ud_1, u_{k-1}v$ are bicoloured edges, and for each i, $1 \le i \le k-2$, we have (1) both $u_i u_{i+1}$ and $d_i d_{i+1}$ are edges of $\widehat{H}$ while $d_i u_{i+1}$ is not an edge of $\widehat{H}$, or (2) both $u_i u_{i+1}$ and $d_i d_{i+1}$ are bicoloured edges of $\widehat{H}$ while $d_i u_{i+1}$ is not a bicoloured edge of $\widehat{H}$.

Theorem 3 [1]. *If a signed graph $\widehat{H}$ contains a chain, then the list homomorphism problem for $\widehat{H}$ is NP-complete.*

Figure 1 shows some important signed trees with a chain.

An *invertible pair* in an undirected graph H is a pair of vertices a, b, with two walks U, D of the same length, where U has vertices $a = u_0, u_1, \ldots, u_k = b, u_{k+1}, \ldots, u_t = a$, and D has vertices $b = d_0, d_1, \ldots, d_k = a, d_{k+1}, \ldots, d_t = b$,

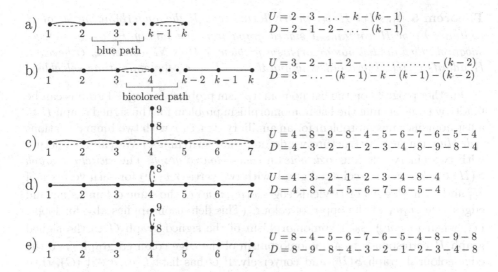

Fig. 1. The family $\mathcal{F}$ of signed graphs yielding NP-complete problems, and a chain in each. (The figure appeared first in [1].) (Color figure online)

such that for each i, $1 \leq i \leq k-2$, both $u_i u_{i+1}$ and $d_i d_{i+1}$ are edges of H, while $d_i u_{i+1}$ is not an edge of $\widehat{H}$. For simplicity we say that a signed graph has an invertible pair if its underlying graph has an invertible pair. It follows from [1,9] that we have the following observation.

Theorem 4. *If $\widehat{H}$ has an invertible pair, then the list homomorphism problem for $\widehat{H}$ is NP-complete.*

Figure 2 shows the graph F_1, with an invertible pair $1, 10$. The walks U, D begin as indicated, then continue from $7, 10$ to $7, 1$ in a similar manner, and then to $10, 1$, and similarly for the second half, from $10, 1$ to $1, 10$.

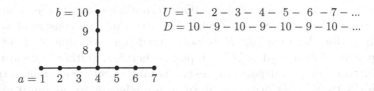

Fig. 2. The graph F_1, with an invertible pair.

The following result from [1] therefore implies that for irreflexive signed trees, the only NP-complete cases have a chain or an invertible pair. (We will make the same conclusion about the irreflexive signed graphs discussed in this paper.)

Theorem 5 [1]. *Let $\widehat{H}$ be an irreflexive tree. If the underlying graph of $\widehat{H}$ contains F_1 or $\widehat{H}$ contains a signed graph from the family $\mathcal{F}$, as an induced subgraph, then the list homomorphism problem to $\widehat{H}$ is NP-complete. Otherwise, $\widehat{H}$ admits a special min ordering and the problem is polynomial-time solvable.*

Further progress on the list homomorphism problem for signed graphs can be made by transforming the list homomorphism problem for the signed graph $\widehat{H}$ to a list homomorphism problem for an auxiliary structure with two binary relations *red, blue*. (In such a structure we do not allow switchings.) We call a structure with two binary relations *red, blue* an *edge-coloured graph*. The *switching graph* $S(\widehat{H})$ of $\widehat{H}$ is an edge-coloured graph with two vertices v_1, v_2 for each vertex v of $\widehat{H}$, and each edge vw of $\widehat{H}$ yields edges v_1w_1, v_2w_2 of the same colour as vw and edges v_1w_2, v_2w_1 of the opposite colour. (This definition applies also for loops, i.e., when $v = w$.) Each homomorphism of the signed graph $\widehat{G}$ to the signed graph $\widehat{H}$ corresponds to a homomorphism of the edge-coloured graph $\widehat{G}$ to the edge-coloured graph $S(\widehat{H})$ and conversely. If $\widehat{G}$ has lists $L(v), v \in V(G)$, then the new lists $L^+(v), v \in V(G)$, for $S(\widehat{H})$ are defined as follows: for any $x \in L(v)$ with $v \in V(G)$, we place both x_1 and x_2 in $L^+(v)$. It is easy to see that the signed graph $\widehat{G}$ has a list homomorphism to the signed graph $S(\widehat{H})$ with respect to the lists L if and only if the edge-coloured graph $\widehat{G}$ has a list homomorphism to the edge-coloured graph $S(\widehat{H})$ with respect to the lists L^+. The new lists L^+ are *symmetric sets in H^+*, meaning that for any $x \in V(H), v \in V(G)$, we have $x_1 \in L^+(v)$ if and only if we have $x_2 \in L^+(v)$. Thus we obtain the list homomorphism problem for the edge-coloured graph $S(\widehat{H})$, restricted to input instances $\widehat{G}$ with lists L that are symmetric in $S(\widehat{H})$. We shall call the corresponding vertices x_1, x_2 *mates*.

A *polymorphism* of an edge-coloured graph $\widehat{H}$ is a homomorphism f of some power $\widehat{H}^t$ to $\widehat{H}$, i.e., a function f that assigns to each ordered t-tuple $(v_1, v_2, \ldots, v_t)$ of vertices of $\widehat{H}$ a vertex $f(v_1, v_2, \ldots, v_t)$ such that two coordinate-wise tuples adjacent in blue (red) obtain images adjacent in blue (red). A polymorphism of order $t = 3$ is a *majority* if $f(v, v, w) = f(v, w, v) = f(w, v, v) = v$ for all v, w. A *Siggers polymorphism* is a polymorphism of order $t = 4$, if $f(a, r, e, a) = f(r, a, r, e)$ for all a, r, e. One formulation of the dichotomy theorem proved by Bulatov [8] and Zhuk [25] states that the constraint satisfaction problem for the template H is polynomial-time solvable if H admits a Siggers polymorphism, and is NP-complete otherwise. (Other equivalent versions refer to other useful polymorphisms, notably *weak near-unanimity polymorphisms* [8,18,25].) Majority polymorphisms are less powerful, but it is known (see [12,17]) that if H admits a majority then the constraint satisfaction problem for the template H is polynomial-time solvable. We say that a polymorphism is *conservative* if $f(v_1, v_2, \ldots, v_t)$ is always one of $v_1, v_2, \ldots, v_t$, and we say that a polymorphism of $S(\widehat{H})$ is *semi-conservative* if $f(v_1, v_2, \ldots, v_t)$ is always one of $v_1, v_2, \ldots, v_t$ or their mates.

To distinguish the two parts of a bipartite graph we speak of *black* and *white* vertices. A *min ordering* of a bipartite edge-coloured graph H is a pair $<_b, <_w$, where $<_b$ is a linear ordering of the black vertices and $<_w$ is a linear ordering of the white vertices, such that for white vertices $x <_w x'$ and black vertices $y <_b y'$, if $xy', x'y$ are both red (blue) edges in H, then xy is also a red (blue) edge in H. It is known [12] that if a bipartite graph H has a min ordering, then the list homomorphism problem for H can be solved in polynomial time. In fact, min ordering can be viewed as a polymorphism of order $t = 2$ [12]. We call a bipartite min ordering of the signed irreflexive tree $\widehat{H}$ *special* if for black vertices x, x' and white vertices y, y', if xy is bicoloured and xy' is unicoloured, then $y <_w y'$, and if xy is bicoloured and $x'y$ is unicoloured, then $x <_b x'$. In other words, the bicoloured neighbours of any vertex appear before its unicoloured neighbours.

For weakly balanced irreflexive signed graphs, [18] suggests the following.

Conjecture. *For a weakly balanced irreflexive signed graph $\widehat{H}$, the list homomorphism problem is polynomial-time solvable if $\widehat{H}$ has a special min ordering; otherwise $\widehat{H}$ contains a chain or an invertible pair and the problem is NP-complete.*

We note that in [18], authors prove that the existence of a special min ordering implies the existence of a semi-conservative majority which means that the problem is polynomial-time solvable; so to confirm their conjecture it remains only to prove the remaining cases are NP-complete.

Theorem 5 from [1] confirms the above conjecture, when $\widehat{H}$ is a signed tree.

We say that an irreflexive signed graph $\widehat{H}$ is *path-separable (cycle-separable)* if the unicoloured edges of $\widehat{H}$ form a spanning path (cycle) in the underlying graph of $\widehat{H}$. For brevity we also say a signed graph is *separable* if it is path-separable or cycle-separable. In this paper we explicitly classify the complexity of the list homomorphism problem for separable signed graphs $\widehat{H}$, see Theorems 6 and 7. The descriptions suggest that the polynomial cases are rather rare and very nicely structured.

In particular, we confirm the above conjecture in the special case of separable signed graphs. Moreover, in our results we do not assume that $\widehat{H}$ is weakly balanced.

2 Path-Separable Signed Graphs

Irreflexive signed graphs are in a sense the core of the problem. By Theorem 1, the list homomorphism problem for H is NP-complete unless the underlying graph H is bipartite. There is a natural transformation of each general problem to a problem for a bipartite irreflexive signed graph, akin to what is done for unsigned graphs in [11]; this is nicely explained in [18].

However, for bipartite H, we don't have a combinatorial classification beyond the case of trees H, except in the case $\widehat{H}$ has no bicoloured edges or loops (when Theorem 2 applies), or when $\widehat{H}$ has no unicoloured edges or loops (when the problem essentially concerns unsigned graphs and thus is solved by [11]).

Therefore we may assume that both bicoloured and unicoloured edges or loops are present. We focus in this paper on those bipartite irreflexive signed graphs $\widehat{H}$ in which the unicoloured edges form simple structures, such as paths and cycles. In this section, we consider irreflexive signed graphs in which the unicoloured edges form a spanning path.

Recall that an irreflexive signed graph $\widehat{H}$ is path-separable if the unicoloured edges of $\widehat{H}$ form a hamiltonian path P in the underlying graph H. We may assume the edges of P are all blue. In other words, all the edges of the hamiltonian path P are blue, and all the other edges of $\widehat{H}$ are bicoloured. Recall that the distinction between unicoloured and bicoloured edges is independent of switching, thus such a hamiltonian path $P = v_1 v_2 \ldots v_n$ is unique, if it exists.

We first observe that for any irreflexive signed graph $\widehat{H}$, the list homomorphism problem for $\widehat{H}$ is NP-complete if the underlying graph H contains an odd cycle, since then the s-core of $\widehat{H}$ has at least three edges. Moreover, we now show that the list homomorphism problem for $\widehat{H}$ is also NP-complete if H contains an induced cycle of length greater than four. Indeed, it suffices to prove this if H is an even cycle of length $k > 4$. If all edges of H are unicoloured, then the problem is NP-complete by Theorem 2, since an irreflexive cycle of length $k > 4$ is not a bi-arc graph. If all edges of the cycle H are bicoloured, then we can easily reduce from the previous case. If H contains both unicoloured and bicoloured edges, then $\widehat{H}$ contains an induced subgraph of type a) or b) in the family $\mathcal{F}$ in Fig. 1, and the problem is NP-complete by Theorem 3. (There are cases when the subgraphs are not induced, but the chains from the proof of Theorem 3 are still applicable.)

We further identify two additional cases of $\widehat{H}$ with NP-complete list homomorphism problems. An *alternating 4-cycle* is a 4-cycle $v_1 v_2 v_3 v_4$ in which the edges $v_1 v_2, v_3 v_4$ are bicoloured and the edges $v_2 v_3, v_4 v_1$ unicoloured. A *4-cycle pair* consists of 4-cycles $v_1 v_2 v_3 v_4$ and $v_1 v_5 v_6 v_7$, sharing the vertex v_1, in which the edges $v_1 v_2, v_1 v_5$ are bicoloured, and all other edges are unicoloured. An alternating 4-cycle has the chain $U = v_1, v_4, v_3; D = v_1, v_2, v_3$, and a 4-cycle pair has the chain $U = v_1, v_4, v_3, v_2, v_1; D = v_1, v_5, v_6, v_7, v_1$. Therefore, if a signed graph $\widehat{H}$ contains an alternating 4-cycle or a 4-cycle pair as an induced subgraph, then the list homomorphism problem for $\widehat{H}$ is NP-complete. Note that the latter chain requires only $v_2 v_6$ and $v_3 v_5$ to be non-edges. The problem remains NP-complete as long as these edges are absent; all other edges with endpoints in different 4-cycles can be present. If both $v_2 v_6$ and $v_3 v_5$ are bicoloured edges, then there is an alternating 4-cycle $v_2 v_3 v_5 v_6$. Thus we conclude that the problem is NP-complete if $\widehat{H}$ contains a 4-cycle pair as a subgraph (not necessarily induced), unless exactly one of $v_3 v_5$ or $v_2 v_6$ is a bicoloured edge.

From now on we will assume that $\widehat{H}$ is a path-separable signed graph with the unicoloured edges (all blue) forming a hamiltonian path $P = v_1, \ldots, v_n$. We will assume further that the list homomorphism problem for $\widehat{H}$ is not NP-complete, and derive information on the structure of $\widehat{H}$. In particular, the underlying graph H is bipartite and does not contain any induced cycles of length greater than 4, and $\widehat{H}$ does not contain an alternating 4-cycle or a 4-cycle pair; more generally,

$\widehat{H}$ does not contain a chain. If $\widehat{H}$ has no bicoloured edges (and hence no edges not on P), then the list homomorphism problem for $\widehat{H}$ is polynomially solvable by Theorem 2, since a path is a bi-arc graph. If there is a bicoloured edge in $\widehat{H}$, then we may assume there is an edge $v_i v_{i+3}$, otherwise there is an induced cycle of length greater than 4.

A *block* in a path-separable signed graph $\widehat{H}$ is a subpath $v_i v_{i+1} v_{i+2} v_{i+3}$ of P, with the bicoloured edge $v_i v_{i+3}$. The previous paragraph concluded that $\widehat{H}$ must contain a block. Note that if $v_i v_{i+1} v_{i+2} v_{i+3}$ is a block, then $v_{i+1} v_{i+2} v_{i+3} v_{i+4}$ cannot be a block: in fact, $v_{i+1} v_{i+4}$ cannot be a bicoloured edge, otherwise $\widehat{H}$ would contain an alternating 4-cycle. However, $v_{i+2} v_{i+3} v_{i+4} v_{i+5}$ can again be a block, and so can $v_{i+4} v_{i+5} v_{i+6} v_{i+7}$, etc. If both $v_i v_{i+1} v_{i+2} v_{i+3}$ and $v_{i+2} v_{i+3} v_{i+4} v_{i+5}$ are blocks then $v_i v_{i+5}$ must be a bicoloured edge, otherwise $v_i v_{i+3} v_{i+2} v_{i+5}$ would induce a signed graph of type a) in family $\mathcal{F}$ from Fig. 1. A *segment* in $\widehat{H}$ is a maximal subpath $v_i v_{i+1} \ldots v_{i+2j+1}$ of P with $j \geq 1$ that has all bicoloured edges $v_{i+e} v_{i+e+3}$, where e is even, $0 \leq e \leq 2j - 2$. (A *maximal* subpath is not properly contained in another such subpath.) Thus each subpath $v_{i+e} v_{i+e+1} v_{i+e+2} v_{i+e+3}$ of the segment is a block, and the segment is a consecutively intersecting sequence of blocks; note that it can consist of just one block. Two segments can touch as the second and third segment in Fig. 3, or leave a gap as the first and second segment in the same figure.

In a segment $v_i v_{i+1} \ldots v_{i+2j+1}$ we call each vertex v_{i+e} with $0 \leq e \leq 2j - 2$ a *forward source*, and each vertex v_{i+o} with $3 \leq o \leq 2j + 1$ a *backward source*. Thus forward sources are the beginning vertices of blocks in the segment, and the backward sources are the ends of blocks in the segment. If $u < b$, we say the edge $v_a v_b$ is a *forward edge* from v_a and a *backward edge* from v_b. In this terminology, each forward source has a forward edge to its corresponding backward source. Because of the absence of a signed graph of type a) in family $\mathcal{F}$ from Fig. 1, we can in fact conclude, by the same argument as in the previous paragraph, that each forward source in a segment has forward edges to all backward sources in the segment.

We say that a segment $v_i v_{i+1} \ldots v_{i+2j+1}$ is *right-leaning* if $v_{i+e} v_{i+e+o}$ is a bicoloured edge for all e is even, $0 \leq e \leq 2j - 2$, and *all* odd $o \geq 3$; and we say it is *left-leaning* if $v_{i+2j+1-e} v_{i+2j+1-e-o}$ is a bicoloured edge for all e even, $0 \leq e \leq 2j - 2$ and all odd $o \geq 3$. Thus a in a right-leaning segment each forward source has *all* possible forward edges (that is, all edges to vertices of opposite colour in the bipartition, including vertices with subscripts greater than $i + 2j + 1$). The concepts of left-leaning segments, backward sources and backward edges are defined similarly.

We say that a path-separable signed graph $\widehat{H}$ is *right-segmented* if all segments are right-leaning, and there are no edges other than those mandated by this fact. In other words, each forward source has all possible forward edges, and each vertex which is not a forward source has no forward edges. Similarly, we say that a path-separable signed graph $\widehat{H}$ is *left-segmented* if all segments are left-leaning, and there are no edges other than those mandated by this fact. In other words, each backward source has all possible backward edges, and each

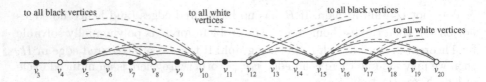

Fig. 3. An example of a left-right-segmented signed graph. The additional bicoloured edges from all white vertices before v_{12} to all black vertices after v_{15} are not shown. (The figure appeared first in [1].) (Color figure online)

vertex which is not a backward source has no backward edges. Finally, $\widehat{H}$ is left-right-segmented if there is a unique segment $v_i v_{i+1} \ldots v_{i+2j+1}$ that is both left-leaning and right-leaning, all segments preceding it are left-leaning, all segments following it are right-leaning, and moreover there are *additional* bicoloured edges $v_{i-e} v_{i+2j+o}$ for all even $e \geq 2$ and all odd $o \geq 3$, but no other edges. In other words, vertices $v_1, v_2, \ldots, v_{i+2j+1}$ induce a left-segmented graph, vertices $v_i, v_{i+1}, \ldots, v_n$ induce a right-segmented graph, and in addition to the edges this requires there are all the edges joining v_{i-e} from $v_1, \ldots, v_{i-1}$ to v_{i+o} from $v_{i+2j+2}, \ldots, v_n$, with even e and odd o. A *segmented graph* is a path-separable signed graph that is right-segmented or left-segmented or left-right-segmented.

See Fig. 3 there are three segments, the left-leaning segment $v_5 v_6 v_7 v_8 v_9 v_{10}$, the left- and right-leaning segment $v_{12} v_{13} v_{14} v_{15}$, and the right-leaning segment $v_{15} v_{16} v_{17} v_{18} v_{19} v_{20}$. Thus this is a left-right-segmented signed graph.

Theorem 6. *Let $\widehat{H}$ be a path-separable signed graph. Then the list homomorphism problem for $\widehat{H}$ is polynomial-time solvable if $\widehat{H}$ is switching equivalent to a segmented signed graph $\widehat{H}$. Otherwise, the problem is NP-complete.*

The NP-completeness is proved in the journal version. To show that the problem is polynomial when $\widehat{H}$ is a segmented signed graph, we use the result from [18] which asserts that the existence of a special min ordering ensures the existence of a polynomial-time algorithm. (We provide an alternate proof in [2].)

We now describe a special min ordering of the vertices for the case of a right-segmented signed graph. Consider two vertices v_x and v_y with even subscripts $x < y$, and each with a forward bicoloured edge. Then all forward neighbours of v_y are also forward neighbours of v_x, and all backward neighbours of v_x are also backward neighbours of v_y. Vertices v_z with no forward bicoloured edges have backward edges to all vertices with forward bicoloured edges from vertices v_t with t odd and $t < z$. We now order the vertices with even subscripts as follows: first we take vertices with forward bicoloured edges in increasing order of subscripts, then we take the remaining vertices in the decreasing order of subscripts. The same ordering is applied on the vertices with odd subscripts. It now follows from our observations that this is a min ordering, and for every vertex the bicoloured edges come before the unicoloured edges.

For left-segmented graphs the ordering is similar; for left-right-segmented graphs the ordering is described in the journal version.

3   Cycle-Separable Signed Graphs

As an application of Theorem 6, we now consider irreflexive signed graphs in which the unicoloured edges form a spanning cycle. We say that an irreflexive signed graph $\widehat{H}$ is *cycle-separable* if the unicoloured edges of $\widehat{H}$ form a hamiltonian cycle C in the underlying graph H. In other words, we have a hamiltonian cycle C whose edges are all unicoloured, and all the other edges of $\widehat{H}$ are bicoloured. In contrast to the path-separable signed graphs, we cannot assume the edges of C are all blue, see below.

We first introduce three cycle-separable signed graphs for which the list homomorphism problem will turn out to be polynomial-time solvable. The signed graph $\widehat{H_0}$ is the 4-cycle with all edges unicoloured blue. The signed graph $\widehat{H_1}$ consists of a blue path $b = t_0, t_1, t_2, t_3 = w$, a red path b, s_1, s_2, w, together with a bicoloured edge bw. The signed graph $\widehat{H_\ell}$ consists of a blue path b, s_1, s_2, w, a blue path $b = t_0, t_1, t_2, \ldots, t_\ell = w$ (with $\ell \geq 3$ odd), and all bicoloured edges $t_i t_j$ with even i and odd $j, j > i + 1$. (Note that this includes the edge bw.) These three cycle separable signed graphs are illustrated in Fig. 4. Note that if the subscript ℓ is greater than 0, then it is odd. Moreover, both H_1 and H_3 have 6 vertices and differ only in the colours of the unicoloured edges forming the hamiltonian cycle C: H_1 has the cycle C unbalanced, and H_3 has the cycle C balanced.

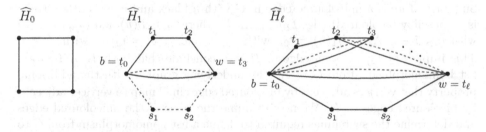

Fig. 4. The cycle-separable signed graphs $\widehat{H_0}$, $\widehat{H_1}$, and $\widehat{H_\ell}$ with $\ell \geq 3$ odd.

Theorem 7. *Let $\widehat{H}$ be a cycle-separable signed graph. Then the the list homomorphism problem for $\widehat{H}$ is polynomial-time solvable if $\widehat{H}$ is switching equivalent to $\widehat{H_0}$, or to $\widehat{H_1}$, or to $\widehat{H_\ell}$ for some odd $\ell \geq 3$. Otherwise, the problem is NP-complete.*

The NP-complete cases are again found in the journal version. Here we show that the list homomorphism problem for $\widehat{H}$ can be solved in polynomial time for all remaining cycle-separable signed graphs.

If $\widehat{H}$ is switching equivalent to $\widehat{H_0}$, then the list homomorphism problem for $\widehat{H}$ is polynomial-time solvable by Theorem 2.

Let $\widehat{G}$ together with lists L be an instance of LIST-S-HOM($\widehat{H}$). We may assume G is connected and bipartite. We will call the vertices of parts of bipartition in $\widehat{G}$ black and white as well. First, we try mapping the black vertices of $\widehat{G}$ to the black vertices of $\widehat{H}$. If that fails, we try mapping the white vertices of $\widehat{G}$ to the black vertices of $\widehat{H}$. In the former case we remove all white (respectively black) vertices from the lists of the black (respectively white) vertices in $\widehat{G}$. The latter case is analogous.

First, we preform the arc consistency procedure (cf. [9]) and also the arc consistency procedure for bicoloured edges. If there is a vertex with empty list after this step, then no suitable list homomorphism exists. Otherwise, we define two mappings f_1 and f_2 as follows.

- $f_1(v) = \min\{t_i \ : \ t_i \in L(v)\}$,
- $f_2(v) = \min\{s_i \ : \ s_i \in L(v)\}$.

(Observe when $L(v)$ consists of black vertices $f_j(v)$ is the vertex, t_i or s_i for $j = 1$ or $j = 2$, with the smallest index, and conversely for the case $L(v)$ consists of white vertices $f_j(v)$ is the vertex with the largest index.) Let uv be a bicoloured edge of $\widehat{G}$ with u black and v white. Then by arc consistency there is a bicoloured edge between a vertex from $L(u)$ and a vertex from $L(v)$. By the labelling of $\widehat{H}$, it has the form $t_{2i}t_{2j+3}$, where $i \leq j$. By our observation, $f_1(u) = t_{2i'}$ where $i' \leq i$. Similarly, $f_1(v) = t_{2j'+3}$ where $j' \geq j$. This implies $i' \leq j'$ and consequently, $f_1(u)f_2(v)$ is a bicoloured edge. A similar argument applies for f_2. Similarly, if uv is a unicoloured edge in $\widehat{G}$ with u black and v white, then there is a (possibly bicoloured) edge $t_{2i}t_{2j+1}$ in $\widehat{H}$ where $t_{2i} \in L(u)$ and $t_{2j+1} \in L(v)$ with $i \leq j - 1$. Again $f_1(u) = t_{2i'}$ with $i' \leq i$ and $f_1(v) = t_{2j'+1}$ with $j' \geq j$. This implies $t_{2i'}t_{2j'+1}$ is an edge of $\widehat{H}$. We conclude that both f_1 and f_2 are list homomorphisms from G to H (the underlying graphs) with the additional property that vertices adjacent by bicoloured edges in $\widehat{G}$ map to vertices adjacent by bicoloured edges in $\widehat{H}$. We now examine the signs of the unicoloured edges and determine the switchings required to define a list homomorphism from $\widehat{G}$ to $\widehat{H}$. We make the following key observation. Due to the ordering on the vertices if, for example, $f_1(u)f_1(v)$ is a unicoloured edge in $\widehat{H}$ (again u is black and v is white), then under no list homomorphism of $\widehat{G}$ to $\widehat{H}$ (with lists L) does uv map to a bicoloured edge. (If such a mapping did exist, then the bicoloured edge would remain as a possible image during the consistency check, and a bicoloured edge would have end points occurring first in the ordering $<$.)

If $b \in L(v)$ for some black vertex v, then $f_1(v) = f_2(v) = b$. Analogously, if $w \in L(v)$ for some white vertex v, then $f_1(v) = f_2(v) = w$. That is, any vertex that can map to b (respectively w) will be mapped to b (respectively w). Moreover, when examining the resigning of vertices (below), if there is no resigning that works when v maps to b, then there is no homomorphism at all, as b dominates all white vertices (in $\widehat{H}$). Similarly w dominates all black vertices.

Consequently, we can partition the vertices of $\widehat{G}$ into those mapped to b or w under f_1 and under f_2 and those vertices that can only map to interior vertices of the two segments, i.e. to $t_1, \ldots, t_{\ell-1}$ and s_1, s_2. The vertices in the pre-images $f_1^{-1}(b) = f_2^{-1}(b)$ and $f_1^{-1}(w) = f_2^{-1}(w)$ are called *boundary vertices*. Removing the boundary vertices from $\widehat{G}$ leaves a union of components. Consider such a component K. The subgraph of $\widehat{G}$ induced by K is called a *region* (similar to [1]). For each region, either its vertices all map to s-vertices or all to t-vertices.

We now examine how to test if there is a switching of the boundary vertices of $\widehat{G}$ so that each region maps to $\widehat{H}$.

First suppose that $\widehat{H}$ is switching equivalent to $\widehat{H}_\ell$ with odd $\ell \geq 3$. Let K be some region of $\widehat{G}$. If the lists of vertices of K contain only s-vertices or only t-vertices, then there is no choice and we will use mappings f_2 or f_1, respectively. Now suppose that the lists of vertices of K contains both s-vertices and t-vertices. We claim we can use f_1 to map K to $\widehat{H}$. If there is any list homomorphism $\widehat{G} \to \widehat{H}$ under which K maps to s-vertices, then there is a switching of $\widehat{G}$ such that K, together with its boundary vertices, induces a subgraph having only blue edges. (Recall b, s_1, s_2, w is a blue path.) The mapping f_1 restricted to K and its boundary vertices is a homomorphism of G to H. As each edge in the segment containing the t-vertices is at least blue, f_1 is a homomorphism of the induced subgraph to $\widehat{H}$. Thus for any region that has both s-vertices and t-vertices in its lists after the consistency checks, we may assume if is mapped $\widehat{H}$ under f_1. The remaining regions must map to s-vertices and we may assume they are mapped using f_2. Moreover, by our key observation above, the edges mapping to unicoloured edges under these mappings must map to unicoloured edges under any mapping. In particular, the discovery of a cycle consisting of unicoloured edges in $\mathcal{G}$ whose sign does not agree with the sign of its image under our use of f_1 and f_2 certifies that $\mathcal{G}$ is a no-instance of the problem.

It now remains to determine the switching of boundary vertices to ensure the signs of all unicoloured edges are positive. We describe this in detail in the journal version.

4 Conclusions

It seems difficult to give a full combinatorial classification of the complexity of list homomorphism problems for general signed graphs. For irreflexive signed graphs, which are in a sense the core of the problem, there is a conjectured classification in [18]. We have obtained a full dichotomy classification in the special case of separable irreflexive signed graphs. The classification confirms the dichotomy conjecture of [18] for this case, and also confirms that the only polynomial cases enjoy a special min ordering and the only NP-complete cases have chains or invertible pairs, as also conjectured in [18].

Acknowledgements. The first author received funding from the European Union's Horizon 2020 project H2020-MSCA-RISE-2018: Research and Innovation Staff Exchange. The second author was supported by his NSERC Canada Discovery Grant. The fourth and fifth author were also partially supported by the fourth author's NSERC Canada Discovery Grant. The first and the fifth author were also supported by the Charles University Grant Agency project 1198419 and by SVV–2020–260578.

References

1. Bok, J., Brewster, R., Feder, T., Hell, P., Jedličková, N.: List homomorphism problems for signed graphs. In Esparza, J., Král, D. (eds.) 45th International Symposium on Mathematical Foundations of Computer Science (MFCS 2020), volume 170 of Leibniz International Proceedings in Informatics (LIPIcs), Dagstuhl, Germany, pp. 20:1–20:14. Schloss Dagstuhl-Leibniz-Zentrum für Informatik (2020). https://doi.org/10.4230/LIPIcs.MFCS.2020.20, https://drops.dagstuhl.de/opus/volltexte/2020/12688
2. Bok, J., Brewster, R., Feder, T., Hell, P., Jedličková, N.: List homomorphism problems for signed graphs. arXiv:2005.05547 (2021)
3. Bok, J., Brewster, R.C., Hell, P., Jedličková, N.: List homomorphisms of signed graphs. In: Bordeaux Graph Workshop, pp. 81–84 (2019)
4. Brewster, R.C., Foucaud, F., Hell, P., Naserasr, R.: The complexity of signed graph and edge-coloured graph homomorphisms. Discrete Math. **340**(2), 223–235 (2017)
5. Brewster, R.C., Graves, T.: Edge-switching homomorphisms of edge-coloured graphs. Discrete Math. **309**(18), 5540–5546 (2009)
6. Brewster, R.C., Siggers, M.: A complexity dichotomy for signed h-colouring. Discrete Math. **341**(10), 2768–2773 (2018)
7. Bulatov, A.A.: Complexity of conservative constraint satisfaction problems. ACM Trans. Comput. Logic (TOCL) **12**(4), 1–66 (2011)
8. Bulatov, A.A.: A dichotomy theorem for nonuniform CSPs. In: 58th Annual IEEE Symposium on Foundations of Computer Science–FOCS 2017, Los Alamitos, CA, pp. 319–330. IEEE Computer Society (2017)
9. Feder, T., Hell, P.: List homomorphisms to reflexive graphs. J. Comb. Theory Ser. B **72**(2), 236–250 (1998)
10. Feder, T., Hell, P., Huang, J.: List homomorphisms and circular arc graphs. Combinatorica **19**(4), 487–505 (1999)
11. Feder, T., Hell, P., Huang, J.: Bi-arc graphs and the complexity of list homomorphisms. J. Graph Theory **42**(1), 61–80 (2003)
12. Feder, T., Vardi, M.Y.: The computational structure of monotone monadic SNP and constraint satisfaction: a study through Datalog and group theory. In: STOC, pp. 612–622 (1993)
13. Guenin, B.: Packing odd circuit covers: a conjecture. Manuscript (2005)
14. Harary, F.: On the notion of balance of a signed graph. Michigan Math. J. **2**, 143–146 (1955). 1953/54
15. Harary, F., Kabell, J.A.: A simple algorithm to detect balance in signed graphs. Math. Social Sci. **1**(1), 131–136 (1980/81)
16. Hell, P., Nešetřil, J.: On the complexity of H-coloring. J. Comb. Theory Ser. B **48**(1), 92–110 (1990)
17. Jeavons, P.: On the algebraic structure of combinatorial problems. Theor. Comput. Sci. **200**(1–2), 185–204 (1998)
18. Kim, H., Siggers, M.: Towards a dichotomy for the switch list homomorphism problem for signed graphs. arXiv:2104.077646 (2021)
19. Naserasr, R., Rollová, E., Sopena, É.: Homomorphisms of signed graphs. J. Graph Theory **79**(3), 178–212 (2015)
20. Zaslavsky, T.: Characterizations of signed graphs. J. Graph Theory **5**(4), 401–406 (1981)
21. Zaslavsky, T.: Signed graph coloring. Discrete Math. **39**(2), 215–228 (1982)
22. Zaslavsky, T.: Signed graphs. Discrete Appl. Math. **4**(1), 47–74 (1982)

23. Zaslavsky, T.: Is there a matroid theory of signed graph embedding? Ars Comb.
    45, 129–141 (1997)
24. Zaslavsky, T.: A mathematical bibliography of signed and gain graphs and allied
    areas. Electron. J. Combin. **5**, 124 (1998). Dynamic Surveys 8. Manuscript prepared
    with Marge Pratt
25. Zhuk, D.: A proof of CSP dichotomy conjecture. In: 58th Annual IEEE Symposium
    on Foundations of Computer Science–FOCS 2017, Los Alamitos, CA, pp. 331–342.
    IEEE Computer Society (2017)

Some Position Problems for Graphs

James Tuite[1(✉)], Elias John Thomas[2], and Ullas Chandran S. V.[3]

[1] Department of Mathematics and Statistics, Open University,
Walton Hall, Milton Keynes, UK
`james.tuite@open.ac.uk`
[2] Department of Mathematics, Mar Ivanios College, University of Kerala,
Thiruvananthapuram 695015, Kerala, India
[3] Department of Mathematics, Mahatma Gandhi College, University of Kerala,
Thiruvananthapuram 695004, Kerala, India

Abstract. The general position problem for graphs stems from a puzzle of Dudeney and the general position problem from discrete geometry. The general position number of a graph G is the size of the largest set of vertices S such that no geodesic of G contains more than two elements of S. The monophonic position number of a graph is defined similarly, but with 'induced path' in place of 'geodesic'. In this abstract we discuss the smallest possible order of a graph with given general and monophonic position numbers, determine the asymptotic order of the largest size of a graph with given order and position numbers and finally determine the possible diameters of a graph with given order and monophonic position number.

Keywords: General position · Monophonic position · Turán problems · Size · Diameter · Induced path

1 Introduction

In this paper all graphs will be taken to be simple and undirected. The order of the graph G will be denoted by n and its size by m. The *clique number* $\omega(G)$ of a graph G is the order of the largest clique in G. An *independent union of cliques* in a graph G is an induced subgraph H of G such that every component of H is a clique. The *independent clique number* $\alpha^\omega(G)$ is the order of a largest independent union of cliques in G. The distance $d(u, v)$ between two vertices u and v in a graph G is the length of the shortest path in G from u to v and a shortest path is called a *geodesic*. An *induced* or *monophonic* path is a path without any chords. The *join* of two graphs G and H is the graph $G \vee H$ obtained from the disjoint union of G and H by joining every vertex of G to every vertex of H.

The general position problem for graphs can be traced back to one of the many puzzles of Dudeney [7]. This problem was introduced in the context of

The research of the first author was supported by an LMS Early Career Fellowship, Project ECF-2021-27. The second author was supported by a Junior Research Fellowship from the University of Kerala.

© Springer Nature Switzerland AG 2022
N. Balachandran and R. Inkulu (Eds.): CALDAM 2022, LNCS 13179, pp. 36–47, 2022.
https://doi.org/10.1007/978-3-030-95018-7_4

graph theory independently in [4,13]. A set S of vertices of a graph G is in *general position* if no geodesic of G contains more than two points of S; in this case S is a *general position set*, or a *gp-set*. The general position problem asks for the largest possible size of a gp-set for a given graph G; this number is denoted by $\mathrm{gp}(G)$. A characterisation of the structure of a gp-set is derived in [2]. Other recent papers on the general position problem include [8–10,14].

In [15] the authors introduced the *monophonic position number* of a graph, or *mp-number* for short. A set S of vertices in a graph G is in *monophonic position* if there is no monophonic path in G that contains more than two elements of S. A set satisfying this condition is called a *monophonic position set* or simply an *mp-set*. The size of a largest mp-set in a graph G is the mp-number $\mathrm{mp}(G)$ of G. The mp- and gp-numbers of trees have a particularly simple form.

Lemma 1 [4,15]. *For any tree T with $\ell(T)$ leaves we have $\mathrm{mp}(T) = \mathrm{gp}(T) = \ell(T)$.*

In Sect. 2 we discuss the problem of finding the smallest possible order of a graph with given mp- and gp-numbers. In Sect. 3 we introduce the Turán-type problem of the largest possible size of a graph with given order and mp-number. We solve this problem asymptotically and present exact values for mp-number two, along with a classification of the extremal graphs. Finally in Sect. 4 we consider the problem of the possible diameters of graphs with given order and mp-number.

2   The Smallest Graph with Given mp- and gp-Numbers

In a previous paper the authors characterised the values of $a, b \in \mathbb{N}$ such that there exists a graph with mp-number a and gp-number b.

Theorem 1 [15]. *For all $a, b \in \mathbb{N}$ there exists a graph with mp-number a and gp-number b if and only if $2 \leq a \leq b$ or $a = b = 1$.*

This raises the question: for $2 \leq a \leq b$ what are the possible values of the order of a graph with mp-number a and gp-number b? In particular, what is the smallest such order? We give strong bounds on this order and solve the problem for a certain range of a and b. For all $a, b \in \mathbb{N}$ such that $2 \leq a \leq b$ the order of the smallest graph G with $\mathrm{mp}(G) = a$ and $\mathrm{gp}(G) = b$ will be denoted by $\mu(a, b)$. Trivially for $a \geq 2$ we have $\mu(a, a) = a$. The following lower bound on $\mu(a, b)$ for $a < b$ can be derived by considering the points lying on a longest induced path that is not a geodesic.

Lemma 2. *For $2 \leq a < b$ we have $\mu(a, b) \geq b + 2$.*

For $r \geq 3$ we define the *pagoda graph* $\mathrm{Pag}(r)$ as follows. The vertex set of $\mathrm{Pag}(r)$ consists of three sets $A = \{a_1, a_2, \ldots, a_r\}$, $B = \{b_1, \ldots, b_r\}$, $C = \{c_1, \ldots, c_r\}$ of size r and an additional vertex x. For $1 \leq i \leq r$ we set b_i to be adjacent to a_j and c_j for $j \neq i$ and also add an edge from x to every vertex of C. $\mathrm{Pag}(4)$ is illustrated in Fig. 1.

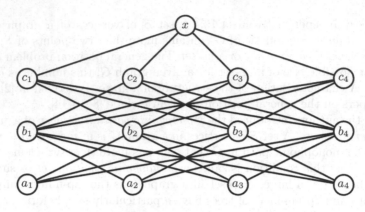

Fig. 1. Pag(4)

Lemma 3. *For $r \geq 3$ we have* $\mathrm{mp}(\mathrm{Pag}(r)) = 2$ *and* $\mathrm{gp}(\mathrm{Pag}(r)) = 2r$.

Proof. The order of $\mathrm{Pag}(r)$ is $n = 3r + 1$. We now show that for $r \geq 3$ we have $\mathrm{mp}(\mathrm{Pag}(r)) = 2$ and $\mathrm{gp}(\mathrm{Pag}(r)) = 2r$. Trivially any two vertices constitute an mp-set and $A \cup C$ is a gp-set of size $2r$. It therefore suffices to show that $\mathrm{mp}(\mathrm{Pag}(r)) \leq 2$ and $\mathrm{gp}(\mathrm{Pag}(r)) \leq 2r$.

For a contradiction, let M be an mp-set in $\mathrm{Pag}(r)$ with size ≥ 3. For $1 \leq i, j \leq r$ and $i \neq j$ let $P_{i,j}$ be the monophonic path $a_i, b_j, c_i, x, c_j, b_i, a_j$. The existence of this path shows that if $x \in M$, then M cannot contain two other vertices of $\mathrm{Pag}(r)$, so we can assume that $x \notin M$. Suppose that M contains two vertices from the same 'layer' A, B or C. For the sake of argument say $a_1, a_2 \in M$; the other cases are similar. The path $P_{1,2}$ shows that $b_1, b_2, c_1, c_2 \notin M$. If another element of A, say a_3, belonged to M, then we would have the monophonic path a_1, b_2, a_3, b_1, a_2, a contradiction. For $3 \leq i \leq r$ the path a_1, b_i, a_2 is trivially monophonic, so $M \cap B = \varnothing$. It follows that there must be a point $c_i \in M$ for some $3 \leq i \leq r$. However a_1, b_2, c_i, b_1, a_2 is a monophonic path, another contradiction. As M cannot contain ≥ 3 points of $\mathrm{Pag}(r)$ we obtain the necessary inequality.

Now assume that K is any gp-set in $\mathrm{Pag}(r)$ with size $\geq 2r$. For $1 \leq i, j, k \leq r$, $j \notin \{i, k\}$, let $Q_{i,j,k}$ be the geodesic a_i, b_j, c_k, x. Suppose that $x \in K$. If also $K \cap C \neq \varnothing$, say $c_1 \in K$, then as c_1, x, c_i is a geodesic for $2 \leq i \leq r$, it follows that $K \cap \{c_2, \ldots, c_r\} = \varnothing$. Also, letting $j \notin \{1, i\}$ in the path $Q_{i,j,1}$ shows that $K \cap A = K \cap (B - \{b_1\}) = \varnothing$, so that we would have $|K| \leq 3 < 2r$. Hence $K \cap C = \varnothing$. Furthermore if some b_j lies in K, then for $1 \leq i, k \leq r$ and $j \notin \{i, k\}$ the geodesic $Q_{i,j,k}$ contains x, b_j and a_i, so that we would have $K \subseteq B \cup \{a_i, x\}$ and $|K| \leq r + 2 < 2r$. Therefore $K \subseteq A \cup \{x\}$ and $|K| \leq r + 1 < 2r$. Therefore x is not contained in any gp-set of $\mathrm{Pag}(r)$ of size $\geq 2r$.

Suppose now that $K \cap B \neq \varnothing$, say $b_1 \in K$. For $2 \leq i, k \leq r$ the existence of the geodesic $Q_{i,1,k}$ shows that K cannot intersect both $A - \{a_1\}$ and $C - \{c_1\}$. Therefore if $|K| \geq 2r + 1$, K must either have the form $A \cup B \cup \{c_1\}$ or $\{a_1\} \cup B \cup C$; however, a_1, b_2, c_1 is a geodesic that contains three points from both of these sets. It follows that $|K| \leq 2r$. Furthermore, if $|K \cap B| \geq 2$, say $b_2 \in K$, then

$K \cap ((A - \{a_1, a_2\}) \cup (C - \{c_1, c_2\})) = \varnothing$ and also K cannot contain both a_i and c_i for $i = 1, 2$, so that $|K|$ would be bounded above by $r + 2$, which is strictly less than $2r$, whereas if K contains a unique vertex of B, then again $|K| \leq r + 2$; therefore $A \cup C$ is the unique gp-set in $\mathrm{Pag}(r)$ with size $2r$.

In a similar fashion it can be shown that the graphs $\mathrm{Pag}'(r) = \mathrm{Pag}(r) - \{a_r\}$ have order $3r$, mp-number 2 and gp-number $2r - 1$ for $r \geq 3$.

We now define a second family of graphs. We need the following result on the position numbers of the join of graphs.

Lemma 4 [15]. *The mp- and gp-numbers of the join $G \vee H$ of graphs G and H satisfies*

$$\mathrm{mp}(G \vee H) = \max\{\omega(G) + \omega(H), \mathrm{mp}(G), \mathrm{mp}(H)\}$$

and

$$\mathrm{gp}(G \vee H) = \max\{\omega(G) + \omega(H), \alpha^\omega(G), \alpha^\omega(H)\}.$$

It follows from Lemmas 1 and 4 that if T is any tree with $\ell(T) \geq 3$ leaves, then $\mathrm{mp}(K_1 \vee T) = \ell(T)$, whilst $\mathrm{gp}(K_1 \vee T) = \alpha^\omega(T)$. If T is a starlike tree with r branches of length one and s branches of length two, then $K_1 \vee T$ has order $r + 2s + 2$, mp-number $r + s$ and gp-number $r + 2s$; the gp-number of this graph matches the lower bound in Lemma 2. It follows that if $3 \leq a < b$ and $\frac{b}{2} \leq a$, then $\mu(a, b) = b + 2$; furthermore for this range there exists a graph with mp-number a, gp-number b and order n if and only if $n \geq b + 2$.

More generally, if we allow one branch of the starlike tree to have length longer than two, then our constructions yield the following upper bounds.

Theorem 2. $- \mu(2, 3) = 5$ *and for $b \geq 4$ we have $\mu(2, b) \leq \lceil \frac{3b}{2} \rceil + 1$, with equality for $4 \leq b \leq 8$.*
$- $ *For $3 \leq a < b$ and $\frac{b}{2} \leq a$ we have $\mu(a, b) = b + 2$.*
$- $ *For $3 \leq a < \frac{b}{2}$ we have $\mu(a, b) < b - a + 2 + \lceil \frac{b}{2} \rceil$.*

We conjecture the bounds in Theorem 2 to be best possible. This has been confirmed by computation for most pairs (a, b) between 2 and 11 [11].

3   The Largest Size of Graphs with Given Order and Position Numbers

A Turán-type problem asks for the largest possible size of a graph with order n that contains no subgraph isomorphic to a graph from a family $\mathcal{F}$ of forbidden subgraphs. The first such result was proved by Mantel, who showed in [12] that the largest possible size of a triangle-free graph with order n is $\lfloor \frac{n^2}{4} \rfloor$, the unique extremal graph being the complete bipartite graph $K_{\lfloor \frac{n}{2} \rfloor, \lceil \frac{n}{2} \rceil}$. This result was generalised by Turán as follows.

Theorem 3 [16]. *The number of edges of a K_{a+1}-free graph H is at most*

$$\binom{n-r}{2} + (a-1)\binom{r+1}{2},$$

where $r = \lfloor \frac{n}{a} \rfloor$. Equality holds if and only if H is isomorphic to the Turán graph $T_{n,a}$, which is the complete a-partite graph with every partite set of size $\lfloor \frac{n}{a} \rfloor$ or $\lceil \frac{n}{a} \rceil$.

We will denote the size of the Turán graph $T_{n,a}$ by $t_{n,a}$. We now discuss some Turán-type problems for the general and monophonic position numbers. For $a \geq 2$ and $n \geq a$ we define $\mathrm{mex}(n; a)$ (respectively $\mathrm{gex}(n; a)$) to be the largest possible size of a graph with order n and mp-number (resp. gp-number) a. Both of these numbers are defined for $n \geq a$ by Theorem 1. The only graphs with gp-number two are the cycle C_4 and paths, so for $n \geq 5$ we have $\mathrm{gex}(n; 2) = n-1$. We can derive a quadratic upper bound for $\mathrm{mex}(n; a)$ by an elementary application of Turán's Theorem. To reduce the bound slightly we will need a lemma on the mp-numbers of complete multipartite graphs that extends the result on complete bipartite graphs from [15].

Lemma 5. *For integers $r_1 \geq r_2 \geq \cdots \geq r_t$ the mp-number and gp-number of the complete multipartite graph $K_{r_1,r_2,\ldots,r_t}$ are given by*

$$\mathrm{gp}(K_{r_1,r_2,\ldots,r_t}) = \mathrm{mp}(K_{r_1,r_2,\ldots,r_t}) = \max\{r_1, t\}.$$

Proof. Let the partite sets of $K_{r_1,r_2,\ldots,r_t}$ be $W_1, W_2, \ldots, W_t$ and let M be a maximum mp-set of $K_{r_1,r_2,\ldots,r_t}$. Suppose that M contains two vertices u_1, u_2 in the same partite set W. Then M cannot contain any vertex v in any other partite set, for u_1, v, u_2 is a monophonic path. Hence in this case $|M| \leq |W| \leq r_1$. If M contains at most one vertex from every partite set then $|M| \leq t$.

For the converse, observe that $K_{r_1,r_2,\ldots,r_t}$ contains a clique of size t, so that $|M| \geq t$. Each partite set is also an mp-set, so that $|M| \geq r_1$. The proof for gp-sets is identical.

Lemma 6. *For $a \leq n \leq a^2$*

$$\mathrm{mex}(n; a) = t_{n,a},$$

but for $n \geq a^2+1$ we have the strict inequalities $\mathrm{mex}(n; a) < t_{n,a}$ and $\mathrm{gex}(n; a) < t_{n,a}$.

Proof. Any clique is in monophonic position. Thus if $\mathrm{mp}(G) = a$, then G is K_{a+1}-free and the conclusion follows from Turán's Theorem and Lemma 5.

We now show that the upper bound given in Lemma 6 is asymptotically tight.

Theorem 4. *For $a \geq 2$ and $n \geq a^2 + 1$ we have*

$$t_{n,a} - \lfloor \frac{n}{a} \rfloor \binom{a}{2} \leq \mathrm{mex}(n; a) \leq t_{n,a} - 1.$$

Proof. Take the Turán graph $T_{n,a}$ and label the partite sets $T_1, T_2, \ldots, T_a$, where $|T_i| = \lceil \frac{n}{a} \rceil$ for $1 \leq i \leq s$ and $|T_i| = \lfloor \frac{n}{a} \rfloor$ for $s + 1 \leq i \leq a$, where $s = n - \lfloor \frac{n}{a} \rfloor a$. For $1 \leq i \leq a$ we denote the vertices of T_i by u_{ij}, where $1 \leq j \leq \lceil \frac{n}{a} \rceil$ if $i \leq s$ and $1 \leq j \leq \lfloor \frac{n}{a} \rfloor$ if $s + 1 \leq i \leq a$.

For each j in the range $1 \leq j \leq \lfloor \frac{n}{a} \rfloor$ delete the edges of the clique of order a on the vertices u_{ij}, $1 \leq i \leq a$, from $T_{n,a}$. This yields a new graph $T^*_{n,a}$ with size $t_{n,a} - \lfloor \frac{n}{a} \rfloor \binom{a}{2}$.

Considering the vertices u_{ii} for $1 \leq i \leq a$, we see that $T^*_{n,a}$ contains a clique of size a, so certainly $\mathrm{mp}(T^*_{n,a}) \geq a$. For the converse, let M be a largest mp-set of $T^*_{n,a}$. Suppose that M contains two vertices u_{ij} and u_{ik} from the same partite set T_i. Then M cannot contain a vertex $u_{i'j'}$ from a different partite set $T_{i'}$ where $j' \notin \{j, k\}$, as $u_{ij}, u_{i'j'}, u_{ik}$ is a monophonic path. We have $\mathrm{mex}(5; 2) = 5$, so the lower bound holds for $n = 5$ and $a = 2$; otherwise for $n \geq a^2 + 1$ we have $\lfloor \frac{n}{a} \rfloor \geq 3$. Hence let $1 \leq l \leq \lfloor \frac{n}{a} \rfloor$ and $l \notin \{j, k\}$. Then for any $i' \in \{1, 2, \ldots, a\} - \{i\}$ the path $P = u_{ij}, u_{i'k}, u_{il}, u_{i'j}, u_{ik}$ is monophonic, so $M \subseteq V(T_i)$. However the path P shows that no three points of the partite set T_i can all lie in M, so that $|M| \leq 2$. Therefore we can assume that any optimal mp-set has at most one point in each partite set, so that $\mathrm{mp}(T^*_{n,a}) \leq a$, completing the proof.

Theorem 4 shows that the asymptotic order of $\mathrm{mex}(n; a)$ is $\frac{1}{2}(1 - \frac{1}{a})n^2 + O(n)$. For $a = 2$ we were able to push our result further to give an exact formula for $\mathrm{mex}(n; 2)$ and classify the extremal graphs. We omit the lengthy proof, which makes use of Turán stability results from [3] and the classification of non-bipartite triangle-free graphs with largest size from [1].

Theorem 5. *For $n \geq 6$ we have $\mathrm{mex}(n; 2) = \lceil \frac{(n-1)^2}{4} \rceil$, with the unique extremal graph given by $T^*_{n,a}$ for odd n and $T^*_{n,a}$ with one edge added between the partite sets for even n.*

In marked contrast to the quadratic size of extremal graphs with given monophonic position number, the function $\mathrm{gex}(n; a)$ is $O(n)$. A graph with order $n = 10$, general position number 3 and largest size is shown in Fig. 2. This can be shown by a simple upper bound on the maximum degree of such a graph that comes from Ramsey theory. The Ramsey number $R(s, t)$ is the smallest value of n such that any graph with order n contains either a clique of size s or an independent set of size t; taking the converse of the extremal graphs we trivially have the symmetry $R(s, t) = R(t, s)$.

Theorem 6. *For $a \geq 3$ the function $\mathrm{gex}(n; a)$ is bounded above in terms of the Ramsey number $R(a, a + 1)$ by $\mathrm{gex}(n; a) \leq \frac{R(a, a+1) - 1}{2} n$.*

Proof. Let G be a graph with order n, gp-number a, size $\mathrm{gex}(n; a)$ and maximum degree Δ. Suppose that there exists a vertex x with degree $d(x) \geq R(a, a + 1)$ and let X be the subgraph induced by $N(x)$. Then X contains either a clique of order a, which together with x would give a clique of size $a + 1$, or else X has an independent set of size $a + 1$; either of these sets constitutes a general position set with more than a vertices. Thus $\Delta \leq R(a, a+1) - 1$ and $\mathrm{gex}(n; a) \leq \frac{R(a, a+1) - 1}{2} n$.

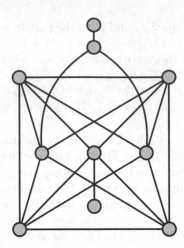

Fig. 2. A graph with order 10, gp-number 3 and largest size

Another interesting question is to find the smallest possible size of a graph with order n, mp-number a and gp-number b; we denote this number by $ex^-(n; a; b)$. It follows from the results of Sect. 2 that this number exists for $a \leq b$ and sufficiently large n. Again we conjecture the following constructions to be extremal.

Theorem 7. *If a and b have the same parity, then for $n \geq \frac{5b-3a}{2} + 4$ we have $ex^-(n; a; b) \leq n + \frac{b-a}{2} + 1$. If a and b have opposite parities, then for $n \geq \frac{5b-3a+11}{2}$ we have $ex^-(n; a; b) \leq n + \frac{b-a+3}{2}$.*

Proof. For $r \geq 2$ and $t \geq 0$ we define a graph $S(r,t)$ as follows. Take a cycle C_{5r+1} of length $5r + 1$ and identify its vertex set with $\mathbb{Z}_{5r+1}$ in the natural way. Join the vertex 0 to the vertices $3 + 5s$ for all $s \in \mathbb{N}$ in the range $0 \leq s \leq r - 1$. Finally append a set $W = \{w_1, \ldots, w_t\}$ of t pendant vertices to the vertex 0. An example is shown in Fig. 3. We claim that $S(r,t)$ has $\text{mp}(S(r,t)) = t + 2$ and $\text{gp}(S(r,t)) = 2r + t$.

The set $W \cup \{1, -1\}$ is obviously in monophonic position, so $\text{mp}(S(r,t)) \geq t + 2$. Let M be any optimal mp-set of $S(r,t)$. By a result of [15] on triangle-free graphs any set in monophonic position in $S(r,t)$ is an independent set. The path $1, 2, 3, \ldots, 5r - 1, 5r$ in C_{5r+1} is monophonic in $S(r,t)$ and hence contains at most two points of M, so if the mp-number of $S(r,t)$ is any greater than $t + 2$, then M contains three vertices of C_{5r+1}, one of which is 0, so that $M \cap W = \varnothing$ and $t = 0$. A simple argument shows that the mp-number of $S(r, 0)$ is two. Thus $\text{mp}(S(r,t)) = t + 2$.

Consider now the set $\{2 + 5s, 4 + 5s : 0 \leq s \leq r - 1\} \cup W$. The vertices of this set are at distance at most four from each other and it is easily verified that none of the geodesics between them pass through other vertices of the set. Thus $\text{gp}(S(r,t)) \geq 2r + t$. Let K be a gp-set of $S(r,t)$ that contains $\geq 2r + t + 1$ vertices. For $0 \leq s \leq r - 1$ the set $S[s] = \{1 + 5s, 2 + 5s, 3 + 5s, 4 + 5s, 5 + 5s\}$ on C_{5r+1} contains at most two vertices of K. It follows that K must contain the vertex 0, two vertices in each of the aforementioned sets and every vertex of W. As the vertices of W have shortest paths to the vertices of K in $S[0]$, we must have $t = 0$. For $r \geq 3$, if $0 \leq s < s' \leq r - 1$ and $s + 2 \leq s'$, then vertices in $S[s]$ have shortest paths to the vertices in $S[s']$ passing through 0, so $0 \notin K$ and $gp(S(r, t)) = 2r + t$. Also $gp(S(2, 0)) = 4$. If a and b have the same parity and $b > a$ the graph $S(\frac{b-a}{2} + 1, a - 2)$ therefore has the required parameters. It is shown in [15] that adding a pendant vertex to an extreme vertex (i.e. a vertex with neighbourhood that induces a clique) preserves the mp-number. As this also holds for the gp-number if $a \geq 3$ we can add a path to a vertex of W to give a graph with any larger order $n' \geq n$ and the same mp- and gp-numbers. If $a = 2$ then lengthening one of the sections of length five on C_{5r+1} accomplishes the same aim. If a and b have opposite parities, then shortening one of the sections of length five on C_{5r+1} yields a graph with order $n = 5r + t$, size $m = n + r$, mp-number $a = t + 2$ and gp-number $b = 2r + t - 1$; solving for r and t and substituting yields the desired bounds.

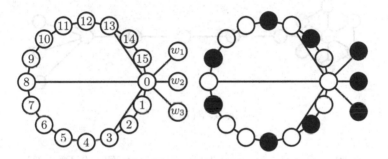

Fig. 3. $S(3, 3)$ (left) with an optimal gp-set in black (right)

4   The Diameters of Graphs with Given Order and mp-Number

The possible diameters of graphs with given geodetic number or hull number are classified in [5,6]. This raises the following question: what are the possible diameters of a graph with given order n and monophonic position number k? In particular, what are the largest and smallest possible diameters of such a graph? We now solve this problem. We split our analysis into two parts, beginning with mp-numbers $k \geq 3$.

Theorem 8. *For any integers k and n with $3 \leq k \leq n - 1$, there exists a connected graph G with order n, monophonic position number k and diameter D if and only if $2 \leq D \leq n - k + 1$.*

Proof. It was shown in [15] that the monophonic position number of a graph with order n and longest monophonic path with length L is bounded above by

$n - L + 1$; rearranging, it follows that $L \leq n - k + 1$. As any geodesic in G is induced, it follows that $n - k + 1$ is the largest possible diameter of a graph G with order n and $\mathrm{mp}(G) = k$. Therefore it remains only to show existence of the required graphs for the remaining values of the parameters. For $k = n - 1$ and $D = 2$ this follows easily by considering the star graph $K_{1,n-1}$, so we can assume that $k \leq n - 2$.

For $n \geq 2$, Theorem 1 shows that any caterpillar graph formed by adding $k - 2$ leaves to the internal vertices of a path of length $n - k + 1$ has order n, mp-number k and diameter $D = n - k + 1$. For $k \geq 3$ we can construct a graph $F(n, k, n - k)$ with order n, mp-number k and diameter $D = n - k$ as follows. Take a path P of length $n - k$; let $V(P) = \{u_0, u_1, \ldots, u_{n-k}\}$, where $u_i \sim u_{i+1}$ for $0 \leq i \leq n - k - 1$. Introduce a set Q of $k - 1$ new vertices $v_1, v_2, \ldots, v_{k-1}$ and join each of them to u_0 and u_1. A straightforward argument shows that this graph has the required parameters.

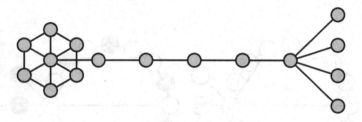

Fig. 4. $F(16, 6, 7)$

Finally for $2 \leq D \leq n - k - 1$ we define the graph $F(n, k, D)$ as follows. Take a path P of length $D - 2$ with vertices $\{x_0, x_1, \ldots, x_{D-2}\}$, where $x_i \sim x_{i+1}$ for $0 \leq i \leq D - 3$. Let C_s be a cycle with length $s = n - D - k + 3$ and vertex set $\{u_0, u_1, \ldots, u_{s-1}\}$, where $u_i \sim u_{i+1}$ for $0 \leq i \leq s - 1$ and addition is carried out modulo s. Join x_0 to every vertex of C_s, so that $C_s \cup \{x_0\}$ induces a wheel. Finally append a set $Q = \{v_1, v_2, \ldots, v_{k-2}\}$ of $k - 2$ pendant edges to x_{D-2}. An example of this construction is given in Fig. 4. This graph has order n and diameter D. It is simple to verify that the set $Q \cup \{u_0, u_1\}$ is an mp-set and that this is largest possible, so that $\mathrm{mp}(F(n, k, D)) = k$.

It is more challenging to determine the possible diameters of graphs with mp-number two.

Theorem 9. *There exists a graph with order n, monophonic position number $k = 2$ and diameter $D \geq 3$ if and only if $D = n - 1$ or $3 \leq D \leq \lfloor \frac{n}{2} \rfloor$.*

Proof. For $n \geq 2$ the path with length $n - 1$ is a graph with order n, mp-number $k = 2$ and diameter $n - 1$. Theorem 4 shows that for $n \geq 3$ the complete bipartite graph $K_{\lceil \frac{n}{2} \rceil, \lfloor \frac{n}{2} \rfloor}$ minus a matching of size $\lfloor \frac{n}{2} \rfloor$ has mp-number $k = 2$; this graph has diameter $D = 3$. The existence of graphs with other values of the diameter in the claimed range follow by altering the number of spokes in the 'half-wheel graphs' defined in [15].

We now show that there is no graph with order n, mp-number $k = 2$ and diameter D, where $\lfloor \frac{n}{2} \rfloor < D < n - 1$. Suppose that G is such a graph and let u and v be vertices at distance D. Assume that G is 2-connected. Then u and v are joined by internally disjoint paths of length $\geq D$, so that $n \geq 2D > n$, which is impossible. Hence G contains a cut-vertex w. Suppose that $d(w) \geq 3$. Then choose a set M of three neighbours of w such that the vertices of M are not all contained in the same component of $G - w$; it is easily seen that M is an mp-set. Hence we must have $d(w) = 2$. Each neighbour of w is either a leaf of G or a cut-vertex, so repeating this reasoning shows that G is a path, which contradicts $D < n - 1$.

We now introduce a special graph operation to clarify when there exists a graph with diameter two and mp-number two.

Lemma 7. *If there exists a graph with order $n \geq 4$, monophonic position number $k = 2$ and diameter $D = 2$, then there exist graphs with monophonic position number $k = 2$, diameter $D = 2$ and orders $3n$, $3n + 1$ and $3n + 2$.*

Proof. Let H be a graph with order $n \geq 4$, monophonic position number $k = 2$ and diameter $D = 2$. Label the vertices of H as $h_1, h_2, \ldots, h_n$. We will construct new graphs with orders $3n$, $3n + 1$ and $3n + 2$ with mp-number $k = 2$ and diameter $D = 2$ from H as follows.

First we define the graph $G(H)$ with order $3n + 2$. Let $X = \{x_1, x_2, \ldots, x_n\}$ and $Y = \{y_1, y_2, \ldots, y_n\}$ be two new sets of vertices disjoint from $V(H)$. On $X \cup Y$ draw a complete bipartite graph with partite sets X and Y and then delete the perfect matching $x_i y_i$, $1 \leq i \leq n$. For $1 \leq i \leq n$ join both x_i and y_i to the vertex h_i by an edge. Finally add two new vertices z_1 and z_2, join z_1 to each vertex of X by an edge, join z_2 to each vertex of Y and lastly add the edge $z_1 z_2$ between the two new vertices. An example of this construction for $H = C_4$ is displayed in Fig. 5. It is easily seen that G has diameter $D = 2$. To show that the mp-number of G is $k = 2$ it is sufficient to show that for any set M of three vertices of G there is an induced path containing each vertex of M. It is evident that for any vertex $v \in V(G) - \{z_1, z_2\}$ there is an induced path in G containing z_1, z_2 and v, so we can assume that $|M \cap \{z_1, z_2\}| \leq 1$.

The map fixing every element of H, interchanging x_i and y_i for $1 \leq i \leq n$ and swapping z_1 and z_2 is an automorphism of G, which reduces the number of cases that we need to check. Suppose that M is a set of three vertices of $G(H)$ containing one of z_1, z_2; say $z_1 \in M$. Without loss of generality we have the following nine possibilities for $M' = M - \{z_1\}$: i) $M' = \{x_1, x_2\}$, ii) $M' = \{x_1, h_1\}$, iii) $M' = \{x_1, h_2\}$, iv) $M' = \{x_1, y_1\}$, v) $M' = \{x_1, y_2\}$, vi) $M' = \{h_1, h_2\}$, vii) $M' = \{h_1, y_1\}$, viii) $M' = \{h_1, y_2\}$ and ix) $M' = \{y_1, y_2\}$.

Consider the following two cycles. For $1 \leq i, j \leq n$, where $i \neq j$, we define $C(i, j)$ to be the cycle $z_1, x_i, y_j, h_j, x_j, z_1$ and, if P is a shortest path in H from h_i to h_j, then $D(i, j)$ is the cycle formed from the path P from h_i to h_j, followed by the path h_j, x_j, z_1, x_i, h_i. Both of these cycles are induced and so can contain at most two points of M. By varying the parameters i and j we see that the first seven configurations for M' above are not possible.

For viii) let P be a shortest path in H from h_1 to h_2. By assumption $n \geq 4$, so as P has length at most two, there exists a vertex of H, say h_3, not appearing in P. Then the path P, followed by the path h_2, y_2, x_3, z_1 contains all three vertices of M. Finally for case ix) the induced path y_1, x_2, z_1, x_1, y_2 contains all three vertices of $\{z_1, y_1, y_2\}$.

We can now suppose that $M \cap \{z_1, z_2\} = \varnothing$. Observe that the subgraph of G induced by $X \cup Y$ is isomorphic to the graph $T^*_{2n,2}$ from the proof of Theorem 4, so we can also assume that $M \nsubseteq X \cup Y$. Furthermore, as $\mathrm{mp}(H) = 2$, we can take $M \nsubseteq V(H)$. For all $1 \leq i, j \leq n$ and $i \neq j$ there is an induced cycle $x_i, h_i, y_i, x_j, h_j, y_j, x_i$, so M must contain vertices with at least three different subscripts $i \in \{1, \ldots, n\}$. Therefore without loss of generality we are left with the following three cases: i) $M = \{x_1, x_2, h_3\}$, ii) $M = \{x_1, y_2, h_3\}$ and iii) $M = \{x_1, h_2, h_3\}$.

For cases i) and ii), let P' be the shortest path in H from h_3 to $\{h_1, h_2\}$; without loss of generality P' is a h_2, h_3-path that does not pass through h_1. Then in case i) x_1, z_1, x_2, h_2 followed by P' contains all three points of $M = \{x_1, x_2, h_3\}$ and in case ii) the path x_1, y_2, h_2 followed by P' contains all three vertices of $M = \{x_1, h_3, y_2\}$. For case iii), if P' is a shortest h_2, h_3-path in H, then x_1, y_2, h_2 followed by P' contains all three vertices of $M = \{x_1, h_2, h_3\}$ unless $d(h_2, h_3) = 2$ and P' is the path h_2, h_1, h_3, in which case the path h_3, y_3, x_1, y_2, h_2 suffices.

This analysis also shows that the graphs $G'(H) = G(H) - \{z_2\}$ with order $3n + 1$ and the graph $G''(H) = G(H) - \{z_1, z_2\}$ with order $3n$ also have mp-number $k = 2$ and diameter $D = 2$. Hence for any $n \geq 4$ if there is a graph with order n, mp-number $k = 2$ and diameter $D = 2$ there also exists such a graph for orders $3n$, $3n + 1$ and $3n + 2$.

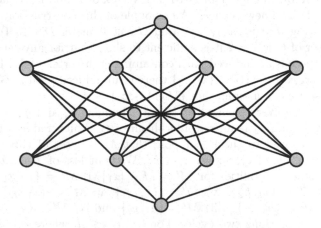

Fig. 5. The graph $G(C_4)$

Theorem 10. *There is a graph with order n, monophonic position number $k = 2$ and diameter $D = 2$ if and only if $n \in \{3, 4, 5, 8\}$ or $n \geq 11$.*

Proof. The statement of the theorem has been verified by computer search for all $n \leq 32$ [11]. Let $n \geq 33$ and assume that the result is true for all orders $< n$. Write $n = 3r + s$, where s is the remainder on division of n by 3. Then $r \geq 11$ and by the induction hypothesis there exists a graph with order r, mp-number $k = 2$ and diameter $D = 2$. Then by Lemma 7 there exists a graph with order n, mp-number $k = 2$ and diameter $D = 2$. The theorem follows by induction.

Acknowledgements. The authors are grateful to the two anonymous reviewers for their suggestions to improve the presentation of this abstract.

References

1. Amin, K., Faudree, J., Gould, R.J., Sidorowicz, E.: On the non-$(p - 1)$-partite K_p-free graphs. Discuss. Math. Graph Theory **33**(1), 9–23 (2013)
2. Anand, B.S., Chandran, S.V.U., Changat, M., Klavžar, S., Thomas, E.J.: Characterization of general position sets and its applications to cographs and bipartite graphs. Appl. Math. Comp. **359**, 84–89 (2019)
3. Brouwer, A.E.: Some lotto numbers from an extension of Turán's theorem. Math. Centr. report ZW152, Amsterdam (1981)
4. Chandran, S.V.U., Parthasarathy, G.J.: The geodesic irredundant sets in graphs. Int. J. Math. Comb. **4**, 135 (2016)
5. Chartrand, G., Harary, F., Zhang, P.: On the geodetic number of a graph. Networks **39**, 1–6 (2002)
6. Chartrand, G., Zhang, P.: Extreme geodesic graphs. Czech. Math. J. **52**(4), 771–780 (2002)
7. Dudeney, H.E.: Amusements in Mathematics, vol. 473. Courier Corporation, North Chelmsford (1958)
8. Klavžar, S., Patkós, B., Rus, G., Yero, I.G.: On general position sets in Cartesian products. Results Math. **76**(3), 1–21 (2021)
9. Klavžar, S., Rus, G.: The general position number of integer lattices. Appl. Math. Comput. **390**, 125664 (2021)
10. Ghorbani, M., Klavžar, S., Maimani, H.R., Momeni, M., Rahimi-Mahid, F., Rus, G.: The general position problem on Kneser graphs and on some graph operations. Discuss. Math. Graph Theory **41**, 1199–1213 (2021)
11. Erskine, G.: Personal communication (2021)
12. Mantel, W.: Problem 28. Wiskundige Opgaven **10**, 60–61 (1907)
13. Manuel, P., Klavžar, S.: A general position problem in graph theory. Bull. Aust. Math. Soc. **98**, 177–187 (2018)
14. Patkós, B.: On the general position problem on Kneser graphs. Ars Math. Contemp. **18**, 273–280 (2020)
15. Thomas, E.J., Chandran, S.V.U., Tuite, J.: On monophonic position sets in graphs. Preprint (2020)
16. Turán, P.: On an extremal problem in graph theory. Math. Fiz. Lapok **48**, 436–452 (1941)

Comparability Graphs Among Cover-Incomparability Graphs

Arun Anil[ID] and Manoj Changat[✉][ID]

Department of Futures Studies, University of Kerala,
Thiruvananthapuram 695581, India
mchangat@keralauniversity.ac.in

Abstract. Comparability graphs and cover-incomparability graphs (C-I graphs) are two interesting classes of graphs from posets. Comparability graph of a poset $P = (V, \leq)$ is a graph with vertex set V and two vertices u and v are adjacent in V if u and v are comparable in P. A C-I graph is a graph from P with vertex set V, and the edge-set is the union of edge sets of the cover graph and the incomparability graph of the poset. C-I graphs have interesting implications on both graphs and posets. In this paper, the C-I graphs, which are also comparability graphs are studied. We identify the class of comparability C-I graphs, which are Ptolemaic graphs, cographs, chordal cographs, distance-hereditary and bisplit graphs. We also determine the posets of these C-I graphs.

Keywords: Comparability graphs · Cover-incomparability graphs · Ptolemaic graph · Cographs · Distance-hereditary graphs · Bisplit graphs

1 Introduction

Comparability graphs form a well-studied class of graphs from posets having many applications and algorithmic interest. Several characterizations are available for comparability graphs [3]. One of them is a characterization in terms of forbidden subgraphs by Gallai in his classic paper [9]. Comparability graphs are also termed as transitively orientable graphs, partially orderable graphs, and containment graphs of the family of sets [3]. Comparability graphs can be recognized in polynomial time [14]. Another important graph from posets is the cover graph, which is the abstract undirected graph behind the Hasse diagram of the poset. It is well known that recognition complexity of a cover graph is NP-complete (Nešetřil and Rödl [15], and Brightwell [7]).

Cover-incomparability graphs of posets, or shortly C-I graphs, were introduced in [4] as underlying graphs of the standard transit function on posets. The C-I graphs are precisely the graphs whose edge set is the union of edge sets of the cover graph and the incomparability graph (complement of a comparability graph) of a poset. Like the cover graph, the recognition complexity of a

© Springer Nature Switzerland AG 2022
N. Balachandran and R. Inkulu (Eds.): CALDAM 2022, LNCS 13179, pp. 48–61, 2022.
https://doi.org/10.1007/978-3-030-95018-7_5

C-I graph is also NP-complete (Maxová et al. [13]) in contrast to the comparability graphs. Hence the problem of characterizing well-known graph families whose C-I graphs have polynomial recognition complexity is interesting. Such C-I graphs studied include the family of split graphs, block graphs [5], cographs [6], Ptolemaic graphs [11], distance-hereditary graphs [11] and k-trees [12]. C-I graphs were recently characterized among the planar graphs and chordal graphs along with new characterizations of Ptolemaic graphs, respectively in [2] and [1]. It is also interesting to note that every C-I graph has a Ptolemaic C-I graph as a spanning subgraph [2].

It is trivial to note that the C-I graphs, which are cover graphs, are precisely the paths, but the same problem for comparability graphs is nontrivial and exciting. This paper identifies the C-I graphs that are comparability graphs among the Ptolemaic graphs, cographs, distance-hereditary graphs and bisplit graphs. Of these graphs, bisplit graphs and cographs are comparability graphs in general. We observe that the class of C-I graphs, which are Ptolemaic graphs and distance-hereditary graphs are comparability graphs. If $\mathcal{G}$ is the class of C-I graphs that are also comparability graphs, then for a graph $G \in \mathcal{G}$, there exists two different partial orders on the vertex set of G, one gives rise to the C-I graph and the other to the comparability graph and both graphs being isomorphic to G. We address this problem also for the families of graphs as mentioned above and determine the posets.

We organize the paper as follows. In the rest of this introductory section, we fix the terminology and notations and discuss some preliminary results on C-I graphs. In Sect. 2, we characterize the posets and graphs whose comparability graphs are Ptolemaic C-I graphs. In Sect. 3, we characterize the posets and graphs whose comparability graphs are cographs and distance hereditary graphs. Similarly we do the same for bisplit graphs in Sect. 4. Finally, in Sect. 5, we study the composition of C-I graphs and observe that the composition of C-I graphs need not be a C-I graph in general and determine some cases when the composition of graphs is a C-I graph as well as a comparability graph.

A *partially ordered set* or *poset* $P = (V, \leq)$ consists of a nonempty set V and a reflexive, anti-symmetric, transitive relation $\leq$ on V, denoted as $P = (V, \leq)$, we call $u \in V$ an element of P. If $u \leq v$ or $v \leq u$ in P, we say u and v are *comparable*, otherwise *incomparable*. If $u \leq v$ but $u \neq v$, then we write $u < v$. If u and v are in V, then v *covers* u in P if $u < v$ and there is no w in V with $u < w < v$, denoted by $u \lhd v$. We write $u \lhd \lhd v$ if $u < v$ but not $u \lhd v$. By $u \| v$, we mean that u and v are incomparable elements of P. A poset P is pictured as a *Hasse diagram* consisting of elements of P and the covering relation between elements denoted as line segments in upward orientation. Let $V' \subseteq V$ and $Q = (V', \leq')$ be a poset, Q is called a *subposet* of P, if $u \leq' v$ if and only if $u \leq v$, for any $u, v \in V'$. The subposet $Q = (V', \leq)$ is a *chain (antichain)* in P, if every pair of elements from V' is comparable (incomparable) in P. A chain of maximum cardinality is named as the *height* of P denoted as $h(P)$. An element u in P is a *minimal (maximal)* if there is no $x \in V$ such that $x \leq u(x \geq u)$ in

P. A poset P is *dual* to a poset Q if for any $x, y \in P$ the following holds: $x \leq y$ in P if and only if $y \leq x$ in Q.

A finite *ranked poset* (also known as *graded poset* [8]) is a poset $P = (V, \leq)$ that is equipped with a rank function $\rho : V \to \mathbb{Z}$ satisfying:

- ρ has value 0 on all minimal elements of P, and
- ρ preserves covering relations: if $a \lessdot b$ then $\rho(b) = \rho(a) + 1$.

A ranked poset P is said to be *complete* if for every i, every element of rank i covers all the elements of rank $i - 1$. For a completely ranked poset $P = (V, \leq)$ we say that element $v \in V$ is on height i, if $\rho(v) = i - 1$. We refer [8], for notions of posets.

Let $G = (V, E)$ be a connected graph, vertex set and edge set of G denoted as $V(G)$ and $E(G)$ respectively, the complement of G is denoted as $\overline{G}$. A graph H is said to be a *subgraph* of G if $V(H) \subseteq V(G)$ and $E(H) \subseteq E(G)$. H is an *induced subgraph* of G if for $u, v \in V(H)$ and $uv \in E(G)$ implies $uv \in E(H)$. A graph G is said to be *H-free*, if G has no induced subgraph isomorphic to H. A *complete graph* is a graph whose vertices are pairwise adjacent, denoted as K_n, a set $S \subseteq V(G)$ is a *clique* if the subgraph of G induced by S is a complete graph, and a *maximal clique* is a clique which is not contained by any other clique. A vertex v is called *simplicial vertex* if its neighborhood induces a complete subgraph. An *independent set* in a graph is a set of pairwise nonadjacent vertices. The *3-fan* is the graph that consists of a path on 4 vertices and a vertex adjacent to all vertices of the path. The distance between u and v in G is the length (*i.e.* the number of edges) of the shortest path from u to v in G. The diameter of G, *diam(G)*, is defined as the maximum distance over all the pair of vertices in G. A graph G is *bipartite* if $V(G)$ is the union of two disjoint independent sets called partite sets of G. A *complete bipartite graph* or *biclique* is a bipartite graph such that two vertices are adjacent if and only if they are in different partite sets. If graphs G_1 and G_2 have disjoint vertex set V_1 and V_2 and edge set E_1 and E_2 respectively, then their *union* $G = G_1 \cup G_2$ has $V = V_1 \cup V_2$ and $E = E_1 \cup E_2$ and their *join*, denoted by $G_1 \vee G_2$, consists of $G = G_1 \cup G_2$ and all edges joining V_1 with V_2.

A graph G is *chordal* if it contains no induced cycles of length more than 3, it is *distance-hereditary* if every induced path is also a shortest path in G. A graph G is *Ptolemaic* if it is distance-hereditary and chordal. Equivalently, G is Ptolemaic if and only if it is 3-fan free chordal graph. P_4- free graphs are called *cographs*. A graph G that is both chordal and cograph is called *chordal cograph*. A graph G is the *comparability graph* of a poset P denoted as C_P, if two vertices are adjacent in C_P if and only if they are comparable in P. Finally the cover-incomparability graph of a poset $P = (V, \leq)$ denoted as G_P is the graph $G = (V, E)$, where $uv \in E(G)$, if either $u \lessdot v$ or $v \lessdot u$ or $u \| v$ in P. A graph is C-I (comparability) graph if it is the C-I (comparability) graph of some poset P.

Now we recall some basic properties of posets and their C-I graphs.

Lemma 1. *[4] Let P be a poset. Then*

(i) the C-I graph of P is connected;

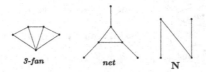

Fig. 1. 3-fan, net graph and the poset **N**

(ii) points of P that are independent in the C-I graph of P lie on a common chain;
(iii) an antichain of P corresponds to a complete subgraph in the C-I graph of P;
(iv) the C-I graph of P contains no induced cycles of length greater than 4.

In the following sections, we discuss some families of graphs, which are comparability graphs, as well as C-I graphs and describe the family of posets whose comparability graphs are precisely the graphs in the family. We begin with the class of Ptolemaic graphs.

2 Ptolemaic Graph

It is proved in [2] that a graph G is a Ptolemaic C-I graph if and only if G is the C-I graph of a completely ranked poset P. In this section, we construct the family of posets whose comparability graphs are Ptolemaic C-I graphs, and prove that every Ptolemaic C-I graph is a comparability graph.

Define a family of posets $\mathscr{P}$ as follows. Consider a sequence of disjoint chains denoted by $L_1, L_3, \ldots, L_{2d+1}$.

Where L_i consists of elements $\{a_{i1}, a_{i2}, \ldots, a_{ik_i}, a_{(i+1)1}, \ldots, a_{(i+1)k_{i+1}}\}$, for $i = 1, 3, \ldots 2d+1, d \geq 0$. Make a covering relation between a_{ik_i} in L_i and $a_{(i-1)1}$ in L_{i-2}; that is, $a_{ik_i} \lhd a_{(i-1)1}$, for $i = 3, 5, \ldots, 2d+1$. (see Fig. 2)

Theorem 1. *A comparability graph C_P of a poset P is a Ptolemaic C-I graph if and only if $P \in \mathscr{P}$. Moreover, every Ptolemaic C-I graph is a comparability graph.*

Proof. Let G be a comparability graph of some poset P. That is, $G \cong C_P$.

Suppose that G is also a Ptolemaic C-I graph of a poset, say P'. Then it follows that P' is a completely ranked poset. Let the elements of rank i be denoted as C_{i+1}. Now every element of C_{i+1} covers every element of C_i. That is, the sets C_i and $C_i \cup C_{i+1}$ form cliques covering all the edges in G and that $V(G) = C_1 \cup C_2 \cup C_3 \cdots \cup C_h$, for some $h > 0$. Also an element x is adjacent to y in G if and only if $x, y \in C_i \cup C_{i+1}$ for some $i \in \{1, 2, \ldots h-1\}$. Let $C_i = \{a_{i1}, a_{i2}, \ldots, a_{ik_i}\}$ for $i = 1, 2, \ldots, h$. The adjacent vertices in a comparability graph must lie on the same chain and non adjacent vertices lie on different chains in P. So the elements in C_i and C_{i+1} lie on a chain and the elements in C_i and C_j lie on different chains for $i = 1, 2, \ldots, h-1$ and $j \neq i-1, i+1$. This implies that $C_i \cup C_{i+1} = L_i$, for $i = 1, 3, \ldots 2d+1$, where $2d+1$ is h or $h-1$ according to h is odd and even respectively. That is, we have proved that $P \in \mathscr{P}$.

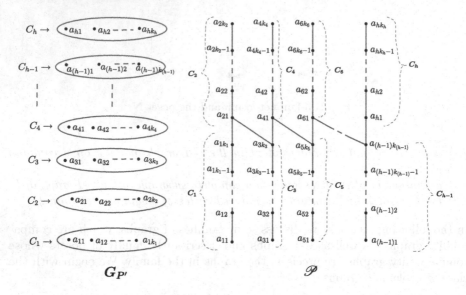

Fig. 2. $G_{P'}$ be the general form of a family of the Ptolemaic C-I graph and $\mathscr{P}$ be the corresponding family of poset whose comparability graphs are Ptolemaic C-I graphs $G_{P'}$.

Conversely, suppose that $P \in \mathscr{P}$. That is, the elements of P form disjoint chains L_i with the covering relation defined between elements a_{ik_i} in L_i and $a_{i-1k_{i-1}}$ in L_{i-2}. Now partition the elements of L_i as $C_i = \{a_{i1}, a_{i2}, \ldots, a_{ik_i}\}$ for $i = 1, 2, \ldots, h$. In C_P, the elements in C_i and $C_i \cup C_{i+1}$ form cliques for $i = 1, 2, \ldots, h$. Also no pair of elements in C_i and C_j, for $|i - j| > 1$ are adjacent in C_P. Let P' be the poset defined such that every element of C_{i+1} cover every element of C_i, $i = 1, \ldots, h$. Now P' is a completely ranked poset and hence $G_{P'}$ is a Ptolemaic C-I graph. It is clear that $G_{P'} \cong C_P$. That is, we have proved that corresponding to every completely ranked poset P', there exists a poset $P \in \mathscr{P}$ and conversely for every $P \in \mathscr{P}$, there exists a completely ranked poset P' such that $C_P \cong G_{P'}$ which completes the proof.   □

3   Cograph and Distance-Hereditary Graph

In this section, we prove that C-I graphs among distance-hereditary graphs are comparability graphs. We first prove that the C-I graph, which is a distance-hereditary graph, is either a Ptolemaic graph or a cograph. We determine the posets whose comparability graphs are distance-hereditary C-I graphs.

Distance-hereditary C-I graphs have been studied in [11]. We quote the following result from [6] and [11].

Theorem 2. *[6] Let G be a chordal cograph. Then G is a C-I graph if and only if G is a connected graph that contains at most two maximal cliques.*

From Theorem 2, a graph G is chordal, C-I graph and cograph (called *chordal C-I cograph*) if and only if there exists three pairwise disjoint set C_1, C_2 and C_3 such that $V(G) = C_1 \cup C_2 \cup C_3$, and $x, y \in V(G)$ are adjacent in G if and only if $x, y \in C_1 \cup C_2$ or $x, y \in C_2 \cup C_3$. That is, $C_1, C_2, C_3, C_1 \cup C_2$ and $C_2 \cup C_3$ forms cliques and the graph has no other edges. It is clear that a chordal C-I cograph is a Ptolemaic C-I graph as it is C-I graph of a completely ranked poset of height 3. From this observation and the posets $\mathscr{P}$ from Sect. 2, we can describe the family of posets whose comparability graphs are chordal C-I cographs. The figure is depicted in Fig. 3(b).

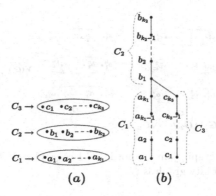

$$(a) \qquad\qquad (b)$$

Fig. 3. (a) General form of a chordal C-I cograph, where $C_1, C_2, C_3, C_1 \cup C_2$ and $C_2 \cup C_3$ form cliques. (b) Corresponding poset whose comparability graph is a chordal C-I cograph.

Now from the fact that a C-I cograph is the join of chordal C-I cographs (which is proved in [6]) and from the family of posets in Fig. 3, we can describe the family of posets whose comparability graphs are also C-I graphs as well as cographs. It is obtained by the ordinal sum of the poset or its dual as shown in Fig. 3(b). (The *ordinal sum* $Z = P \oplus R$ of two disjoint posets P and R is the poset with the underlying set as union of the underlying sets of P and R and the Hasse diagram of Z is obtained by placing the Hasse diagram of R above the Hasse diagram of P and joined by a covering relation from the maximal elements of P to all the minimal elements of R). In Fig. 4(a), one such family of posets is described.

It is known that, in general, a cograph is a comparability graph of a series-parallel partial order. (A poset $P = (V, \le)$ is *series parallel* if and only if the poset $\mathbf{N}$ (shown in Fig. 1) is not a subposet of P.) Now the C-I cographs are the join of chordal C-I cographs whose posets are a particular class of series-parallel partial orders.

Theorem 3. *A graph G is both a C-I graph and a cograph if and only if $\overline{G}$ is a vertex disjoint union of b complete bipartite subgraphs and i isolated vertices and $i \ge b$.*

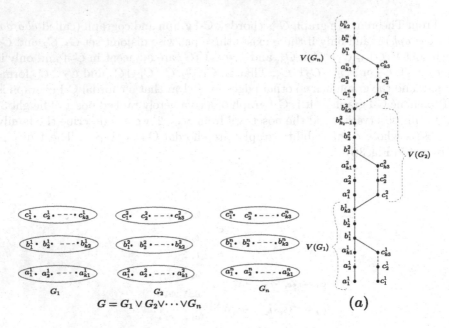

Fig. 4. G be the C-I cograph, join of chordal C-I cographs $G_1, G_2, \ldots, G_n$ and (a) be a poset whose comparability graph is G.

Proof. If G is a complete graph, then the result follows trivially. Now let G is not complete.

Suppose G is a C-I cograph. That is, it is the join of chordal C-I cographs. Let $G = G_1 \vee G_2 \vee \ldots \vee G_n$, where $G_1, G_2, \ldots, G_n$ are chordal C-I cographs. For each G_k, the vertex set consist of pairwise disjoint sets C_1^k, C_2^k and C_3^k such that $V(G_k) = C_1^k \cup C_2^k \cup C_3^k$ and that $C_1^k, C_2^k, C_3^k, C_1^k \cup C_2^k$ and $C_2^k \cup C_3^k$ forms cliques and covers all the edges of G_k. If some G_k's are complete graphs, then we can take C_1^k and C_3^k as empty sets (see the Fig. 3(a)). Now consider $\overline{G}$. In $\overline{G}$, the vertices $C_1^k \cup C_3^k$ induces complete bipartite subgraphs for $k = 1, 2, \ldots, n$ and $\bigcup_{k=1}^{n} C_2^k$ induces isolated vertices. Consider $i = |\bigcup_{k=1}^{n} C_2^k|$, then clearly $i \geq n$.

Conversely, let G be a graph such that $\overline{G}$ is a vertex disjoint union of b complete bipartite subgraphs and i isolated vertices, $i \geq b$. Let $\overline{G}_1, \overline{G}_2, \ldots, \overline{G}_b$ be complete bipartite subgraphs of $\overline{G}$ and for each $\overline{G}_j$, let the vertex set $V(\overline{G}_j)$ be $C_1^j \cup C_3^j$ and since every $\overline{G}_j$ is a complete bipartite graph, every vertex in C_1^j adjacent to every vertex in C_3^j, the vertices in C_1^j and C_3^j are independent for $j = 1, 2, \ldots, b$. Since there are i isolated vertices in $\overline{G}$ with $i \geq b$, we can partition the i isolated vertices into b sets, labeled as C_2^j for $j = 1, 2, \ldots, b$. Now, we consider the graph G and construct an induced subgraph of G as follows. Let G_j be the induced subgraph of G with the vertex set $V(G_j) = C_1^j \cup C_2^j \cup C_3^j$ and by the definition of C_i^j, $i = 1, 2, 3$ and their adjacency relations in $\overline{G}$, it follows that $C_1^j, C_2^j, C_3^j, C_1^j \cup C_2^j$ and $C_2^j \cup C_3^j$ form cliques covering all the edges in G_j, for $j = 1, 2, \ldots, b$. That is, G_j is a chordal C-I cograph. Also every vertex of G_j is adjacent to every vertex of G_k for $k = 1, 2, \ldots, b$, $j \neq k$ in G. Since $V(G) = V(G_1) \cup V(G_2) \cup \ldots \cup V(G_b)$, G is the join of chordal C-I cographs or G is C-I cograph. Hence the Theorem.   □

Theorem 4. *[11] Let G be a distance-hereditary graph. G is a C-I graph if and only if one of the following two conditions holds:*

(1) $diam(G) = 2$ and $\overline{G}$ is a vertex disjoint union of b complete bipartite subgraphs and i isolated vertices and $i \geq b$, or

(2) $diam(G) \geq 3$, G is chordal, and G does not contain a triple of independent simplicial vertices (That is, G is Ptolemaic).

From Theorems 3 and 4 we have,

Theorem 5. *A graph G is a distance-hereditary C-I graph if and only if either*

(i) G is a Ptolemaic C-I graph if $diam(G) > 2$, Or,

(ii) G is a C-I cograph if $diam(G) = 2$.

In general, a distance-hereditary graph need not be a comparability graph. For example, it can be verified easily that the *net graph* (shown in Fig. 1) is a distance-hereditary graph, but it is not a comparability graph. From Theorem 5, it follows that distance-hereditary C-I graph is either a C-I cograph or a Ptolemaic C-I graph. Hence distance-hereditary C-I graphs are comparability graphs. The family of the posets, whose comparability graphs are distance-hereditary graphs are either of the form $\mathscr{P}$ in Fig. 2 or the ordinal sum of posets or the dual posets in Fig. 3(b). In Fig. 4(a), one such family of posets is described.

4   Bisplit Graphs

In this section, we identify C-I graphs that are bisplit graphs, and prove that bisplit C-I graphs are comparability graphs.

A graph G is a *bisplit graph* if its vertex set can be partitioned into three independent sets X, Y and Z such that $Y \cup Z$ induces a complete bipartite subgraph (*bi-clique*) in G. That is, a graph is bisplit if and only if it can be partitioned into an independent set and a bi-clique.

It is trivial to observe that the path graphs are bipartite. Now suppose a C-I graph G_P of a poset P is a bipartite graph. That is, the vertex set V ($|V| = n$) can be partitioned into two independent sets, say $\{u_1, u_2, ..., u_r\}$ and $\{v_1, v_2, ..., v_s\}$, where $n = r + s$. Since $u_1, u_2, ..., u_{r-1}$, and u_r are independent, they lie on a chain alternately. Also $v_1, v_2, ..., v_{s-1}$, and v_s lie on a chain alternately. Therefore, the poset P is a chain, when $|V| > 2$. When $|V| = 2$, the poset is a chain of height 2 or an antichain of size 2. Thus we have the following remark.

Remark 1. A C-I graph G is bipartite if and only if it is a path. Hence a C-I graph is a complete bipartite graph if and only if G is either the path graph P_1, P_2 or P_3.

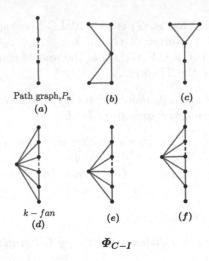

Path graph,P_n
(a)   (b)   (c)

$k - fan$
(d)   (e)   (f)

Φ_{C-I}

Fig. 5. Family of bisplit C-I graphs (Φ_{C-I})

The family of graphs denoted by Φ_{C-I} are depicted in Fig. 5.

Theorem 6. *A graph G is a bisplit C-I graph if and only if G is from Φ_{C-I} (in Fig. 5).*

Proof. Let G be a bisplit C-I graph. So the vertices of G can be partitioned into three independent set X, Y and Z such that $X \cup Y$ induced a bi-clique in G. Since G is a C-I graph, independent elements lie on a chain non consecutively. $X \cup Y$ induces a bi-clique, so every element of X is adjacent to every element of Y. That is, either of the following will occur.

(i) Every element in X is incomparable with every element of Y.
(ii) The maximal element in X covers the minimal element of Y.
(iii) The minimal element of X covers the maximal element of Y.
(iv) Satisfies both conditions (ii) and (iii).

It may be noted that other elements in X and Y are incomparable. Let C_1 be the chain containing the elements in X and C_2 be the chain containing the elements in Y. It is further noted that the elements in X, Y cannot occur consecutively in C_1, respectively C_2. Since X, Y, Z form a partition of V, the remaining elements of C_1 and C_2 must be from Z. Since the elements in Z are also independent, the elements in Z should also lie on a chain. This will happen only if either $|X| \leq 1$ or $|Y| \leq 1$. Without loss of generality assume $|X| \leq 1$. The following cases can happen.

Case 1: $|X| = 0$, $|Y| = 0$ then $Z = 1$. In this case, G is the single vertex graph.
Case 2: $|X| = 0$, $|Y| = n$, $n \geq 1$ then $|Z| = n - 1$, n or $n + 1$.
       In this case G is the path graph P_m, where $m = 2n - 1, 2n$ or $2n + 1$.

Case 3: $|X| = 1$, $|Y| = 1$, then $|Z| = 0, 1$ or 2
(the case $|X| = 1, |Y| = 0$ is similar to Case 2, when $n = 1$).
Let $X = \{x\}$ and $Y = \{y\}$ then
  (i) if $|Z| = 0$, then the graph G is K_2.
  (ii) if $|Z| = 1$, let $Z = \{z\}$, then the possible posets are isomorphic to the posets in Fig. 6 or its duals.
  It is clear that the C-I graph corresponding to the posets in Fig. 6(i) is P_3 and in all the other cases in Fig. 6, the C-I graph is isomorphic to K_3 (special case of Fig. 5(d)).

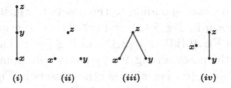

(i)   (ii)   (iii)   (iv)

Fig. 6. Posets whose C-I graphs have a bi-clique of size 2 and independent set of size 1 (*i.e.*, $|X| = |Y| = |Z| = 1$)

  (iii) $|Z| = 2$, let $Z = \{z_1, z_2\}$, then the possible poset are isomorphic to the posets in Fig. 7 or its duals.
  The C-I graph corresponding to the posets in Fig. 7(i),(ii) and (iii) are isomorphic to 2-fan (Fig. 5(d) is the k-fan), the C-I graph corresponding to the poset Fig. 7(iv) is a sub-case of Fig. 5(e) and for the poset Fig. 7(v) is a path graph P_4 (Fig. 5(a)).

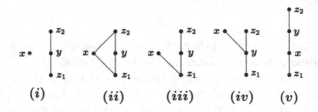

(i)   (ii)   (iii)   (iv)   (v)

Fig. 7. Posets whose C-I graphs have a bi-clique of size 2 and independent set of size 2 (*i.e.*, $|X| = |Y| = 1$ and $|Z| = 2$).

Case 4: $|X| = 1$, $|Y| = 2$, then $|Z| = 0, 1, 2$ or 3.
  (i) $|Z| = 0$, then the graph is just a P_3.
  (ii) $|Z| = 1$, let $Z = \{z\}$, then the possible posets are isomorphic to the posets in Fig. 8 or its duals and they are precisely, the posets isomorphic to those of Case 3(iii).
  (iii) $|Z| = 2$, Let $Z = \{z_1, z_2\}$, then the possible poset are isomorphic to the posets in Fig. 9 or its duals.

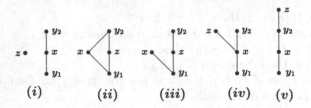

Fig. 8. Posets whose C-I graphs have a bi-clique of size 3 and independent set of size 2 (*i.e.*, $|X| = 1, |Y| = 2$ and $|Z| = 2$).

The C-I graphs corresponding to the posets in Fig. 9 (i) and (ii), respectively are shown in Fig. 5 (*a*) and (*b*). The C-I graph corresponding to the posets in Fig. 9 (iii) is depicted in Fig. 5 (*c*). The C-I graphs corresponding to the posets in Fig. 9 (iv) and (v) are isomorphic to the graph depicted in Fig. 5 (*e*). For all the other posets in Fig. 9, the C-I graphs are isomorphic to the graph shown in Fig. 5 (*d*).

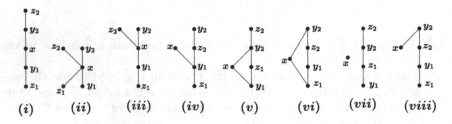

Fig. 9. Posets whose C-I graphs have a bi-clique of size 3 and independent set of size 1 (*i.e.*, $|X| = 1, |Y| = 2$ and $|Z| = 1$)

Case 5: $|X| = 1, |Y| = n, n \geq 3$ then $|Z| = n - 1, n$ or $n + 1$. the possible posets are isomorphic to the posets in Fig. 10 or its duals.

The C-I graphs corresponding to the posets in Fig. 10 (i) (ii) and (iii) are isomorphic and is shown in Fig. 5 (*d*). The C-I graphs corresponding to the posets in Fig. 10 (iv) and (v) are the isomorphic and is depicted in Fig. 5 (*e*). The C-I graph corresponding to the poset Fig. 10 (vi) is shown in Fig. 5 (*f*).

In all the cases, we have shown that the C-I graphs G which are bisplit graphs belong to the family Φ_{C-I} and hence the necessary part follows.

It is easy to verify that the graphs from the family Φ_{C-I} are bisplit C-I graphs. Hence the theorem follows.   □

The posets whose comparability C-I graphs are bisplit graphs are easily constructed, and for each bisplit graph in Fig. 4, the corresponding posets are respectively represented in Fig. 11.

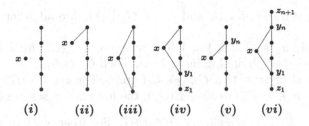

Fig. 10. Posets whose C-I graphs have a bi-clique of size ≥ 4 (*i.e.*, $|X| = 1$, $|Y| = n$, $n \geq 3$ and then $|Z| = n - 1, n$ or $n + 1$).

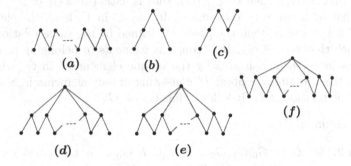

Fig. 11. The posets whose comparability graphs are bisplit C-I graphs (Φ_{C-I})

5   Composition of C-I Graphs

It may be noted that the composition of graphs plays a vital role in the theory of comparability graphs as noted in the Theorem 7 by Martin Golumbic in [10].

We first define the composition of graphs. Let G_0 be a graph with n vertices $v_1, v_2, \ldots, v_n$ and let $G_1, G_2, \ldots, G_n$ be n disjoint graphs. The *composition* graph, $G = G_0[G_1, G_2, \ldots, G_n]$ is formed as follows: For all $1 \leq i, j \leq n$, replace vertex v_i in G_0 with the graph G_i and make each vertex of G_i adjacent to each vertex of G_j whenever v_i is adjacent to v_j in G_0. It may be noted that if G_0 is a complete graph, then the composition becomes the join operation of graphs.

Theorem 7. *[10] Let $G = G_0[G_1, G_2, \ldots, G_n]$, where G_i's are disjoint graphs. Then G is a comparability graph if and only if each G_i ($0 \leq i \leq n$) is a comparability graph.*

This section attempts to study the comparability graphs among C-I graphs using the composition operation.

Theorem 8. *If $G = G_0[G_1, G_2, \ldots, G_n]$ is a C-I graph then G_0 is a C-I graph.*

Proof. Let G be the composition graph which is also a C-I graph of a poset P. Let $G = G_0[G_1, G_2, \ldots, G_n]$. If $u, v \in V(G)$ then $uv \in E(G)$ if and only if either $u, v \in V(G_i)$ with $uv \in E(G_i)$, or $u \in V(G_i)$ and $v \in V(G_j)$ whose corresponding vertex v_i for G_i and v_j for G_j in G_0 are adjacent (*i.e.*, $v_iv_j \in E(G_0)$).

Let $S = \{u_1, u_2, u_3, \ldots, u_n\} \subseteq V(G)$, where $u_i \in V(G_i)$ for $i = 1, 2, \ldots, n$. Let G_S be the subgraph of G induced by S. Then $G_S \cong G_0$.

Now we need to prove that G_S is a C-I graph. For $u, v \in V(G_S)$, depending upon whether $uv \in E(G_S)$ or $uv \notin E(G_s)$, the following cases can occur.

Case 1: If $uv \in E(G_S)$ then clearly $uv \in E(G)$. So either $u \lhd v$ or $v \lhd u$ or $u||v$ in P. Now the subposet P' of P consisting of elements $u, v \in S$ with $u \lhd v$ or $v \lhd u$ or $u||v$ is a poset whose C-I graph is isomorphic to G_S.

Case 2: If $uv \notin E(G_s)$, then $uv \notin E(G)$, that is, either $u \lhd \lhd v$ or $v \lhd \lhd u$ in P. That is, u and v lie on some chain, say C in P. If all the elements of C belongs to S, then the poset P' defined by the same relation in C is such that $G_{P'} \cong G_S$. If C contains elements w belonging to some G_j not in S, then replace w by the unique element w' in G_j which is in S, thus obtaining a poset P' consisting of only elements in S using the same relation in C. It follows that $G_{P'} \cong G_S$.

Hence the theorem. $\square$

Theorem 9. *Let $G = G_0[G_1, G_2, \ldots, G_n]$. If G_0 is a C-I graph and G_i for $i = 1, 2, \ldots, n$ are complete graphs then G is a C-I graph.*

Proof. Let $G = G_0[G_1, G_2, \ldots, G_n]$ be the composition of graphs $G_1, G_2, \ldots, G_n$. By definition of G, the vertices of G consists of vertices in G_i, for $i = 1, \ldots, n$. Let G_0 be the C-I graph of a poset P_0 and G_i for $i = 1, 2, \ldots, n$ be complete graphs. Clearly G_i is the C-I graph of a poset P_i with every element $a, b \in P_i$ being incomparable ($a||b$). Now we construct a poset P from P_0 by replacing the element v_i in P_0 by the vertices of G_i. Corresponding to vertices $a, b \in V(G_i)$ make $a||b$ in P for $i = 1, 2, \ldots, n$. If $v_i \lhd v_j$ in P_0, then make every element $u \in V(G_i)$ and $v \in V(G_j)$ as $u \lhd v$ in P.

Consider the C-I graph G_P of the poset P. Now we need to prove that $G_P \cong G$. Now $ab \in E(G_P)$ if and only if $a \lhd b$ or $b \lhd a$ or $a||b$ in P. This implies that $a \lhd b$ (or $b \lhd a$) in P if and only if for $a \in V(G_i)$ and $b \in V(G_j)$ with $v_i \lhd v_j$ (or $v_j \lhd v_i$)in P_0 and also $a||b$ in P if and only if either a and b are in $V(G_i)$ or $a \in V(G_i)$ and $b \in V(G_j)$ with $v_i||v_j$ in P_0. That is, either $ab \in E(G_i)$ or $a \in V(G_i)$ and $b \in V(G_j)$ such that $v_iv_j \in E(G_0)$. Therefore, $ab \in E(G)$. Hence the result. $\square$

From the above results, we get the following,

Corollary 1. *If $G = G_0[G_1, G_2, \ldots, G_n]$ is a comparability C-I graph then G_0 is a comparability C-I graph.*

Corollary 2. *Let $G = G_0[G_1, G_2, \ldots, G_n]$ and G_i's are complete graphs for $i = 1, 2, \ldots, n$. Then G is a comparability C-I graph if and only if G_0 is a comparability C-I graph.*

Concluding Remarks: We have given some preliminary results of comparability graphs which are also C-I graphs, along with their corresponding posets and obtained some families of such graphs. Using Theorem 9, we may obtain several new classes of these graphs by considering the known C-I graphs as the graph G_0. In particular, we can take the known C-I graphs that we have seen in this paper as the graph G_0 in Theorem 9, and obtain new classes of C-I graphs that are comparability graphs. It will be an interesting problem to characterize comparability C-I graphs. We will address the composition of C-I graphs in detail in a forthcoming paper.

Acknowledgements. A.A. acknowledges the financial support from UGC, Govt. of India, for providing Senior Research Fellowship (1220/(CSIR-UGC NET DEC.2017)). M.C. acknowledges the financial support from SERB, Department of Science & Technology, Govt. of India (research project under MATRICS scheme No. MTR/2017/000238).

References

1. Anil, A., Changat, M.: Ptolemaic and chordal cover-incomparability graphs. Order **38**, 1–15 (2021). https://doi.org/10.1007/s11083-021-09551-w
2. Anil, A., Changat, M., Gologranc, T., Sukumaran, B.: Ptolemaic and planar cover-incomparability graphs. Order **38**(3), 421–439 (2021). https://doi.org/10.1007/s11083-021-09549-4
3. Brandstädt, A., Le, V.B., Spinrad, J.P.: Graph classes - a survey. In: SIAM Monographs on Discrete Mathematics and Applications, ISBN 0-89871-432-X
4. Brešar, B., Changat, M., Klavžar, S., Kovše, M., Mathew, J., Mathews, A.: Cover-incomparability graphs of posets. Order **25**, 335–347 (2008)
5. Brešar, B., Changat, M., Gologranc, T., Mathew, J., Mathews, A.: Cover-incomparability graphs and chordal graphs. Discrete Appl. Math. **158**, 1752–1759 (2010)
6. Brešar, B., Gologranc, T., Changat, M., Sukumaran, B.: Cographs which are cover-incomparability graphs of posets. Order **32**(2), 179–187 (2014). https://doi.org/10.1007/s11083-014-9324-x
7. Brightwell, G.: On the complexity of diagram testing. Order **10**(4), 297–303 (1993)
8. Brightwell, G., West, W.B.: Partially ordered sets. In: Rosen, K.H. (ed.) Handbook of Discrete and Combinatorial Mathematics, Chapter 11, pp. 717–752. CRC Press, Boca Raton (2000)
9. Gallai, T.: Transitiv orientierbare Graphen. Acta Math. Acad. Sci. Hung. **18**, 25–66 (1967)
10. Golumbic, M.C.: Algorithmic Graph Theory and Perfect Graphs, 2nd edn. Annals of Discrete Mathematics, vol. 57. Elsevier (2004)
11. Maxová, J., Turzík, D.: Which distance-hereditary graphs are cover-incomparability graphs? Discrete Appl. Math. **161**, 2095–2100 (2013)
12. Maxová, J., Dubcová, M., Pavlíková, P., Turzík, D.: Which k-trees are cover-incomparability graphs. Discrete Appl. Math. **167**, 222–227 (2014)
13. Maxová, J., Pavlíková, P., Turzík, D.: On the complexity of cover-incomparability graphs of posets. Order **26**(3), 229–236 (2009)
14. McConnell, R.M., Spinrad, J.: Linear-time transitive orientation. In: 8th ACM-SIAM Symposium on Discrete Algorithms, pp. 19–25 (1997)
15. Nešetřil, J., Rödl, V.: Complexity of diagrams. Order **3**(4), 321–330 (1987)

Graph Algorithms

Complexity of Paired Domination
in AT-free and Planar Graphs

Vikash Tripathi[1], Ton Kloks[2], Arti Pandey[1]($\boxtimes$), Kaustav Paul[1],
and Hung-Lung Wang[3]

[1] Department of Mathematics, Indian Institute of Technology Ropar,
Rupnagar 140001, Punjab, India
{2017maz0005,arti,kaustav.20maz0010}@iitrpr.ac.in
[2] Eindhoven, Netherlands
klokston@gmail.com
[3] Department of Computer Science and Information Engineering,
National Taiwan Normal University, Taipei, Taiwan
hlwang@ntnu.edu.tw

Abstract. For a graph $G = (V, E)$, a subset D of vertex set V, is
a dominating set of G if every vertex not in D is adjacent to atleast
one vertex of D. A dominating set D of a graph G with no isolated
vertices is called a *paired dominating set (PD-set)*, if $G[D]$, the subgraph induced by D in G has a perfect matching. The MIN-PD problem requires to compute a PD-set of minimum cardinality. The decision version of the MIN-PD problem remains NP-complete even when
G belongs to restricted graph classes such as bipartite graphs, chordal
graphs etc. On the positive side, the problem is efficiently solvable for
many graph classes including intervals graphs, strongly chordal graphs,
permutation graphs etc. In this paper, we study the complexity of the
problem in AT-free graphs and planar graph. The class of AT-free graphs
contains cocomparability graphs, permutation graphs, trapezoid graphs,
and interval graphs as subclasses. We propose a polynomial-time algorithm to compute a minimum PD-set in AT-free graphs. In addition, we
also present a linear-time 2-approximation algorithm for the problem in
AT-free graphs. Further, we prove that the decision version of the problem is NP-complete for planar graphs, which answers an open question
asked by Lin et al. (in Theor. Comput. Sci., 591(2015) : 99 − 105 and
Algorithmica, 82(2020) : 2809 − 2840).

Keywords: Domination · Paired domination · Planar graphs ·
AT-free graphs · Graph algorithms · NP-completeness · Approximation
algorithm

1 Introduction

Let $G = (V, E)$ be a graph. A vertex $v \in V$ is *adjacent* to another vertex $u \in V$
if uv is an edge of G. In this case, we say u, a neighbour of v. The set of all

© Springer Nature Switzerland AG 2022
N. Balachandran and R. Inkulu (Eds.): CALDAM 2022, LNCS 13179, pp. 65–77, 2022.
https://doi.org/10.1007/978-3-030-95018-7_6

vertices adjacent to $v \in V$, denoted by $N_G(v)$, is known as *open neighbourhood* of v, whereas the set $N_G[v] = N_G(v) \cup \{v\}$ is known as *closed neighbourhood* of v in G.

In a graph $G = (V, E)$, a vertex $v \in V$ *dominates* a vertex $u \in V$ if $u \in N_G[v]$. A subset D of vertex set V, is a *dominating set* of G if every vertex of V is dominated by at least one vertex of D. The *domination number*, symbolized as $\gamma(G)$, is the minimum cardinality of a dominating set. The concept of domination has wide applications and is thoroughly studied by researchers in the literature. A survey of the results, both algorithmic as well as combinatorial, on domination can be found in [7,8]. Due to several applications in the real world problems, numerous variations of domination are introduced by imposing one or more additional condition on dominating set. Many of these variations are thoroughly studied by researchers in the literature. Total domination is one of the important variation of domination. For a graph $G = (V, E)$ without an isolated vertex, a *total dominating set* of G is a subset D of vertex set such that every vertex of the graph is adjacent to at least one vertex in D.

Paired domination is another important variation of domination, introduced by Haynes and Slater in [9]. A detailed survey of results on domination problem and its variations can also be found in a recent book by Haynes et al. [6]. Given a graph $G = (V, E)$ with no isolated vertices, a subset D of vertex set V, is a *paired dominating set(PD-set)* if D is a dominating set and the subgraph induced by D in G has a perfect matching. The *paired domination number*, symbolized as $\gamma_{pr}(G)$, is the cardinality of a minimum PD-set of G. The MIN-PD problem requires to compute a PD-set of a graph G without an isolated vertex. More precisely, the MIN-PD problem and its decision version of the same are defined as follows:

MIN-PD problem

Instance: A graph G with no isolated vertices.
Solution: A PD-set D.
Measure: Size of D.

DECIDE PD-SET problem

Instance: A graph G and an integer $k > 0$, satisfying $k \leq |V|$.
Query: Is there is a PD-set D of G, satisfying $|D| \leq k$?

It is shown that the decision version of the problem is NP-complete for general graphs [9]. Therefore, complexity of the problem is studied for several restricted graph classes. It is proven that, the decision version of the problem is NP-complete when restricted to special graph classes, including bipartite graphs [3], perfect elimination bipartite graphs [16], and split graphs [3]. But, on the good side, the problem is efficiently solvable in several important graph classes, including permutation graphs [12], interval graphs [3], block graphs [3], strongly chordal graphs [4], circular-arc graphs [13] and some others. A detailed survey of the results on paired domination can be found in [5]. In Fig. 1 we show the hierarchy of some important graph classes and the complexity status of the DECIDE PD-SET problem in these graph classes.

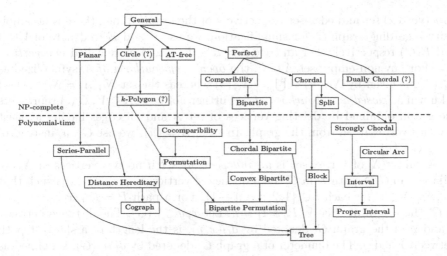

Fig. 1. Complexity status of MIN-PD problem in some well known graph classes.

The computational complexity of the problem is still unknown in some graph classes including planar graphs, AT-free graphs and circle graphs. AT-free graphs is introduced by Corneil et al. in [1]. AT-free graph class includes some important classes of graphs such as interval graphs, permutation graphs and cocomparability graphs as subclasses. A minimum dominating and total dominating set of an AT-free graph can be computed in polynomial-time, see [11]. In this paper, we investigate the computational complexity of the problem on AT-free graph and planar graphs. We show that minimum PD-set of an AT-free graph can be computed in polynomial-time. In addition, we give an approximation algorithm which computes a PD-set of any AT-free graph, within a factor of 2. Lin et al. in [13] and [14] asked to determine the complexity of the problem in planar graphs. In this paper, we prove that DECIDE PD-SET problem remain NP-complete even for planar graphs. The section wise contribution of the paper is outlined as follows:

In Sect. 2, we give insights on some notations and definitions, including properties of AT-free graphs. In Sect. 3, we prove the existence of a linear-time 2-approximation algorithm to compute a PD-set of an AT-free graph. In Sect. 4, we design a polynomial time algorithm to compute a minimum cardinality PD-set of an AT-free graph. In Sect. 5, we show that the problem remains NP-hard for planar graphs. Finally, Sect. 6 wind up the paper with some interesting open questions on the problem.

2 Preliminaries

2.1 Basic Notations and Definitions

In this paper, we consider only simple, connected and finite graphs with no isolated vertices. Let $G = (V, E)$ be a graph. The sets $V(G)$ and $E(G)$ represents node(vertex) set and edge set respectively of the graph. When there is no ambiguity regarding graph G, for simplification, we use V and E to denote of $V(G)$ and $E(G)$ respectively. For an edge $e = uv \in E$, u and v are called *end vertices* of e. For any non-empty set $A \subseteq V$, the *open neighbourhood of A*, symbolized as $N_G(A)$, is given by $N_G(A) = \bigcup_{v \in A} N_G(v)$ whereas the set $N_G[A] = N_G(A) \cup A$ is known as *closed neighbourhood of A*. Further, for a set $A \subseteq V$, $G \backslash A$ represents the graph obtained by deleting vertices of set A and all edges having at least one end vertex in A, from the graph. In case, $A = \{u\}$, we use $G \backslash u$, instead of using $G \backslash \{u\}$.

A subset X of vertex set is an *independent set* if no two vertices of X are adjacent in G. A path P in G is a sequence of vertices $(x_1, x_2, \ldots, x_n)$ such that $(x_i, x_{i+1}) \in E$ for each $i \in \{1, 2, \ldots, n-1\}$. For a path $P = (x_1, x_2, \ldots, x_{n+1})$ in G, the length of P is $|V(P) - 1| = n$. Let $x, y \in V(G)$. The distance between x and y in the graph G, denoted by $d_G(x, y)$, is the length of a shortest path between x and y. The diameter of a graph G, denoted by $diam(G)$, is defined as $diam(G) = \max\{d_G(x, y) \mid x, y \in V(G)\}$. We use the standard notation $[n]$ to denote the set $\{1, 2, \ldots, n\}$.

2.2   AT-free Graphs

Let $G = (V, E)$ be a graph. A set $T = \{p, q, r\}$ of three vertices, is called an *asteroidal tripe*(in short AT) if T is an independent set and for any two vertices in the set T there exits a path $\mathcal{P}$ between them such that $V(\mathcal{P})$ does not contain any vertex from the closed neighbourhood of third. A graph is *AT free* if it does not contain an asteroidal tripe. A path on six vertices is an example of an AT-free graph.

Definition 1. *In a graph $G = (V, E)$, a pair of vertices (x, y) is called a dominating pair, if the vertex set of any path between x and y in G is a dominating set of G. A dominating shortest path is a shortest path connecting x and y in G.*

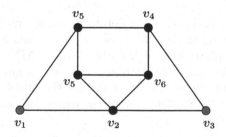

Fig. 2. An AT-free graph G

An asteroidal triple free graph is shown in Fig. 2. For the graph G in Fig. 2, (v_1, v_3) is a dominating pair, and $P = (v_1, v_2, v_3)$ is a dominating shortest path. We have the following result for a connected AT-free graph in the literature.

Theorem 1. *[1, 2] A dominating pair exists in every AT-free graph which can be computed in linear time.*

3 Approximation Algorithm

In this section, we show that a PD-set of an AT-free graph G, can be computed in linear time whose cardinality is at most twice of $\gamma_{pr}(G)$. Let G is an AT-free graph. Using Theorem 1, we note that there exists a dominating pair (x, y) in G. Assume that P is a dominating shortest path between x and y in G, and the number of vertices in P are t. Note that any vertex that is not in P is adjacent to some vertex of $V(P)$, as the set $V(P)$ is a dominating set of G. We may also conclude that any vertex not in P has at most three neighbours in P, since otherwise P will not be a shortest path. By a similar argument we note that any two adjacent vertices in G dominate at most the vertices of a P_4 in P. Consequently, $\frac{\gamma_{pr}}{2} \geq \lceil \frac{t}{4} \rceil$, that is, $\gamma_{pr} \geq 2 \cdot \lceil \frac{t}{4} \rceil$. Before proving the Theorem 2, which is the main result of this section, we notice that the following lemma is true.

Lemma 1. *For any odd positive integer n, $\lceil \frac{n}{4} \rceil \geq \frac{n+1}{4}$.*

Proof. The proof is easy, and hence is omitted. □

Theorem 2. *Given an AT-free graph G, a PD-set D of G can be computed in linear time, satisfying $|D| \leq 2 \cdot \gamma_{pr}(G)$.*

Proof. Given an AT-free graph $G = (V, E)$, there is a linear-time algorithm to find a dominating pair (x, y) of G (by Theorem 1). Let $P = (x = v_1, v_2 \ldots v_{t-1}, v_t = y)$ be a shortest path between x and y, and $D = V(P)$. We have already observed that $\gamma_{pr}(G) \geq 2 \cdot \lceil \frac{t}{4} \rceil$. We prove the result under the following assumptions:

Case 1: If t is even.
Here, we note that the set D is a PD-set and $|D| = t \leq 4 \cdot \lceil \frac{t}{4} \rceil \leq 2 \cdot \gamma_{pr}(G)$.

Case 2: If t is odd.
In this case, we construct a PD-set of the graph G by adding at most one vertex in D. Clearly, D is a dominating set. For pairing, we pair v_i with v_{i+1} for $i \in [t-2]$. Now we need to pair v_t. Note that if $N(v_t) \subseteq D$ then $D \setminus \{v_t\}$ is a PD-set of G, otherwise if there exists a vertex $u \in N(v_t) \setminus D$ then the updated set $D = D \cup \{u\}$ is a PD-set of g. Therefore, we can always construct a PD-set D of G, where $|D| \leq t + 1$. Using Lemma 1, we have $t + 1 \leq 4 \cdot \lceil \frac{t}{4} \rceil$. Hence, $|D| \leq t + 1 \leq 2\gamma_{pr}(G)$.

In both the cases, we can obtain a PD-set D satisfying, $|D| \leq 2\gamma_{pr}(G)$. Hence, we have an efficient 2-approximation algorithm to computes a PD-set of an AT-free graph. □

4 Exact Polynomial-Time Algorithm

The main purpose this section is to establish a polynomial time algorithm that outputs a minimum cardinality PD-set, when the input graph is an AT-free graph. For this, we first present a theorem, which will be useful in designing our algorithm. In this theorem, we show that there exists a BFS-tree T of G and a minimum PD-set D of G such that the number of vertices of D in some consecutive levels of T are bounded. We will use the notation L_i to denote the vertices, which are at i^{th} level in the tree T, that is, the set of vertices which are at distance i from the root node in tree T. The following result is already known in literature.

Theorem 3. *[10] Let G be an AT-free graph with dominating pair (x, y) and T be a BFS-tree of G rooted at x. Let $L_0, L_1, L_2, \ldots, L_l$ are the BFS-levels of the BFS-tree T. Then there exists a linear-time algorithm which computes a path $P = (x = x_0, x_1, x_2, \ldots, x_d = y)$ such that $x_i \in L_i$ for each $0 \leq i \leq d$ and every vertex $w \in L_i$ for $i \in \{1, 2, \ldots, l\}$ is adjacent to either x_{i-1} or x_i.*

Theorem 4. *Let $G = (V, E)$ be an AT-free graph and (x, y) be a dominating pair of G. If $L_0, L_1, L_2, \ldots, L_l$ are the BFS-levels of the BFS-tree T rooted at x then there exists a minimum cardinality PD-set D_p of G such that $|D_p \cap \bigcup_{k=i}^{i+j} L_k| \leq j + 4$ for all $i \in \{0, 1, \ldots l\}$ and $j \in \{0, 1, \ldots l - i\}$.*

Proof. Let $G = (V, E)$ be an AT-free graph and D_p be a minimum cardinality PD-set of the graph G. Suppose that the set D_p does not satisfy the given property, that is, there is at least one pair (i, j) such that $|D_p \cap \bigcup_{k=i}^{i+j} L_k| > j + 4$ where $i \in \{0, 1, \ldots l\}$ and $j \in \{0, 1, \ldots l - i\}$. Let $B = \{(i, j) : |D_p \cap \bigcup_{k=i}^{i+j} L_k| \geq j + 5\}$. Note that $B \neq \emptyset$. Now we choose pair (i', j') such that $i' = \min\{i | (i, j) \in B\}$ and $j' = \max\{j \mid (i', j) \in B\}$. By the choice of the pair (i', j'), note that $D_p \cap L_{i'-1} = \emptyset$ and $D_p \cap L_{i'+j'+1} = \emptyset$. Using the properties of a BFS-tree, we note that for any vertex $v \in (D_p \cap \bigcup_{k=i'}^{i'+j'} L_k)$, any neighbor of v belongs to one of the levels $L_{i'-1}, L_{i'}, \cdots L_{i'+j'+1}$. Let $A = \{x_{i'-2}, x_{i'-1}, \ldots x_{i'+j'+1}\}$. Note that, $|A| = j' + 4$. Since $V(P)$ is a dominating set of G and each vertex $z \in L_i$ is adjacent to either x_{i-1} or x_i, $\bigcup_{k=i'-1}^{i'+j'+1} L_k \subseteq N[A]$. Now by updating D_p we will find another minimum PD-set D_p' such that $A \subseteq D_p'$.

Case 1: If $x_{i'-2} \notin D_p$ and $|A|$ is even.

Since $D_p \cap L_{i'+j'+1} = \emptyset$, $x_{i'+j'+1} \notin D_p$. If $x_{i'-2} \notin D_p$ and $|A|$ is even then the set $D_p' = (D_p \setminus \bigcup_{k=i'}^{i'+j'} L_k) \cup A$ is a PD-set of G with $|D_p'| < |D_p|$, a contradiction to the choice of D_p.

Case 2: If $x_{i'-2} \notin D_p$ and $|A|$ is odd.

Note that $|A|$ is odd and $G[A]$ is a path, if we include A in a PD-set we can pair all the vertices in A except one. We pair $(x_{i'-2}, x_{i'-1})$, $(x_{i'}, x_{i'+1}), \ldots, (x_{i'+j'-1}, x_{i'+j'})$. Now we need to pair $x_{i'+j'+1}$. If $(N_G(x_{i'+j'+1}) \setminus \{x_{i'+j'}\}) \subseteq D_p$ and $(N_G(x_{i'+j'+1}) \setminus \{x_{i'+j'}\}) \cap (L_{x_{i'+j'+1}} \cup L_{x_{i'+j'}}) = \emptyset$. In this case using the property of path P, note that all the vertices in $L_{i'+j'+1}$ is adjacent to $x_{i'+j'}$. Hence the set $D'_p = (D_p \setminus \bigcup_{k=i'}^{i'+j'} L_k) \cup (A \setminus \{x_{i'+j'+1}\})$ is a PD-set of G with $|D'_p| < |D_p|$, a contradiction. If there is a vertex $u \in N_G(x_{i'+j'+1}) \setminus \{x_{i'+j'}\}$ such that $u \notin D_p$ or if $N_G(x_{i'+j'+1}) \setminus \{x_{i'+j'}\} \subseteq D_p$ but there is a vertex $u \in N_G(x_{i'+j'+1}) \setminus \{x_{i'+j'}\}$ such that $u \in (L_{x_{i'+j'+1}} \cup L_{x_{i'+j'}})$ then take $A' = A \cup \{u\}$. Note that the set $D'_p = (D_p \setminus \bigcup_{k=i'}^{i'+j'} L_k) \cup A'$ is a PD-set of G, implying that $|D'_p| \geq |D_p|$. Also we have $|D_p \cap \bigcup_{k=i'}^{i'+j'} L_k| \geq j'+5$ and $|A'| = j'+5$ implying that $|D'_p| \leq |D_p|$. Hence D'_p is also a minimum PD-set of G.

Case 3: If $x_{i'-2} \in D_p$ and $|A|$ is even.
Proof is omitted due to space constraints.

Case 4: If $x_{i'-2} \in D_p$ and $|A|$ is odd.
Proof is omitted due to space constraints.

Further, note that if i' is 0 or 1, we can choose $A = \{x_0, x_1, \ldots, x_{i'+j'+1}\}$ if $|\{x_0, x_1, \ldots, x_{i'+j'+1}\}|$ is even, otherwise we can choose $A = \{x_0, x_1, \ldots, x_{i'+j'+1}, u\}$, where $u \in N(x_{i'+j'+1})$. We can show the existence of u as we did above. In both the cases $|A| \leq j' + 4$ implying that $D'_p = (D_p \setminus \bigcup_{k=i'}^{i'+j'} L_k) \cup A$ is a PD-set of G having cardinality less than the minimum cardinality PD-set D_p of G, a contradiction. Hence $i' \notin \{0,1\}$. Similarly we can claim that $i' + j' \notin \{l-1, l\}$.

We call this replacement of D_p with D'_p an exchange step. Now, if $|D'_p \cap \bigcup_{k=i}^{i+j} L_k| \leq j + 4$ for all $i \in \{0, 1, \ldots l\}$ and $j \in \{0, 1, \ldots l - i\}$ then G has a minimum paired dominating D'_p satisfying the condition given in Theorem 4. Otherwise, let $B' = \{(i,j) : |D'_p \cap \bigcup_{k=i}^{i+j} L_k| \geq j+5\}$. Suppose $(i,j) \in B'$. Now we will show that $i > i'$. By contradiction suppose, $i \leq i'$. In this case note that $i + j \geq i' - 2$ otherwise, $(i,j) \in B$, contradicting the choice of i'. Also, $|D'_p \cap L_t| \geq 1$ for all $t \in \{i'-2, i'-1, \ldots, i'+j'+1\}$. Hence for $(i,j) \in B'$ with $i < i'$ and $i+j \geq i'-2$ there exits a j' such that $(i, j') \in B'$ and $i+j' \geq i'+j'+1$. By construction of D'_p, we note that $|D_p \cap \bigcup_{k=i}^{i+j'} L_k| \geq |D'_p \cap \bigcup_{k=i}^{i+j'} L_k| \geq j' + 5$ implying that $(i,j') \in B$, a contradiction to the choice of i' or j'. Hence $i > i'$. Therefore, if $i'' = \min \{i \mid (i,j) \in B'\}$ then $i'' > i'$.

This implies that, at every exchange step, we replace a minimum cardinality PD-set D_p with an updated minimum cardinality PD-set D'_p. After each exchange step, we note that the smallest value of i for which there was a $j \in \{0, 1, \ldots, l - i\}$ satisfying $|D_p \cap \bigcup_{k=i}^{i+j} L_k| \geq j + 5$, for the minimum cardinality PD-set D_p, will increase. Therefore, we conclude that, if we start with any minimum cardinality PD-set D_p, we obtain a minimum cardinality PD-set D'_p, such that $|D'_p \cap \bigcup_{k=i}^{i+j} L_k| \leq j+4$ for all $i \in \{0, 1, \ldots l\}$ and $j \in \{0, 1, \ldots l-i\}$, by executing at most d exchange steps. $\qquad \square$

Now we are ready to present an algorithm to compute a minimum cardinality PD-set of an AT-free G. Using Theorem 4, we may conclude that there is a minimum PD-set of G that contains at most 6 vertices from any three consecutive BFS-levels of x, where (x,y) is a dominating pair of G. The idea behind our algorithm is the following:

In our algorithm, we explore a BFS-level of x in each iteration. In the ith-iteration of the algorithm, we do the following:

- store all the possible sets $X' \subseteq \bigcup_{j=0}^{i+1} L_j$ such that X' dominates all the vertices till i^{th}-level.
- ensure that all the vertices in $X' \cap (\bigcup_{j=0}^{i} L_j)$ are paired as these vertices can not be paired with a vertex at level $i + 2$ or above.
- for every possible set X', store another set $X = X' \cap (L_i \cup L_{i+1})$

The set X helps in extending a partial solution X' to the next level as we are restricted to select at most 6 vertices from any three consecutive levels in a minimum PD-set. Below, we have provided the detailed algorithm for computing a minimum cardinality PD-set D_p of an AT-free graph G. The set D_p maintains the property that it contains at most 6 vertices from any three consecutive BFS-levels of x.

Algorithm 1: Minimum Paired Domination in AT-free Graphs

Input: A connected AT-free graph $G = (V, E)$ with a dominating pair (x, y);
Output: A PD-set D_p of G;
Compute the BFS-levels of x;
For $0 \leq i \leq l$, let $L_i = \{w \in V \mid d_G(x, w) = i\}$ denote the set of vertices at level i in the BFS of G rooted at u.
In particular, $L_0 = \{x\}$.
Initialize the queue Q_1 which contains an ordered tuple $(X, X, size(X))$ for all non-empty $X \subseteq N[x]$ such that $size(X) = |X| \leq 6$;
Initialize $i = 1$;
while $(Q_i \neq \emptyset$ and $i < l)$ **do**

    Update $i = i + 1$;

    for (*each element* $(X, X', size(X'))$ *of the queue* Q_{i-1}) **do**

        for (*every* $U \subseteq L_i$ *with* $|X \cup U| \leq 6$) **do**

            if ($L_{i-1} \subseteq N[X \cup U]$ *and there exists a set* $U' \subseteq U$ *such that* $G[X' \cup U']$ *has a perfect matching*) **then**

                $Y = (X \cup U) \setminus L_{i-2}$;
                $Y' = X' \cup U$;
                $size(Y') = size(X') + |U|$;
                if (*for all element* $(X, X', size(X'))$ *of* Q_i, $X \neq Y$) **then**
                    insert $(Y, Y', size(Y'))$ in the queue Q_i;
                if (*there is a tuple* $(Z, Z', size(Z'))$ *in* Q_i *such that* $Z = Y$ *and* $size(Y') < size(Z')$) **then**
                    delete $(Z, Z', size(Z'))$ form Q_i;
                    insert $(Y, Y', size(Y'))$ in Q_i;

Among all the triples $(X, X', size(X'))$ in the queue Q_l that satisfy $L_l \subseteq N[X]$ and $G[X']$ has a perfect matching, find one such that $size(X')$ is minimum, say $(D, D', size(D'))$;
$D_p = D'$;
return D_p;

Now we prove the following theorem to show that the Algorithm 1 returns a minimum PD-set. We also analyse the running time of the algorithm.

Theorem 5. *Let $G = (V, E)$ be an AT-free graph such that $|V| = n$ and $|E| = m$. Algorithm 1 computes a minimum cardinality PD-set of G in $O(n^{8.5})$-time.*

Proof. Proof is omitted due to space constraints. $\square$

5   Paired Domination in Planar Graphs

In this section we show that the DECIDE PD-SET problem is NP-complete even when restricted to planar graph. For this purpose, we will give a polynomial reduction from the MINIMUM VERTEX COVER(MIN-VC) problem to the MIN-PD problem. In a graph $G = (V, E)$, a *vertex cover* is a set $C \subseteq V$ such that C has at least one end point of every edge $e \in E$. The MIN-VC problem require to compute a minimum cardinality vertex cover of a given graph G. The following theorem is already proved for the MIN-VC problem.

Theorem 6. *[15]   The MIN-VC problem is NP-hard for the planar cubic graphs.*

Now, we prove the main result of this section.

Theorem 7. *The DECIDE PD-SET problem is NP-complete for planar graphs with maximum degree 5.*

Proof. Clearly, the DECIDE PD-SET problem is in NP. To show the hardness of the problem, we give a reduction from MIN-VC problem which is NP-hard for planar cubic graphs, by Theorem 6. Let $G = (V, E)$ be a planar cubic graph with $V = \{v_1, v_2, \ldots, v_n\}$. We transform the graph G into a graph $G' = (V', E')$ as follows:

- replace each vertex $v_i \in V$ with the gadget G_{v_i} as shown in the Fig. 3
- If three edges e_j, e_k, e_l were incident on v_i in G, then in G', we make e_j incident on v_i^1, e_k incident on v_i^2 and e_l incident on v_i^3.

We note that the graph G' is a planar graph with maximum degree 5, and G' can be computed from G in polynomial time. Now, to prove the result we only need to prove the following claim:

Claim. If $\beta(G)$ denotes the cardinality of a minimum vertex cover of G, then $\gamma_{pr}(G') = 4n + 2\beta(G)$, where n denotes the number of vertices in G.

Proof. Let V^c be a minimum cardinality vertex cover of G. Let $D_p = \{v_i^1, y_i^1, v_i^2, y_i^4, v_i^3, z_i^2 \mid v_i \in V^c\} \cup \{y_i^2, z_i^1, y_i^3, z_i^3 \mid v_i \notin V^c\}$ where $i \in [n]$. Note that if $v_i \notin V^c$, then all the three vertices adjacent to v_i in G must be present in V^c. Using this fact, it can be easily verified that D_p is a PD-set of G', and

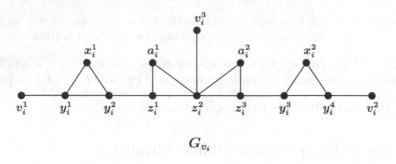

Fig. 3. Gadget G_{v_i} used in the construction of graph G' from G in Theorem 7.

$|D_p| = 6 \cdot \beta(G) + 4 \cdot (n - \beta(G)) = 4n + 2\beta(G)$. Therefore, if D_p^* is a minimum cardinality PD-set of G' then $|D_p^*| \leq 4n + 2\beta(G)$. Hence, we have

$$\gamma_{pr}(G') \leq 4n + 2\beta(G) \tag{1}$$

Conversely, suppose D_p is a minimum cardinality PD-set of G'. Then, to dominate the vertex x_i^1, $D_p \cap \{x_i^1, y_i^1, y_i^2\}$ must be non-empty. Further, a vertex $u \in \{x_i^1, y_i^1, y_i^2\} \cap D_p$ can only be paired with a vertex in the set $\{v_i^1, x_i^1, y_i^1, y_i^2, z_i^1\} \setminus \{u\}$. Hence, $|D_p \cap \{v_i^1, x_i^1, y_i^1, y_i^2, z_i^1\}| \geq 2$. Similarly, we have $|D_p \cap \{v_i^3, x_i^3, y_i^3, y_i^4, z_i^3\}| \geq 2$. Therefore, for each $i \in [n]$, we have $|D_p \cap V(G_{v_i})| \geq 4$. Note that to dominate x_i^1, $D_p \cap \{x_i^1, y_i^1, y_i^2\} \neq \emptyset$. Further, to dominate a_i^1, $D_p \cap \{z_i^1, z_i^2, a_i^1\} \neq \emptyset$. Similarly, to dominate x_i^2 and a_i^2, $D_p \cap \{x_i^2, y_i^3, y_i^4\} \neq \emptyset$ and $D_p \cap \{z_i^3, z_i^2, a_i^2\} \neq \emptyset$ respectively. Therefore, we observe that, if $|D_p \cap V(G_{v_i})| = 4$, then $D_p \cap V(G_{v_i}) = \{y_i^2, z_i^1, y_i^3, z_i^3\}$.

Now, we prove that we can update D_p such that D_p remains a minimum cardinality PD-set of G' and for each $i \in [n]$, $|D_p \cap V(G_{v_i})| = 4$ or $|D_p \cap V(G_{v_i})| \geq 6$. Suppose $|D_p \cap V(G_{v_i})| = 5$ for some $i \in [n]$. As we observed, the vertices dominating x_i^1 and x_i^2 are paired with the vertices of $V(G_{v_i})$, and $|D_p \cap V(G_{v_i})| \geq 4$. Hence if $|D_p \cap V(G_{v_i})| = 5$ then $D_p \cap \{v_i^1, v_i^2, v_i^3\} \neq \emptyset$, as only these vertices of the gadget G_{v_i} can be paired with a vertex of another gadget.

Case 1: Suppose $v_i^1 \in D_p$.

In this case, first we show that $D_p \cap \{v_i^2, v_i^3\} = \emptyset$. Note that v_i^1 is paired with a vertex of some other gadget, and v_i^1 is not dominating x_i^1. Further, if u is the vertex dominating vertex x_i^1 then u can only be paired with a vertex in the set $\{x_i^1, y_i^1, y_i^2, z_i^1\} \setminus \{u\}$. Therefore, $|D_p \cap \{v_i^1, x_i^1, y_i^1, y_i^2, z_i^1\}| \geq 3$. Also as $|D_p \cap \{v_i^3, x_i^2, y_i^3, y_i^4, z_i^3\}| \geq 2$ and $|D_p \cap V(G_{v_i})| = 5$, we have $v_i^3 \notin D_p$. Further, as $|D_p \cap \{v_i^1, x_i^1, y_i^1, y_i^2, z_i^1\}| \geq 3$ therefore, $|D_p \cap \{v_i^2, x_i^2, y_i^3, y_i^4, z_i^3\}| = 2$. Now, if $v_i^2 \in D_p$ then v_i^2 is paired with y_i^4 this leaves the vertex z_i^3 undominated, a contradiction. Therefore, $v_i^2 \notin D_p$. This concludes that $D_p \cap \{v_i^2, v_i^3\} = \emptyset$.

Now, let v_i^1 is paired with a vertex u of another gadget, say G_{v_j} where $i \neq j$. Note that $u \in \{v_j^1, v_j^2, v_j^3\}$. It is easy to observe that $|D_p \cap V(G_{v_j})| \geq 5$. Suppose v_i^1 is paired with v_j^1. Now if $y_j^1 \notin D_p$ then update D_p as follows: $D_p = D_p \setminus V(G_{v_i}) \cup \{y_i^2, y_i^3, z_i^1, z_i^3, y_j^1\}$ and pair v_j^1 with y_j^1. Now, suppose that

y_j^1 already belongs to D_p. Note that y_j^1 is paired with either y_j^2 or x_j^1. If both y_j^2 and $x_j^1 \in D_p$ then y_j^1 must be paired with x_j^1. In this case, the set $D_p' = D_p \setminus (V(G_{v_i}) \cup \{x_j^1\}) \cup \{y_i^2, y_i^3, z_i^1, z_i^3, y_j^1\}$ where v_j^1 is paired with y_j^1 is a PD-set of G' and $|D_p'| < |D_p|$, a contradiction. Therefore, in this case either $y_j^2 \notin D_p$ or $x_j^1 \notin D_p$. If $x_j^1 \notin D_p$ then y_j^1 is paired with y_j^2 and in this case, we update D_p as follows: $D_p = D_p \setminus V(G_{v_i}) \cup \{y_i^2, y_i^3, z_i^1, z_i^3, x_j^1\}$, pair v_j^1 with y_j^1 and y_j^2 with x_j^1. We can update D_p in a similar way if $y_j^2 \notin D_p$.

Similarly we can update D_p if v_i^1 is paired with v_j^2. Now suppose v_i^1 is paired with v_j^3. If $z_j^2 \notin D_p$ then update D_p as follows: $D_p = D_p \setminus V(G_{v_i}) \cup \{y_i^2, y_i^3, z_i^1, z_i^3, z_j^2\}$ and pair v_j^3 with z_j^2. But, if $z_j^2 \in D_p$, we may observe that it is possible to update D_p by giving similar arguments as above with suitable modifications, such that v_j^3 is paired with z_j^2. After update in each case, we may note that $|D_p \cap V(G_{v_i})| = 4$ and $|D_p \cap V(G_{v_j})| \geq 6$.

Case 2: Suppose $v_i^2 \in D_p$.
The arguments are similar to Case 1.

Case 3: Suppose $v_i^3 \in D_p$.
Let v_i^3 is paired with a vertex u of another gadget G_{v_j}. Since, $|D_p \cap V(G_{v_i})| = 5$ we have $|D_p \cap \{v_i^1, x_i^1, y_i^1, y_i^2, z_i^1\}| = 2$ and $|D_p \cap \{v_i^3, x_i^3, y_i^3, y_i^4, z_i^3\}| = 2$. Now, if $v_i^1 \in D_p$ then v_i^1 is paired with y_i^1 this leaves the vertex z_i^1 undominated, a contradiction. Similarly, if $v_i^2 \in D_p$ then v_i^2 is paired with y_i^4 this leaves the vertex z_i^3 undominated, a contradiction. Hence, $D_p \cap \{v_i^1, v_i^2\} = \emptyset$. Now we can give similar arguments as Case 1, to show that D_p can be updated such that $|D_p \cap V(G_{v_i})| = 4$ and $|D_p \cap V(G_{v_j})| \geq 6$.

Now, without loss of generality, we may assume that there exists a minimum cardinality PD-set of G' such that for each $i \in [n]$, $|D_p \cap V(G_{v_i})| = 4$ or $|D_p \cap V(G_{v_i})| \geq 6$

Define $V^c = \{v_i \in V \mid |D_p \cap V(G_{v_i})| \geq 6\}$. Next, we claim that V^c is a vertex cover of G. Consider any two distinct vertices v_i and v_j in G such that $v_i v_j \in E(G)$. We prove that either $|D_p \cap V(G_{v_i})| \geq 6$ or $|D_p \cap V(G_{v_j})| \geq 6$. Let v_i^k is made adjacent to $v_j^{k'}$, where $k, k' \in [3]$. Note that if $|D_p \cap V(G_{v_i})| = 4$ and $|D_p \cap V(G_{v_i})| = 4$ then from above observation, we have $D_p \cap V(G_{v_i}) = \{y_i^2, z_i^1, y_i^3, z_i^3\}$ and $D_p \cap V(G_{v_j}) = \{y_j^2, z_j^1, y_j^3, z_j^3\}$, this leaves the vertices v_i^k and $v_i^{k'}$ undominated, a contradiction. Therefore, V^c is a vertex cover of G. Also, $\gamma_{pr}(G') \geq 6|V^c| + 4(n - |V^c|)$. So, we have $2|V^c| \leq \gamma_{pr}(G') - 4n$. Hence,

$$2\beta(G) \leq \gamma_{pr}(G') - 4n \qquad (2)$$

Therefore, using Eqs. 1 and 2, we have $\gamma_{pr}(G') = 4n + 2\beta(G)$. This proves the claim. $\qquad \square$

Since, the MIN-VC problem is NP-hard for cubic planar graphs, from above claim we conclude that the DECIDE PD-SET problem is NP-complete for planar graphs with maximum degree 5. $\qquad \square$

6   Concluding Remarks

In this paper, we resolve the complexity of the MIN-PD problem for planar graphs and AT-free graphs. We proposed a polynomial time algorithm for MIN-PD problem in AT-free graphs. We also proposed a 2-approximation algorithm to compute a PD-set in AT-free graphs. Since the class of AT-free graphs include the class of cocomparability graphs, the results and algorithms presented for paired domination in AT-free graphs, also holds for cocomparability graphs. We further investigated the computational complexity of the problem in planar graphs and proved that the problem is NP-hard. The complexity of the problem is still not known in circle graphs. One may be interested in investigating the complexity status of the MIN-PD problem in circle graph. Further, it is interesting to design more efficient algorithm for the problem in AT-free graphs and cocomparability graphs.

References

1. Corneil, D.G., Olariu, S., Stewart, L.: Asteroidal triple-free graphs. SIAM J. Discrete Math. **10**, 399–430 (1997)
2. Corneil, D.G., Olariu, S., Stewart, L.: Linear time algorithms for dominating pairs in asteroidal triple-free graphs. In: Fülöp, Z., Gécseg, F. (eds.) ICALP 1995. LNCS, vol. 944, pp. 292–302. Springer, Heidelberg (1995). https://doi.org/10.1007/3-540-60084-1_82
3. Chen, L., Lu, C., Zeng, Z.: Labelling algorithms for paired domination problems in block and interval graphs. J. Comb. Optim. **19**(4), 457–470 (2008)
4. Chen, L., Lu, C., Zeng, Z.: A linear-time algorithm for paired-domination in strongly chordal graphs. Inf. Process. Lett. **110**(1), 20–23 (2009)
5. Desormeaux, W., Henning, M.A.: Paired domination in graphs: a survey and recent results. Util. Math. **94**, 101–166 (2014)
6. Haynes, T.W., Hedetniemi, S.T., Henning, M.A. (eds.): Topics in Domination in Graphs. DM, vol. 64. Springer, Cham (2020). https://doi.org/10.1007/978-3-030-51117-3
7. Haynes, T.W., Hedetniemi, S.T., Slater, P.J.: Fundamentals of Domination in Graphs, vol. 208. Marcel Dekker Inc., New York (1998)
8. Haynes, T.W., Hedetniemi, S.T., Slater, P.J.: Domination in Graphs: Advanced Topics, vol. 209. Marcel Dekker Inc., New York (1998)
9. Haynes, T.W., Slater, P.J.: Paired domination in graphs. Networks **32**, 199–206 (1998)
10. Kloks, T., Kratsch, D., Müller, H.: Approximating the bandwidth for asteroidal triple-free graphs. In: Spirakis, P. (ed.) ESA 1995. LNCS, vol. 979, pp. 434–447. Springer, Heidelberg (1995). https://doi.org/10.1007/3-540-60313-1_161
11. Kratsch, D.: Domination and total domination in asteroidal triple-free graphs. Discrete Appl. Math. **99**, 111–123 (2000)
12. Lappas, E., Nikolopoulos, S.D., Palios, L.: An $O(n)$-time algorithm for paired domination problem on permutation graphs. Eur. J. Combin. **34**(3), 593–608 (2013)
13. Lin, C.C., Tu, H.L.: A linear-time algorithm for paired-domination on circular arc graphs. Theor. Comput. Sci. **591**, 99–105 (2015)

14. Lin, C.-C., Ku, K.-C., Hsu, C.-H.: Paired-domination problem on distance-hereditary graphs. Algorithmica **82**(10), 2809–2840 (2020). https://doi.org/10.1007/s00453-020-00705-7
15. Mohar, B.: Face covers and the genus problem for apex graphs. J. Comb. Theory Ser. A. **82**, 102–117 (2001)
16. Panda, B.S., Pradhan, D.: Minimum paired-dominating set in chordal bipartite graphs and perfect elimination bipartite graphs. J. Comb. Optim. **26**(4), 770–785 (2012). https://doi.org/10.1007/s10878-012-9483-x

The Complexity of Star Colouring in Bounded Degree Graphs and Regular Graphs

M. A. Shalu⬤ and Cyriac Antony(✉)⬤

Indian Institute of Information Technology,
Design and Manufacturing (IIITDM) Kancheepuram, Chennai, India
{shalu,mat17d001}@iiitdm.ac.in

Abstract. A k-star colouring of a graph G is a function $f : V(G) \to \{0, 1, \dots, k-1\}$ such that $f(u) \neq f(v)$ for every edge uv of G, and G does not contain a 4-vertex path bicoloured by f as a subgraph. For $k \in \mathbb{N}$, the problem k-STAR COLOURABILITY takes a graph G as input and asks whether G is k-star colourable. By the construction of Coleman and Moré (SIAM J. Numer. Anal., 1983), for all $k \geq 3$, k-STAR COLOURABILITY is NP-complete for graphs of maximum degree $d = k(k - 1 + \lceil \sqrt{k} \rceil)$. For $k = 4$ and $k = 5$, the maximum degree in this NP-completeness result is $d = 20$ and $d = 35$ respectively. We reduce the maximum degree to $d = 4$ in both cases: i.e., 4-STAR COLOURABILITY and 5-STAR COLOURABILITY are NP-complete for graphs of maximum degree four. We also show that for all $k \geq 3$ and $d < k$, the time complexity of k-STAR COLOURABILITY is the same for graphs of maximum degree d and d-regular graphs (i.e., the problem is either in P for both classes or NP-complete for both classes).

Keywords: Graph coloring · Vertex coloring · Star coloring · Complexity

1 Introduction

The star colouring is a well-known variant of (vertex) colouring introduced by Grünbaum [7] in the 1970s. The scientific computing community independently discovered star colouring in the 1980s and used it for lossless compression of symmetric sparse matrices, which is in turn used in the estimation of sparse Hessian matrices (see the survey [6]). A k-colouring f of a graph G, say $f : V(G) \to \{0, 1, \dots, k - 1\}$, is a k-star colouring of G if G does not contain a 4-vertex path bicoloured by f as a subgraph. For every positive integer k, the problem k-STAR COLOURABILITY takes a graph G as input and asks whether G is k-star colourable. The problem k-COLOURABILITY is defined likewise. The problem STAR COLOURABILITY takes a graph G and a positive integer k as input and asks whether G is k-star colourable.

M. A. Shalu—Supported by SERB(DST), MATRICS scheme MTR/2018/000086.

N. Balachandran and R. Inkulu (Eds.): CALDAM 2022, LNCS 13179, pp. 78–90, 2022.
https://doi.org/10.1007/978-3-030-95018-7_7

The complexity of star colouring is studied in various graph classes. STAR COLOURABILITY is polynomial-time solvable for cographs [10] and line graphs of trees [12]. For the class of co-bipartite graphs, STAR COLOURABILITY is NP-complete eventhough k-STAR COLOURABILITY is polynomial-time solvable for every $k \in \mathbb{N}$ [2,13]. Coleman and Moré [3] proved that for all $k \geq 3$, k-STAR COLOURABILITY is NP-complete for bipartite graphs. The problem 3-STAR COLOURABILITY is NP-complete for planar bipartite graphs [1], line graphs of sub-cubic graphs [9] and graphs of arbitrarily large girth [2]. Gebremedhin et al. [5] produced an inapproximation result on star colouring of bipartite graphs. In this paper, we focus on the classes of bounded degree graphs and regular graphs.

For motivation, let us look at the complexity of colouring in bounded degree graphs. It is well-known that for all $k \geq 3$, k-COLOURABILITY is NP-complete. Emden-Weinert et al. [4] proved that k-COLOURABILITY remains NP-complete when restricted to graphs of maximum degree $d = k - 1 + \lceil \sqrt{k} \rceil$. For sufficiently large k, the maximum degree in this NP-completeness result is the minimum possible (because the problem is in P for $d < k - 1 + \lceil \sqrt{k} \rceil$ [11, Theorem 43]). Coleman and Moré [3] proved that for $k \geq 3$, k-STAR COLOURABILITY is NP-complete for bipartite graphs. From their construction, it follows that for all $k \geq 3$, k-STAR COLOURABILITY is NP-complete for graphs of maximum degree $d = k(k-1+\lceil \sqrt{k} \rceil)$. We seek to reduce the maximum degree in this NP-completeness result to the minimum possible. For $k = 3$, Lei et al. [9] reduced the maximum degree to $d = 4$ (because 3-STAR COLOURABILITY is NP-complete for line graphs of subcubic graphs). No similar result exists for $k > 3$. For $k = 4$ and $k = 5$, the maximum degree of the output graph in Coleman and Moré's construction is $d = 20$ and $d = 35$ respectively. We reduce the maximum degree to $d = 4$ in both cases: i.e., 4-STAR COLOURABILITY and 5-STAR COLOURABILITY are NP-complete for graphs of maximum degree four.

It is easy to show that for all k and d, the time complexity of k-COLOURABILITY is the same for graphs of maximum degree d and d-regular graphs (i.e., the problem is either in P for both classes or NP-complete for both classes). Such a property does not hold in general for star colouring: 3-STAR COLOURABILITY is NP-complete for graphs of maximum degree four [9] whereas it is in P for 4-regular graphs [15]. We show that for $k \geq 3$ and $d < k$, the time complexity of k-STAR COLOURABILITY is the same for graphs of maximum degree d and d-regular graphs. We briefly discuss the consequences of our results to hardness transitions of the problem k-STAR COLOURABILITY restricted to the class of graphs of maximum degree d (resp. d-regular graphs).

The paper is organized as follows. See Sect. 2 for definitions. The hardness results on bounded degree graphs and regular graphs appear in Sects. 3 and 4, respectively.

2   Definitions

All graphs considered in this paper are finite, simple and undirected. We follow West [14] for graph theory terminology and notation. The *girth* of a graph G is the length of a shortest cycle in G. A k-colouring of a graph G is a function f from the vertex set of G to a set of k colours, say $\{0, 1, \ldots, k-1\}$, such that f maps every pair of adjacent vertices to different colours. A k-colouring f of G is a k-*star colouring* if G does not contain a 4-vertex path bicoloured by f as a subgraph.

For every positive integer k, the decision problem k-COLOURABILITY takes a graph G as input and asks whether G is k-colourable. The problem k-STAR COLOURABILITY is defined likewise. To denote the restriction of a decision problem, we write the conditions in parenthesis. For instance, 4-STAR COLOURABILITY($\Delta = 4$, girth= 5) denotes the problem 4-STAR COLOURABILITY restricted to the class of graphs G with the maximum degree $\Delta(G) = 4$ and girth(G) = 5. For every construction in this paper, the output graph is made up of gadgets. For every gadget, only some of the vertices in it are allowed to have edges to vertices outside the gadget; we call these vertices as *terminals*. In diagrams, we draw a *circle around each terminal*. A decision problem is said to have a *hardness transition* with respect to a parameter d at a point $d = x$ if either (i) the problem is in P for $d = x - 1$ and it is NP-complete for $d = x$, or (ii) the problem is NP-complete for $d = x - 1$ and it is in P for $d = x$.

3   Bounded Degree Graphs

In this section, we prove that 4-STAR COLOURABILITY and 5-STAR COLOURABILITY are NP-complete for graphs of maximum degree four. To this end, we employ two similar constructions named Construction 1 and Construction 2. Detailed proofs are omitted from Sect. 3.1 due to space constraints.

3.1   4-Star Colouring

We use Petersen graph minus one vertex as the gadget component to build gadgets in Construction 1.

Construction 1.

Input: A 4-regular graph G.
Output: A graph G' of maximum degree four and girth five.
Guarantee: G is 3-colourable if and only if G' is 4-star colourable.
Steps:

Let $v_1, v_2, \ldots, v_n$ be the vertices in G. First, replace each vertex of G by a vertex gadget as shown in Fig. 1. The vertex gadget for v_i has five terminals, and the terminals $v_{i,1}, v_{i,2}, v_{i,3}, v_{i,4}$ accommodate the four edges incident on v_i in G in a one-to-one fashion (order does not matter). So, corresponding to each edge $v_i v_j$ in G, there is an edge $v_{i,k} v_{j,\ell}$ in G' for some $k, \ell \in \{1, 2, 3, 4\}$. Finally, introduce the chain gadget displayed in Fig. 2, and join $v_{i,0}$ to v_i^* for $i = 1, 2, \ldots, n$.

Proof of Guarantee (Overview). The following claims are pivotal to the proof.

Claim 1: Every 4-star colouring of the gadget component (i.e., Petersen graph minus one vertex) must assign the same colour on all three degree-2 vertices of the gadget component.

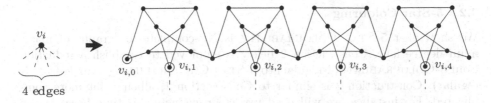

Fig. 1. Replacement of vertex by vertex gadget.

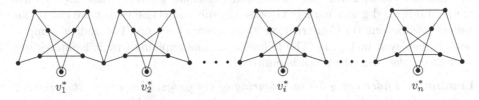

Fig. 2. Chain gadget in Construction 1.

Claim 2: For every 4-star colouring of the vertex gadget (resp. chain gadget), all terminals of the gadget get the same colour.

Claim 3: For every $v \in V(G)$ and every colour $c \neq 0$, the vertex gadget for v admits a 4-star colouring such that all terminals of the gadget have colour c, no neighbour of $v_{i,0}$ is coloured 0, and neighbours of $v_{i,1}, v_{i,2}, v_{i,3}, v_{i,4}$ are all coloured 0.

Claim 4: The chain gadget admits a 4-star colouring such that all terminals of the gadget get colour 0.

The proofs of Claims 1, 3 and 4 are omitted. Claim 2 follows from Claim 1.

Suppose that G admits a 3-colouring $f : V(G) \rightarrow \{1, 2, 3\}$. The following steps give a 4-star colouring of G': (i) for each $v \in V(G)$, choose $c = f(v)$ and colour the vertex gadget for v by the colouring guaranteed in Claim 3, and (ii) colour the chain gadget by the colouring guaranteed in Claim 4.

Conversely, suppose that G' admits a 4-star colouring f'. By Claim 2, all terminals of a vertex gadget (resp. chain gadget) get the same colour under f'. Without loss of generality, assume that $f'(v_i^*) = 0$ for $i = 1, 2, \ldots, n$. Since $v_{i,0} v_i^*$ is an edge for $1 \leq i \leq n$, the chain gadget forbids colour 0 at vertices $v_{i,j}$ for $1 \leq i \leq n$ and $1 \leq j \leq 4$. Therefore, the function $f : V(G) \rightarrow \{1, 2, 3\}$ defined as $f(v_i) = f'(v_{i,0})$ for $1 \leq i \leq n$ is a 3-colouring of G. $\qquad \square$

Construction 1 establishes a reduction from 3-COLOURABILITY(4-regular) to 4-STAR COLOURABILITY($\Delta = 4$, girth = 5). Note that Construction 1 requires only time polynomial in $m + n$ because $|E(G')| = 61n + m$ and $|V(G')| = 41n + 1$ (where $m = |E(G)|$ and $n = |V(G)|$). Thus, we have the following theorem.

Theorem 1. 4-STAR COLOURABILITY *is NP-complete for graphs of maximum degree four and girth five.* $\qquad \square$

3.2   5-Star Colouring

We show that 5-STAR COLOURABILITY is NP-complete for graphs of maximum degree four. Construction 2 below is employed to establish a reduction from 3-COLOURABILITY(4-regular) to 5-STAR COLOURABILITY(triangle-free, 4-regular). Construction 2 is similar to Construction 1, albeit a bit more complicated. For instance, we will need two chain gadgets this time because two colours should be forbidden. The gadgets used in the construction are made of two gadgets called 2-in-2-out gadget and not-equal gadget. These are in turn made of one fixed graph namely Grötzsch graph minus one vertex; we call it the gadget component (in Construction 2) for obvious reason. The gadget component is displayed in Fig. 3a. The following lemma explains why it is interesting for 5-star colouring (the proof is omitted).

Lemma 1. *Under every 5-star colouring of the gadget component, the degree-2 vertices of the graph should get pair-wise distinct colours. Moreover, every 5-star colouring of the gadget component must be of the form displayed in Fig. 3b or Fig. 3c upto colour swaps.*

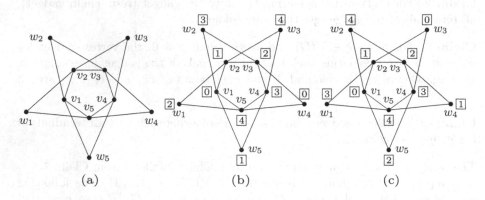

Fig. 3. (a) Gadget component, (b, c) General form of 5-star colouring of it.

The 2-in-2-out gadget is displayed in Fig. 4. Observe that two copies of the gadget component are part of this gadget. The following lemma shows why 5-star colouring of this gadget is interesting (the proof is omitted).

Lemma 2. *For every 5-star colouring f of the 2-in-2-out gadget, there exist two distinct colours c_1 and c_2 such that $f(y_1) = f(y_2) = f(z_1^*) = f(z_2^*) = c_1$ and $f(y_1^*) = f(y_2^*) = f(z_1) = f(z_2) = c_2$. Moreover, every 3-vertex path containing one of the pendant edges of the gadget is tricoloured by f.*

The not-equal gadget is the graph displayed in Fig. 5. The not-equal gadget is made from one 2-in-2-out gadget by identifying vertex y_1^* of the 2-in-2-out gadget with vertex y_2^* and identifying vertex z_1^* with vertex z_2^*. Hence, the next lemma follows from Lemma 2 (note that $c_1 \neq c_2$ in Lemma 2).

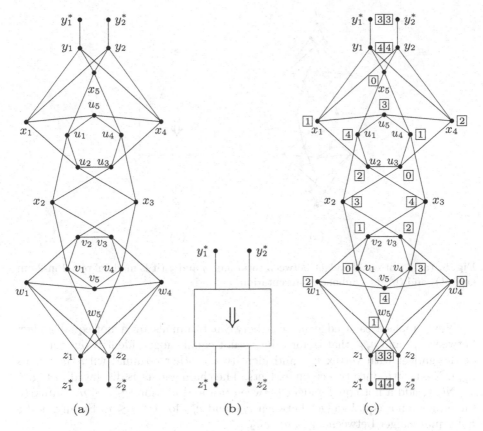

Fig. 4. (a) The 2-in-2-out gadget, (b) its symbolic representation, and (c) a 5-star colouring of the gadget.

Lemma 3. *The terminals of the not-equal gadget should get different colours under each 5-star colouring f. Moreover, every 3-vertex path within the gadget with a terminal as one endpoint is tricoloured by f.*   □

We are now ready to present the construction.

Construction 2.

Input: A 4-regular graph G.
Output: A triangle-free graph G' of maximum degree four.
Guarantee: G is 3-colourable if and only if G' is 5-star colourable.
Steps:

Let $v_1, v_2, \ldots, v_n$ be the vertices in G. First, replace each vertex v_i of G by a vertex gadget as shown in Fig. 6. The vertex gadget for v_i has six terminals namely $v_{i,0}, v_{i,1}, v_{i,2}, v_{i,3}, v_{i,4}$ and $v_{i,5}$. The terminals $v_{i,1}, v_{i,2}, v_{i,3}, v_{i,4}$ accommodate the edges incident on v_i in G. The replacement of vertices by vertex gadgets converts each edge $v_i v_j$ of G to an edge between terminals $v_{i,k}$ and $v_{j,\ell}$ for some $k, \ell \in \{1, 2, 3, 4\}$.

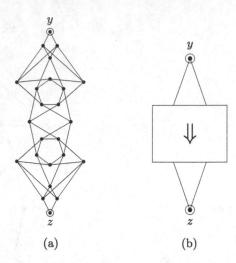

Fig. 5. (a) A not-equal gadget between terminals y and z (it is made of one 2-in-2-out gadget), and (b) its symbolic representation.

Next, replace each edge $v_{i,k} v_{j,\ell}$ between terminals by a not-equal gadget between $v_{i,k}$ and $v_{j,\ell}$ (that is, introduce a not-equal gadget, identify one terminal of the gadget with vertex $v_{i,k}$ and identify the other terminal with the vertex $v_{j,\ell}$). Next, introduce two chain gadgets. The chain gadget is displayed in Fig. 7.

Next, add a not-equal gadget between $v_{i,0}$ and $v_{i,1}^*$ for $1 \leq i \leq n$. Similarly, introduce a not-equal gadget between $v_{i,5}$ and $v_{i,2}^*$ for $1 \leq i \leq n$. Finally, add a not-equal gadget between $x_{1,1}$ and $x_{1,2}$.

Proof of Guarantee. For convenience, let us call the edges $y_1 y_1^*, y_2 y_2^*$ of a 2-in-2-out gadget (see Fig. 4) as in-edges of the 2-in-2-out gadget, edges $z_1 z_1^*, z_2 z_2^*$ as out-edges of the 2-in-2-out-gadget, vertices y_1^*, y_2^* as in-vertices of the 2-in-2-out gadget, and vertices z_1^*, z_2^* as out-vertices of the 2-in-2-out gadget. The next claim follows from Lemma 2.

Claim 1: If an in-edge of a 2-in-2-out gadget is an out-edge of another 2-in-2-out gadget, the colour of the out-vertices of both gadgets must be the same.

Next, we point out a property of the vertex gadget and the chain gadget.

Claim 2: All terminals of a vertex gadget (resp. chain gadget) should get the same colour under a 5-star colouring.

By Claim 1, if an in-edge of a 2-in-2-out gadget is an out-edge of another 2-in-2-out gadget, the colour of out-vertices of both gadgets must be the same. Repeated application of this idea proves Claim 2.

We are now ready to prove the guarantee. Suppose that G admits a 3-colouring $f : V(G) \to \{2, 3, 4\}$. A 5-colouring $f' : V(G') \to \{0, 1, 2, 3, 4\}$ of G' is

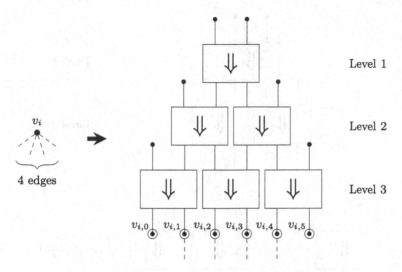

Fig. 6. Replacement of vertex by vertex gadget.

constructed as follows. First, assign $f'(v_{i,j}) = f(v_i)$ for $1 \leq i \leq n$ and $0 \leq j \leq 5$. Extend this into a 5-star colouring of the vertex gadget by using the scheme in Fig. 4c on each 2-in-2-out-gadget within the vertex gadget (use the scheme in Fig. 4c if $f'(v_{i,j}) = 4$; suitably swap colours in other cases). To colour the first chain gadget, colour each 2-in-2-out gadget within this chain gadget using the scheme obtained from Fig. 4c by swapping colour 4 with colour 0. Similarly, for the second chain gadget, colour each 2-in-2-out gadget within the chain gadget using the scheme obtained from Fig. 4c by swapping colour 4 with colour 1. To complete the colouring, it suffices to extend the partial colouring to not-equal gadgets. For each not-equal gadget between two terminals, say terminal y and terminal z, colour the 2-in-2-out gadget within the not-equal gadget using the scheme obtained from Fig. 4c by swapping colour 3 with colour $f'(y)$ and swapping colour 4 with colour $f'(z)$.

By Lemma 2 and Lemma 3 (see the second statements in both lemmas), every 3-vertex path in any gadget in G' containing a terminal of the gadget as an endpoint is tricoloured by f'. In addition, the construction of the graph G' is merely glueing together terminals of different gadgets. Therefore, there is no P_4 in G' bicoloured by f'; that is, f' is a 5-star colouring of G'.

Conversely, suppose that G' admits a 5-star colouring $f' : V(G') \rightarrow \{0, 1, 2, 3, 4\}$. By Claim 2, all terminals of a vertex/chain gadget should have the same colour under f'. As there is a not-equal gadget between $x_{1,1}$ and $x_{1,2}$, $f'(x_{1,1}) \neq f'(x_{1,2})$ (by Lemma 3). Without loss of generality, assume that $f'(x_{1,1}) = 0$ and $f'(x_{1,2}) = 1$. By Claim 2, all terminals of the first chain gadget have colour 0; that is, $f'(x_{1,1}) = f(x_{2,1}) = f'(v_{i,1}^*) = 0$ for $1 \leq i \leq n$. Similarly, all terminals of the second chain gadget have colour 1; that is, $f'(x_{1,2}) = f'(x_{2,2}) = f'(v_{i,2}^*) = 1$ for $1 \leq i \leq n$. By Claim 2,

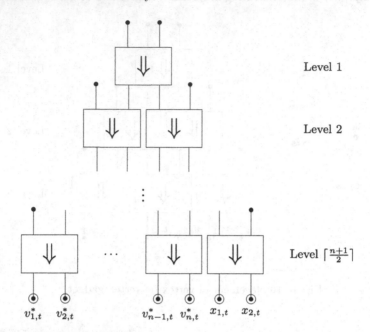

Fig. 7. t-th chain gadget in Construction 2 if n is even where $t = 1$ or 2. If n is odd, the t-th chain gadget is the same except that it has only $n + 1$ terminals $v_{1,t}^*, v_{2,t}^*, \ldots, v_{n,t}^*$ and $x_{1,t}$. A chain gadget is similar to a vertex gadget; the only difference is that it has more levels and terminals.

all terminals of the vertex gadget for v_1 have the same colour under f', say colour c. Since there is a not-equal gadget between $v_{1,0}$ and $v_{1,1}^*$, we have $c = f'(v_{1,0}) \neq f'(v_{1,1}^*) = 0$. Since there is a not-equal gadget between $v_{1,5}$ and $v_{1,2}^*$, we have $c = f'(v_{1,5}) \neq f'(v_{1,2}^*) = 1$. So, $c \in \{2, 3, 4\}$. Hence, for $0 \leq j \leq 5$, $f'(v_{1,j}) \in \{2, 3, 4\}$. Similarly, for $1 \leq i \leq n$ and $0 \leq j \leq 5$, $f'(v_{i,j}) \in \{2, 3, 4\}$. Moreover, whenever $v_i v_j$ is an edge in G, there is a not-equal gadget between terminals $v_{i,k}$ and $v_{j,\ell}$ in G' for some $k, \ell \in \{1, 2, 3, 4\}$ and hence $f'(v_{i,k}) \neq f'(v_{j,\ell})$. Therefore, the function $f : V(G) \to \{2, 3, 4\}$ defined as $f(v_i) = f'(v_{i,0})$ is indeed a 3-colouring of G. This proves the converse part and thus the guarantee. □

Theorem 2. 5-STAR COLOURABILITY *is NP-complete for triangle-free graphs of maximum degree four.*

Proof. We employ Construction 2 to establish a reduction from 3-COLOURABILTY(4-regular) to 5-STAR COLOURABILITY(triangle-free, $\Delta = 4$). Let G be an instance of 3-COLOURABILTY(4-regular). From G, construct an instance G' of 5-STAR COLOURABILITY(triangle-free, $\Delta = 4$) by Construction 2. Let $m = |E(G)|$ and $n = |V(G)|$. In G', there are at most $6n + m + 2(1 + 2 + \cdots + \lceil (n+1)/2 \rceil + n) + 1 \leq \frac{1}{4}(n^2 + 46n + 12)$ 2-in-2-out gadgets and in addition at most $16n + 8$ vertices and $32n + 12$ edges. So, G' can be constructed in time polynomial in n. By the guarantee in Construction 2, G is 3-colourable if and only if G' is 5-star colourable. □

4   Regular Graphs

We prove that for all $k \geq 3$ and $d < k$, the complexity of k-STAR COLOURABILITY is the same for graphs of maximum degree d and d-regular graphs. That is, for all $k \geq 3$ and $d < k$, k-STAR COLOURABILITY restricted to graphs of maximum degree d is in P (resp. NP-complete) if and only if k-STAR COLOURABILITY restricted to d-regular graphs is in P (resp. NP-complete). First, we show that for all $k \geq 3$, the complexity of k-STAR COLOURABILITY is the same for graphs of maximum degree $k - 1$ and $(k - 1)$-regular graphs.

Construction 3.
Parameter: An integer $k \geq 3$.
Input: A graph G of maximum degree $k - 1$.
Output: A $(k - 1)$-regular graph G'.
Guarantee 1: G is k-star colourable if and only if G' is k-star colourable.
Guarantee 2: If G is triangle-free (resp. bipartite), then G' is triangle-free (resp. bipartite).
Steps:
Introduce two copies of G. For each vertex v of G, introduce $(k - 1) - deg_G(v)$ filler gadgets (see Fig. 8) between the two copies of v.

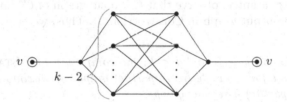

Fig. 8. A filler gadget for $v \in V(G)$.

It is easy to show that each k-star colouring of G can be extended into a k-star colouring of G' (detailed proofs of the guarantees in Construction 3 are omitted). Thanks to Construction 3, we have the following theorem.

Theorem 3. *For all $k \geq 3$, the complexity of k-STAR COLOURABILITY is the same for graphs of maximum degree $k - 1$ and $(k - 1)$-regular graphs. In addition, for $k \geq 3$, the complexity of k-STAR COLOURABILITY is the same for triangle-free (resp. bipartite) graphs of maximum degree $k - 1$ and triangle-free (resp. bipartite) $(k - 1)$-regular graphs.* □

By Theorem 3, the complexity of 5-STAR COLOURABILITY is the same for triangle-free graphs of maximum degree four and triangle-free 4-regular graphs. Thus, by Theorem 2, we have the following.

Theorem 4. *5-STAR COLOURABILITY is NP-complete for triangle-free 4-regular graphs.* □

Construction 4.

Parameters: Integers $k \geq 3$ and $d < k$.
Input: A graph G of maximum degree d.
Output: A d-regular graph G^*.
Guarantee: G is k-star colourable if and only if G^* is k-star colourable.
Steps:

Introduce two copies of G. For each vertex v of G, introduce $d - deg_G(v)$ filler gadgets (see Fig. 9) between the two copies of v.

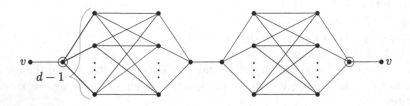

Fig. 9. A filler gadget for $v \in V(G)$.

To prove the guarantee, observe that G is a subgraph of G^* and G^* is a subgraph of G' (the output graph in Construction 3). Thus, we have the following.

Theorem 5. *For all $k \geq 3$ and $d < k$, the complexity of k-STAR COLOURA-BILITY is the same for (triangle-free/bipartite) graphs of maximum degree d and (triangle-free/bipartite) d-regular graphs.* □

5   Conclusion

A decision problem has a *hardness transition* with respect to a parameter d at a point $d = x$ if either (i) the problem is in P for $d = x-1$ and it is NP-complete for $d = x$, or (ii) the problem is NP-complete for $d = x-1$ and it is in P for $d = x$ (see [8]). For all $k \geq 3$, the problem k-COLOURABILITY restricted to graphs of maximum degree d has exactly one point of hardness transition with respect to d (to produce a reduction, add a disjoint copy of $K_{1,d+1}$). By the same reasoning, star colouring displays similar behaviour when restricted to bounded degree graphs. That is, the problem k-STAR COLOURABILITY restricted to graphs of maximum degree d has exactly one point of hardness transition (w.r.t. d), say $d = T_1^{(k)}$. For sufficiently large k, $d = k - 1 + \lceil\sqrt{k}\rceil$ is the unique point of hardness transition (w.r.t. d) for the problem k-COLOURABILITY restricted to graphs of maximum degree d (see [4] and [11, Theorem 43]). In contrast, for the problem k-STAR COLOURABILITY in graphs of maximum degree d, the unique point of hardness transition (namely $T_1^{(k)}$) is unknown even for sufficiently large k. Since 4-STAR

COLOURABILITY and 5-STAR COLOURABILITY are NP-complete for graphs of maximum degree four (see Theorems 1 and 2), $T_1^{(4)} \leq 4$ and $T_1^{(5)} \leq 4$.

In a forthcoming (unpublished) paper, we prove that for all $k \geq 4$, $T_1^{(k)} \leq k$. We suspect that for all $k \geq 5$, $T_1^{(k)} \leq k - 1$.

When it comes to the class of regular graphs, colouring shows the same behaviour as in the class of bounded degree graphs because k-COLOURABILITY is NP-complete for d-regular graphs if and only if k-COLOURABILITY is NP-complete for graphs of maximum degree d. Such a property does not hold in general for star colouring; for example, 3-STAR COLOURABILITY is NP-complete for graphs of maximum degree four [9] whereas it is in P for 4-regular graphs [15]. We show that for all $k \geq 3$ and $d < k$, the complexity of k-STAR COLOURABILITY is the same for graphs of maximum degree d and d-regular graphs. The following observation is a consequence of this (the proof is omitted).

Observation 1. *If $k \geq 3$ and $T_1^{(k)} \leq k - 1$, then between $d = 1$ and $d = k - 1$, the problem k-STAR COLOURABILITY in d-regular graphs has exactly one point of hardness transition (w.r.t. d) namely $d = T_1^{(k)}$.*

In a forthcoming (unpublished) paper, we show that for all $k \geq 4$, the problem k-STAR COLOURABILITY is NP-complete for graphs of maximum degree k and polynomial-time solvable for d-regular graphs for each $d > 2k-4$. In particular, 5-STAR COLOURABILITY is polynomial-time solvable for d-regular graphs for each $d > 6$. Since 5-STAR COLOURABILITY in d-regular graphs is NP-complete for $d = 4$ (see Theorem 4) and in P for $d \leq 2$ as well as $d > 6$, 5-STAR COLOURABILITY in d-regular graphs has at least two points of hardness transition. In general, for $k \geq 5$, k-STAR COLOURABILITY in d-regular graphs has either zero or at least two points of hardness transition (the latter is more likely). We conjecture that there are exactly two points of hardness transition and the second point of hardness transition is close to $2k - 4$.

Conjecture 1. For $k \geq 5$, the problem k-STAR COLOURABILITY in d-regular graphs has exactly two points of hardness transition $d = T_1^{(k)}$ and $d = T_2^{(k)}$, and the second point of hardness transition $T_2^{(k)}$ satisfies $|T_2^{(k)} - (2k - 4)| \leq 1$.

Acknowledgements. The first author is supported by SERB(DST), MATRICS scheme MTR/2018/000086. We thank Kirubakaran V. K. and three anonymous referees for their suggestions.

References

1. Albertson, M.O., Chappell, G.G., Kierstead, H.A., Kündgen, A., Ramamurthi, R.: Coloring with no 2-colored P_4's. Electron. J. Comb. **11**(1), 26 (2004). https://doi. org/10.37236/1779

2. Bok, J., Jedličková, N., Martin, B., Paulusma, D., Smith, S.: Acyclic, star and injective colouring: a complexity picture for H-free graphs. In: Grandoni, F., Herman, G., Sanders, P. (eds.) 28th Annual European Symposium on Algorithms (ESA 2020). Leibniz International Proceedings in Informatics (LIPIcs), vol. 173, pp. 22:1–22:22. Schloss Dagstuhl-Leibniz-Zentrum für Informatik, Dagstuhl, Germany (2020). https://doi.org/10.4230/LIPIcs.ESA.2020.22

3. Coleman, T.F., Moré, J.J.: Estimation of sparse Jacobian matrices and graph coloring problems. SIAM J. Numer. Anal. **20**(1), 187–209 (1983)

4. Emden-Weinert, T., Hougardy, S., Kreuter, B.: Uniquely colourable graphs and hardness of colouring graphs of large girth. Comb. Probab. Comput. **7**(4), 375–386 (1998). https://doi.org/10.1017/S0963548398003678

5. Gebremedhin, A.H., Tarafdar, A., Manne, F., Pothen, A.: New acyclic and star coloring algorithms with application to computing Hessians. SIAM J. Sci. Comput. **29**(3), 1042–1072 (2007). https://doi.org/10.1137/050639879

6. Gebremedhin, A.H., Manne, F., Pothen, A.: What color is your Jacobian? Graph coloring for computing derivatives. SIAM Rev. **47**(4), 629–705 (2005). https://doi.org/10.1137/S0036144504444711

7. Grünbaum, B.: Acyclic colorings of planar graphs. Israel J. Math. **14**, 390–408 (1973). https://doi.org/10.1007/BF02764716

8. mikero (https://cstheory.stackexchange.com/users/149/mikero): Parameterized complexity from P to NP-hard and back again. Theoretical Computer Science Stack Exchange. https://cstheory.stackexchange.com/q/3473, (version: 13 April 2017)

9. Lei, H., Shi, Y., Song, Z.X.: Star chromatic index of subcubic multigraphs. J. Graph Theory **88**(4), 566–576 (2018). https://doi.org/10.1002/jgt.22230

10. Lyons, A.: Acyclic and star colorings of cographs. Discret. Appl. Math. **159**(16), 1842–1850 (2011). https://doi.org/10.1016/j.dam.2011.04.011

11. Molloy, M., Reed, B.: Colouring graphs when the number of colours is almost the maximum degree. J. Comb. Theory Ser. B **109**, 134–195 (2014). https://doi.org/10.1016/j.jctb.2014.06.004

12. Omoomi, B., Roshanbin, E., Dastjerdi, M.V.: A polynomial time algorithm to find the star chromatic index of trees. Electron. J. Comb. **28**(1) (2021). https://doi.org/10.37236/9202. Article No. 16

13. Shalu, M.A., Antony, C.: The complexity of restricted star colouring. Discret. Appl. Math. (2021, in press). https://doi.org/10.1016/j.dam.2021.05.015. Available online: 31 May 2021

14. West, D.B.: Introduction to graph theory, 2nd edn. Prentice Hall, Upper Saddle River (2001)

15. Xie, D., Xiao, H., Zhao, Z.: Star coloring of cubic graphs. Inf. Process. Lett. **114**(12), 689–691 (2014). https://doi.org/10.1016/j.ipl.2014.05.013

On Conflict-Free Spanning Tree:
Algorithms and Complexity

Bruno José S. Barros[1], Luiz Satoru Ochi[1], Rian Gabriel S. Pinheiro[2],
and Uéverton S. Souza[1($\boxtimes$)]

[1] Instituto de Computação, Universidade Federal Fluminense, Niterói, Brazil
bruno_barros@id.uff.br, {satoru,ueverton}@ic.uff.br
[2] Instituto de Computação, Universidade Federal de Alagoas, Maceió, Brazil
rian@ic.ufal.br

Abstract. A natural constraint in real-world applications is to avoid
conflicting elements in the solution of problems. Given an undirected
graph $G = (V, E)$ where each edge $e \in E$ has a positive integer weight
$\omega(e)$, and a conflict graph $\hat{G} = (\hat{V}, \hat{E})$ such that $\hat{V} \subseteq E$ and each edge
$\hat{e} = (e_1, e_2) \in \hat{E}$ represents a conflict between two edges $e_1, e_2 \in E$, in
the MINIMUM CONFLICT-FREE SPANNING TREE (MCFST) problem we
are asked to find (if any) a spanning tree avoiding pairs of conflicting
edges (conflict-free) with minimum cost, i.e., a minimum solution among
spanning trees T such that $E(T)$ is an independent set of $\hat{G}$. A spanning
tree T of G is a feasible solution for an instance $I = (G, \hat{G})$ of MCFST if
$E(T)$ is an independent set of $\hat{G}$. In contrast to the polynomial-time solvability of MINIMUM SPANNING TREE, to determine whether an instance
$I = (G, \hat{G})$ of MCFST admits a feasible solution is $\mathcal{NP}$-complete. In this
paper, we present a multivariate complexity analysis of MCFST by considering particular classes of graphs G and $\hat{G}$. In particular, we show that
the problem of determining whether an instance $I = (G, \hat{G})$ of MCFST
has a feasible solution is $\mathcal{NP}$-complete even if G is a bipartite planar
subcubic graph, and $\hat{G}$ is a disjoint union of paths of size three (P_3).
Moreover, we show that whether G is a complete graph and $\hat{G}$ is a disjoint union of stars, then a feasible solution for $I = (G, \hat{G})$ can be found
in polynomial time. In addition, we present (in)approximability results
for MCFST on complete graphs G, and an FPT algorithm parameter-
ized by the distance to $\mathcal{F}$ of the conflict graph $\hat{G}$, where $\mathcal{F}$ is a hereditary
graph class such that MCFST on conflict graphs $\hat{G} \in \mathcal{F}$ can be solved
in polynomial time.

Keywords: Conflict-free · Spanning tree · Approximation · FPT

1   Introduction

Given an edge-weighted graph G, the MINIMUM SPANNING TREE (MST) problem consists of finding a spanning tree of G having minimum cost, where the

Supported by CAPES, CNPq, and FAPERJ.

N. Balachandran and R. Inkulu (Eds.): CALDAM 2022, LNCS 13179, pp. 91–102, 2022.
https://doi.org/10.1007/978-3-030-95018-7_8

cost is the sum of the weights of its edges. MST is a very important problem with application in several areas. Besides that, the MST problem can be solved in polynomial time by either Kruskal or Prim algorithms [9,13]. A survey on MST can be found in [8].

Conflict-free variants of classical decision and optimization problems have aroused considerable interest and have been studied in the recent literature. A classical computational problem $\mathcal{X}$ can be turned into a conflict-free version of $\mathcal{X}$ by coupling a conflict graph $\hat{G}$ together with the instances $I_\mathcal{X}$ of $\mathcal{X}$. In such a conflict graph, the vertices represent elements of $I_\mathcal{X}$, and its edges represent pairs of elements that are prohibited from being mutually in the same solution. A solution for an instance $(I_\mathcal{X}, \hat{G})$ of the conflict-free version of $\mathcal{X}$ represents, simultaneously, an independent set in $\hat{G}$ and a solution for the instance $I_\mathcal{X}$ of $\mathcal{X}$.

Several optimization problems have already been studied from the viewpoint of conflict-free versions, such as BIN PACKING [1,7], KNAPSACK [11], MAXIMUM MATCHING and SHORTEST PATH [3]. In this work, we deal with the conflict-free versions of the MINIMUM SPANNING TREE problem. The CONFLICT-FREE SPANNING TREE (CFST) problem was introduced in [2] and consists in determining if a graph G has a conflict-free spanning tree according to a conflict graph $\hat{G}$ representing conflicts between edges. In addition, the MINIMUM CONFLICT-FREE SPANNING TREE (MCFST) problem consists in finding a minimum conflict-free spanning tree (if any) of $(G, \hat{G})$ where G is an edge-weighted simple graph. Below we formally present both problems.

CONFLICT-FREE SPANNING TREE (CFST)
Input: A simple undirected graph $G = (V, E)$, and a conflict graph $\hat{G} = (\hat{V}, \hat{E})$, where $\hat{V} \subseteq E(G)$.
Question: Is there a spanning tree T of G such that T induces an independent set in $\hat{G}$?

MINIMUM CONFLICT-FREE SPANNING TREE (MCFST)
Input: A simple undirected graph $G = (V, E)$ where each edge $e \in E$ has a positive integer weight $\omega(e)$, and a conflict graph $\hat{G} = (\hat{V}, \hat{E})$, where $\hat{V} \subseteq E$.
Goal: Find (if any) a conflict-free spanning tree of $(G, \hat{G})$ which minimizes the sum of its edge weights.

Note that given a simple undirected edge-weighted graph $G = (V, E)$, and a conflict graph $\hat{G} = (\hat{V}, \hat{E})$ of G, the CONFLICT-FREE SPANNING TREE problem consists in determining whether $(G, \hat{G})$ admits a feasible solution, while MCFST is the natural conflict-free minimization version of MINIMUM SPANNING TREE.

In [3], the authors proved that CFST remains NP-hard even when the conflict graph is a disjoint union of paths of size three (P_3's - paths of length 2). Also, they show that if the conflict graph is a disjoint union of paths of size 2 (P_2's -paths of length 1), then MCFST becomes polynomial-time solvable. In addition, when the underlying graph G is a cactus it holds that CFST (the feasibility problem) is polynomial-time solvable, but MCFST (the optimization version) is still $\mathcal{NP}$-hard [16]. Another interesting result obtained by [16] is that if the conflict graph is a cluster graph (disjoint union of cliques), then MCFST can be solved in polynomial-time. In addition, a unifying model for locally constrained spanning tree problems was presented in [5].

In this paper, we present a multivariate complexity analysis of MCFST by considering particular classes of graphs G and $\hat{G}$. In particular, we show that the problem of determining whether an instance $I = (G, \hat{G})$ admits a feasible solution is $\mathcal{NP}$-complete even when G is a bipartite planar subcubic graph, and $\hat{G}$ is a disjoint union of paths of size three. Moreover, we show that whether G is a complete graph and $\hat{G}$ is a disjoint union of stars, then a feasible solution for $I = (G, \hat{G})$ can be found in polynomial time, while the problem of finding an optimum solution still $\mathcal{NP}$-hard. Finally, concerning MCFST, (in)approximability results on complete graphs G, and an FPT algorithm parameterized by the distance to $\mathcal{F}$ of $\hat{G}$ are also presented, where $\mathcal{F}$ is any hereditary graph class such that MCFST on conflict graphs $\hat{G} \in \mathcal{F}$ can be solved in polynomial time.

2   Preliminaries

Graphs

We consider a simple undirected graph $G = (V, E)$ (or just a graph, for short) as a pair of sets such that $V(G)$ is the set of vertices and $E(G)$ is the set of edges, where each edge connects a distinct pair of vertices. When two vertices share an edge they are *"adjacent"* (or *"neighbors"*). The number of neighbors of a vertex v is called the *degree* of v. An *induced subgraph* of a graph G is another graph, formed from a subset X of $V(G)$ and all of the edges of G connecting pairs of vertices in X. For $X \subseteq V(G)$, we denote by $G[X]$ the subgraph of G induced by X. A set $X \subseteq V(G)$ is an *independent set* of G if $G[X]$ does not contain any edge. A *complete graph*, K_n, is a simple graph with n vertices and $\frac{n^2-n}{2}$ edges. A *clique* of a graph G is a subgraph of G that is complete. A *path* is a sequence of vertices that do not repeat any vertex, and each pair of consecutive vertices in the sequence are adjacent in the graph. A *cycle* is a path where the first and last vertex of the sequence are also adjacent. Vertices v_1 and v_2 are connected if there is a path beginning with v_1 and ending with v_2. A graph G is *connected* if every pair of vertices in G is connected. A graph that is not connected is called *disconnected*. A *tree* is a connected graph having no cycle. A *star* is a tree having a vertex v called *center* (or *root*) such that all remaining vertices are adjacent to v. A *connected component* of a graph G is a maximal connected subgraph of G. If each connected component of G is a clique then G is a *cluster* graph. We named $G[X]$ as a *block* of G if $G[X]$ is connected, there is no $v \in X$ such that $G[X \setminus \{v\}]$ is disconnected, and X is maximal with respect to such properties. A graph G is *subcubic* if each vertex of G has degree at most three. An *induced path* of a graph G is a path that is an induced subgraph of G, i.e., nonadjacent vertices in the sequence are not connected by an edge in G. An induced path with n vertices is a P_n. We denote by E_v the set of edges incident to v. A *spanning tree* T of a graph G is a subgraph of G that is a tree and contains all vertices of G.

Given a graph $G = (V, E)$ where each edge $e \in E$ has a positive integer weight $\omega(e)$, the MINIMUM SPANNING TREE problem consists in finding a spanning tree with the minimum cost, where its cost is the sum of its edge weights. The conflict graph of an instance G of MINIMUM SPANNING TREE is denoted by $\hat{G}$ and its vertex set is formed only by edges of G.

SAT and Its Variants

Given a CNF Boolean formula F, the SAT problem consists of determining if there exists an assignment to the variables of F that satisfies the formula. 3SAT is the particular case of SAT where each clause has at most three literals, and its subcase where each variable occurs at most three times is denoted by $3SAT_3$. 3SAT as well as $3SAT_3$ are classical NP-complete problems extensively used in NP-hardness proofs.

Let F be a CNF Boolean formula, and let B_F be the bipartite incidence graph where the vertices represent the variables and clauses of F, and the edges represent the occurrence of variables in clauses. When the instances are restricted to formulas having a planar embedding for their bipartite incidence graphs, these particular cases are called PLANAR SAT, PLANAR 3SAT, and PLANAR $3SAT_3$ respectively, and all remain NP-complete.

It is useful to consider additional constraints concerning the bipartite incidence graph of the formulas to prove the NP-hardness of problems in special graph classes. In addition to the constraint of being planar, if we add a Hamiltonian cycle that goes through all the variable nodes, and the incidence graph is still planar, then we have a *Var-Linked Planar instance*. Besides, if all variable nodes can be drawn as straight line segments in the x axis and all clauses can be drawn as horizontal lines, connected to the variable nodes by at most 3 vertical segments also known as three-legged embedding [6,12], then we have a *Rectilinear Var-Linked Planar* 3SAT instance.

Tippenhauer [14] showed that VAR-LINKED PLANAR 3SAT is equivalent to RECTILINEAR VAR-LINKED PLANAR 3SAT, and from a planar embedding of the incidence graph B_F a rectilinear var-linked planar embedding can be easily drawn for B_F. Thus, whenever convenient, we can assume rectilinear embeddings for instances of VAR-LINKED PLANAR 3SAT.

In [10], it was shown that VAR-LINKED PLANAR 3SAT is $\mathcal{NP}$-complete even when all variables occur exactly three times with two positive and one negative occurrence, and no variable appears more than once in the same clause. This, together with the equivalence provided by [14], implies that RECTILINEAR VAR-LINKED PLANAR $3SAT_3$ is $\mathcal{NP}$-complete.

3   Computational Complexity

In this section, we present some new theoretical results concerning CFST.

NP-completeness

Using a reduction from RECTILINEAR VAR-LINKED PLANAR 3SAT$_3$ we show that CONFLICT-FREE SPANNING TREE (CFST) remains NP-complete even when the input graph G is bipartite planar and subcubic, and the conflict graph $\hat{G}$ is a disjoint union of induced paths of size three.

Theorem 1. CONFLICT-FREE SPANNING TREE *is* $\mathcal{NP}$-*complete even when* G *is a bipartite planar subcubic graph, and the conflict graph* $\hat{G}$ *is a disjoint union of* P_3's *(induced paths with three vertices).*

Proof. It is easy to see that CONFLICT-FREE SPANNING TREE is in $\mathcal{NP}$. Therefore, it is enough to show the $\mathcal{NP}$-hardness. Let F be an instance of RECTILINEAR VAR-LINKED PLANAR 3SAT$_3$ where all variables occur exactly three times with one negative and two positive occurrences. From F we construct an instance $(G, \hat{G})$ of CONFLICT-FREE SPANNING TREE as follows.

Let B_F be the bipartite incidence graph of F. Considering a rectilinear var-linked planar embedding of B_F, we use such an embedding of B_F as support to construct G.

1. For each clause C_j, create a vertex c_j in G and position it in the plane in the position corresponding to C_j in the embedding of B_F;
2. For each variable x_i, create three vertices x_i, x_i' and $\bar{x}_i$ in G. To position these three vertices, we consider the relative position between the edges with an endpoint in $x_i \in B_F$. If, from the left to the right, the edge representing the negative connection appears first, in the middle, or last, we position over the X-axis the vertices in the sequence "$\bar{x}_i, x_i, x_i'$", "$x_i, \bar{x}_i, x_i'$", or "$x_i, x_i', \bar{x}_i$", respectively. The relative position between the edges always must be maintained by the vertices, and we also add an intermediate vertex between any pair of consecutive vertices as in Fig. 1; (The intermediate vertices make the resulting graph bipartite.)

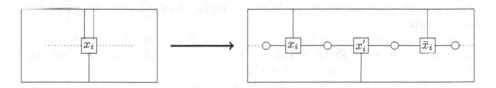

Fig. 1. Vertex gadget

3. If the edge (C_j, x_i) is in B_F, we add in G the equivalent edge, respecting the connection in F and the relative position in the rectilinear var-linked planar embedding of B_F;
4. Create an induced path $\mathcal{P}$ in G connecting all $x_i, x_i', \bar{x}_i$ and intermediate vertices created over the X-axis by adding edges between the consecutive vertices. Note that $\mathcal{P}$ is similar to the straight line crossing the variables in the rectilinear var-linked planar embedding of B_F.

5. The conflict graph $\hat{G}$ is obtained by creating one vertex by each edge representing an occurrence of a literal in a clause. And then, for each i from 1 to n, we add a P_3 representing (x_i, c_j) conflicting with $(\bar{x}_i, c_k)$, and (x'_i, c_l) conflicting with $(\bar{x}_i, c_k)$, where C_k is the clause containing the literal $\bar{x}_i$, and C_j, C_l are the clauses containing the first and second occurrences of the literal x_i, respectively.

Figure 2 shows an instance of CFST obtained by the steps described above. The square and round shapes of the vertices illustrate the bipartition of $V(G)$.

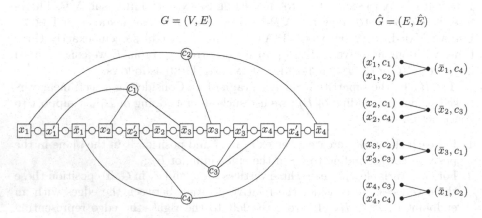

Fig. 2. Instance of CONFLICT-FREE SPANNING TREE obtained from $F = (x_1 + x_2 + \bar{x}_3)(x_1 + x_3 + \bar{x}_4)(\bar{x}_2 + x_3 + x_4)(\bar{x}_1 + x_2 + x_4)$.

Since each clause has at most three variables and each literal occurs at most twice, it is easy to see that the resulting graph G has maximum degree equal to three. In order to observe that G is planar (see step 2), it is enough to note that B_F is planar, and each x_i in B_F has been replaced by a gadget that preserves the planarity.

Now, it remains to show that F is satisfiable if and only if $(G, \hat{G})$ is a "yes"-instance of CONFLICT-FREE SPANNING TREE. Let n be the number of variables of F, and let m be the number of clauses of F.

Let A be a satisfying assignment for F. As G has $6n + m - 1$ vertices, we must select at least $6n + m - 2$ edges forming a connected subgraph of G and an independent set in $\hat{G}$. Note that it is sufficient to find a connected spanning subgraph of G in which the edges induce an independent set in $\hat{G}$ (in this case, the tree can be easily computed). Let X be the set of variables from F, and consider that T initially contains all vertices of G and no edge. From A we obtain the edge set to be added in T as follows.

- Add all edges incident on two vertices over the X-axis in T.
- If $x_i \in X$ has *true* value in A, add the edges incident to x_i and x'_i in T. Otherwise, add in T the remaining edge of $\bar{x}_i$.

Since A is a satisfying assignment, every vertex c_j has at least one edge added to T. Thus, by construction, T is a conflict-free connected spanning subgraph of G, which is enough to our proof.

Now, let T be a conflict-free spanning tree of G. Without loss of generality, we can assume that T contains $\mathcal{P}$. Therefore, we form a satisfying assignment A for F as follows. If the vertex $\bar{x}_i$ in T has a neighbor in $\{c_1, \ldots, c_k\}$, then set x_i equals *false*, otherwise set x_i equals *true*. Since all vertices c_j has some neighbor in T, and edges representing occurrences of x_i and $\bar{x}_i$ are conflicting edges, it holds that every clause of F has at least one literal evaluated as *true* by A, which is a consistent assignment due to the construction of $\hat{G}$.

Therefore, CONFLICT-FREE SPANNING TREE is $\mathcal{NP}$-complete even on instances $(G, \hat{G})$ where G is a bipartite planar subcubic graph, and $\hat{G}$ is a disjoint union of P_3's.   □

Next, we present some remarks regarding complete graphs.

Theorem 2. *Let $\mathcal{F}$ be a family of graphs. If* CONFLICT-FREE SPANNING TREE *is $\mathcal{NP}$-complete when restricted to instances with conflict graph in $\mathcal{F}$, then the* MINIMUM CONFLICT-FREE SPANNING TREE *problem is $\mathcal{NP}$-hard when G is a complete graph and $\hat{G} \in \mathcal{F}$.*

Proof. Let $I = (G, \hat{G})$ be an instance of the CONFLICT-FREE SPANNING TREE problem, where $|V(G)| = n$ and $|E(G)| = m$. From $I = (G, \hat{G})$ we create an instance $I' = (G', \hat{G}')$ of MCFST such that I has a feasible solution if and only if I' has a spanning tree with cost $n - 1$. First, G' starts as a copy of G, so we define a weight function ω on the edges. For each $e \in E(G')$, we set $\omega(e) = 1$, after that, for each edge $f \notin E(G)$ we add f to $E(G')$ with $\omega(f) = 2$. Finally, $\hat{G}'$ is defined as a copy of $\hat{G}$. At this point, it is easy to see that I has a conflict-free solution if and only if I' has an optimal solution of cost $n - 1$. That means the optimal solution of I' does not contain any edge of cost 2. Thus, since $\hat{G}'$ is a copy of $\hat{G}$, the claim holds.   □

Corollary 1. MINIMUM CONFLICT-FREE SPANNING TREE *is strongly $\mathcal{NP}$-hard on complete graphs G even when the conflict graph is a disjoint union of paths of size three.*

Proof. Let $I = (G, \hat{G})$ be an instance of CFST with $\hat{G}$ being a disjoint union of P_3's. By Theorem 1, we know that deciding if I contains a conflict-free spanning tree is an $\mathcal{NP}$-complete problem. Let $\mathcal{F}_3$ be the family of graphs formed by the disjoint union of P_3's. By Theorem 2, we can construct a complete graph G' coupled with $\hat{G}$, such that, $I \in$ CFST if and only if the minimum conflict-free spanning tree of $(G', \hat{G})$ has cost $(|V(G)| - 1)$. Since the instance weights are constants, the strongly $\mathcal{NP}$-hardness holds.   □

Note that Corollary 1 is regarding MCFST. At this point, the reader may be asking about the problem of finding a feasible solution when G is a complete graph and $\hat{G}$ is a disjoint union of paths of size three. Next, in contrast with Corollary 1, we show that finding a feasible solution can be done in polynomial time in this case.

A Polynomial-Time Solvable Case

Recall that a P_3 is a star with three vertices, and CFST is $\mathcal{NP}$-complete even when $\hat{G}$ is a disjoint union of paths of size three. Next, we consider the CFST problem when G is a complete graph and $\hat{G}$ is a disjoint union of stars.

Theorem 3. *If G is a complete graph and $\hat{G}$ is a disjoint union of stars, then $I = (G, \hat{G})$ admits a conflict-free spanning tree.*

Proof. If $n = |V(G)| \leq 2$ then the claim holds.

Now, assume that the claim holds for every possible instance $I = (G, \hat{G})$ of CFST such that G is a complete graph with $n \leq k$ vertices (i.e., $G \cong K_k$) and $\hat{G}$ is a disjoint union of stars.

Let $I = (K_{k+1}, \hat{G})$ be an instance where $V(K_{k+1}) = \{v_1, v_2, \ldots, v_{k+1}\}$ and $\hat{G}$ is a disjoint union of stars. By induction, it remains only to prove that $I = (K_{k+1}, \hat{G})$ also admits a conflict-free spanning tree.

Let $I' = (K_k, \hat{G}')$ be the instance obtained by removing the vertex v_{k+1} and its edges from K_{k+1} and $\hat{G}$. Let $E_{v_{k+1}} = \{(v_1, v_{k+1}), (v_2, v_{k+1}), \ldots, (v_k, v_{k+1})\}$ be the set of edges incident to v_{k+1} in the K_{k+1} graph. Note that $\hat{G}' = \hat{G} - E_x$ remains a disjoint union of stars.

If $E_{v_{k+1}}$ is an independent set in $\hat{G}$ then $T = (\{v_1, v_2, \ldots, v_{k+1}\}, E_{v_{k+1}})$ is a conflict-free spanning tree of I, because K_{k+1} is a complete graph (so, v_{k+1} is a universal vertex). Conversely, if $E_{v_{k+1}}$ contains a pair e_ℓ, e_x of conflicting edges, it holds that either e_ℓ or e_x has degree one in $\hat{G}$, because $\hat{G}$ is a disjoint union of stars. Let e_ℓ be such an edge. By hypothesis, I' admits a conflict-free spanning tree T'. Since e_ℓ has degree one in $\hat{G}$ and it conflicts with e_x, it holds that e_ℓ has no conflict with any $e \in E(T')$ then by adding v_{k+1} and e_ℓ in T' we obtain a conflict-free spanning tree T for $I = (K_{k+1}, \hat{G})$.   □

The constructive proof of Theorem 3 suggests a polynomial-time algorithm to obtain a conflict-free spanning tree when G is a complete graph and $\hat{G}$ is a disjoint union of stars.

Corollary 2. *Let $I = (G, \hat{G})$ be an instance of* MINIMUM CONFLICT-FREE SPANNING TREE *where G is an edge-weighted complete graph and $\hat{G}$ is a disjoint union of stars. It holds that a feasible solution for $I = (G, \hat{G})$ can be found in $\mathcal{O}(n + m)$ time, where $n = |V(G)|$ and $m = |E(G)|$.*

Proof. Let $I = (G, \hat{G})$ be an instance of MCFST where G is an edge-weighted complete graph and $\hat{G}$ is a disjoint union of stars. Let $V(G) = \{v_1, v_2, \ldots, v_n\}$, and let G_k be the subgraph of G induced by $\{v_1, v_2, \ldots, v_k\}$. According to the proof of Theorem 3, for k from 2 to n we obtain a conflict-free spanning tree for G_k by either using the edges incidents to v_k in G_k or adding an edge from v_k in a solution of G_{k-1}.

Since $\hat{G}$ is a disjoint union of stars, in $\mathcal{O}(n + m)$ time, we can identify and store in a list $\mathcal{L}$ the information whether the edges incident to v_k in G_k form an independent set. Note that whenever such a set of edges is not an independent set, there is an edge (v_k, v_j) that is the center of a star S of $\hat{G}$ and is adjacent to an edge (v_k, v_i) that is a leaf of S, where $i, j < k$. Thus, in $\mathcal{O}(n+m)$ time, we can either store the leaf (edge (v_k, v_i)) to be aggregated together with a solution of G_{k-1}, or store a information that $\{(v_1, v_k), (v_2, v_k), \ldots, (v_{k-1}, v_k)\}$ is a solution for G_k. After this preprocessing, We can, in linear time, traverse the stored list $\mathcal{L}$ from position n to 1 recovering the edges of a conflict-free spanning tree of $G = G_n$.

Approximation Results

By Corollary 1, MINIMUM CONFLICT-FREE SPANNING TREE is strongly $\mathcal{NP}$-hard on complete graphs G even when the conflict graph is a disjoint union of paths of size three. However, by Corollary 2, concerning such instances of MCFST, a feasible solution can be found in $\mathcal{O}(n+m)$ time. This motivates the study of the approximability in such a particular case of MCFST.

Theorem 4. MINIMUM CONFLICT-FREE SPANNING TREE *on instances* $I = (G, \hat{G})$ *such that* G *is a complete graph and* $\hat{G}$ *is a disjoint union of stars, admits a c-approximation algorithm where c is the maximum weight among the edges of* G. *In addition, considering the same constraints, it does not admit a* $(c - 1)$*-approximation algorithm unless* $P = NP$.

Proof. Let $I = (K_n, \hat{G})$ be an instance of MCFST where K_n is a complete graph with positive weight on the edges and $\hat{G}$ is a disjoint union of stars.

To prove the first statement, we use the algorithm of Theorem 3. Let $c = max\{\omega(e) \mid e \in E(K_n)\}$. By Theorem 3, we can obtain in polynomial time a feasible solution T. In the worst case, all edges in T have a weight equal to c, so that $\omega(T) = c(n - 1)$, but the optimal solution must have cost at least $n - 1$, because the weights are positive integers. Therefore, T is a c-approximate solution for the problem.

Now, to prove the second claim, suppose that there is a $(c-1)$-approximation algorithm for the problem. Thus, we can run this $(c - 1)$-approximation algorithms in the instances $I = (K_n, \hat{G})$ where K_n is a complete graph with $\omega(e) \in \{1, 2\}, \forall~e \in E(K_n)$ and $\hat{G}$ is a disjoint union of P_3's. Note that on such instances, since $c = 2$, the $(c - 1)$-approximation algorithm must solve them in polynomial time. However, by Corollary 1, MCFST remains $\mathcal{NP}$-hard even restricted on such instances. Therefore, the existence of such a $(c - 1)$-approximation algorithm would imply that $P = NP$.   □

Theorem 5. MINIMUM CONFLICT-FREE SPANNING TREE *on complete graphs* G *with the conflict graph* $\hat{G}$ *being a disjoint union of* P_3's *is a exp-APX-complete problem.*

Proof. To prove the hardness we modify the construction of the proof of Theorem 2. Let $\mathfrak{n}$ be the input size of CFST. Instead of adding edges f with weight $\omega(f) = 2$, we assign $\omega(f) = 2^{\mathfrak{n}}$. Note that encoding $2^{\mathfrak{n}}$ can be done using $\mathfrak{n}$ bits. This implies that, if there is a non-exponential approximation factor algorithm to these modified instances, then the resulting solution does not contain any edge with weight 2^n, solving the original CFST instance in polynomial time, which contradicts the fact that CFST is NP-complete when $\hat{G}$ is a disjoint union of P_3's. Thus, this particular case of MCFST is exp-APX-hard.

Finally, to see that such a particular case is in exp-APX, it is enough to consider the c-approximation algorithm of Theorem 4. $\square$

FPT Algorithm

Throughout this paper, the results show that even if the conflict graph has only isolated P_3's, the CFST problem remains challenging. In addition, P_3's are necessary structures in the conflict graph for the problem to become hard. Recall that a cluster graph is a graph formed from the disjoint union of complete graphs. It is well known that a graph is a cluster graph if and only if it has no induced path with three vertices, i.e., the class of cluster graphs is exactly the class of P_3-free graphs. In [16], Zhang, Kabadi, and Punnen showed that if the conflict graph is a cluster graph (disjoint union of cliques), then MCFST can be solved in polynomial-time. This shows that induced paths with three vertices are key structures for the intractability of the problem.

Besides the existence of P_3's, another structural property necessary in conflict graphs $\hat{G}$ on hard instances of MCFST is the existence of $2K_2$'s. It is well known that $2K_2$-free graphs have a polynomial number of maximal independent sets. In addition, all distinct maximal independent sets of a $2K_2$-free graph can be enumerated in polynomial time [4]. Therefore, it is easy to see that MCFST can be solved in polynomial time on $2K_2$-free conflict graphs $\hat{G}$. Recall that $2K_2$-free graphs generalize interesting graph classes such as the class of split graphs.

At this point, we consider the "distance from trivialities" of the conflict graph as structural parameterizations.

The distance to a graph class $\mathcal{F}$ is the minimum number of vertices to be removed to obtain a graph in $\mathcal{F}$. A $\mathcal{F}$-vertex deletion set K of a graph H is a set of vertices such that $H[V \setminus K]$ is in $\mathcal{F}$. The $\mathcal{F}$-vertex deletion number, also known as distance to $\mathcal{F}$, is the minimum cardinality of a set K such that K is a $\mathcal{F}$-vertex deletion set.

Now, let $\mathcal{F}$ be a hereditary graph class such that MCFST on conflict graphs $\hat{G} \in \mathcal{F}$ can be solved in polynomial time. Given a minimum $\mathcal{F}$-vertex deletion set K of a conflict graph $\hat{G}$, we can design an FPT algorithm to solve the MINIMUM CONFLICT-FREE SPANNING TREE problem parameterized by the distance to $\mathcal{F}$ of the conflict graph $\hat{G}$.

Note that cluster graphs and $2K_2$-free graphs are particular cases of $\mathcal{F}$. Also, notice that CLUSTER VERTEX DELETION and $2K_2$-FREE VERTEX DELETION can be solved in $2^{\mathcal{O}(k)} \cdot n^{\mathcal{O}(1)}$ time when parameterized by the solution size (k) through the bounded search tree technique. In addition, an $1.811^k \cdot n^{\mathcal{O}(1)}$-time algorithm for CLUSTER VERTEX DELETION can be found in [15]. Therefore, we may assume that a minimum $\mathcal{F}$-vertex deletion set of $\hat{G}$ is given whenever it can be computed in FPT time (concerning its size), and the MINIMUM CONFLICT-FREE SPANNING TREE problem is parameterized by the distance to

$\mathcal{F}$ of the conflict graph. Zhang, Kabadi, and Punnen [16] showed that MCFST is polynomial-time solvable if the distance to cluster of the conflict graph is bounded by a constant. Next, we generalize such a result.

Theorem 6. *Let $\mathcal{F}$ be a hereditary graph class such that MCFST on conflict graphs $\hat{G} \in \mathcal{F}$ can be solved in polynomial time, and let $(G, \hat{G})$ be an instance of* MINIMUM CONFLICT-FREE SPANNING TREE. *Given a minimum $\mathcal{F}$-vertex deletion set K of $\hat{G}$, a minimum conflict-free spanning tree of $(G, \hat{G})$ can be found (if any) in $2^k \cdot n^{O(1)}$ time, where $k = |K|$.*

Proof. We assume that a minimum $\mathcal{F}$-vertex deletion set K of $\hat{G}$ is given.

Moreover, we named by $\mathcal{F}$-*Algorithm* a polynomial-time algorithm to solve MCFST on conflict graphs $\hat{G} \in \mathcal{F}$. Next, we show the pseudocode of our FPT-MCFST algorithm. We use $N_G(v)$ to indicate the (open) neighborhood of v in the graph G, and use $N_G[v]$ to denote its closed neighborhood.

```
 1  FPT-MCFST(G, Ĝ, K);
 2  if |K| = 0 then
 3  |   return F-Algorithm(G, Ĝ)
 4  else
 5  |   Take an element e = (u, v) of K
 6  |   K ← K \ {e}
 7  |   (V(G'), E(G')) ← (V(G), E(G) \ {e})   #edge e is not in the solution
 8  |   Ĝ' ← Ĝ − {e}
 9  |   (V(G''), E(G'')) ← (V(G), E(G) \ N_Ĝ(e))   #edge e is in the solution
10  |   K'' ← K \ N_Ĝ(e)
11  |   Ĝ'' ← Ĝ − N_Ĝ[e]
12  |   return min{FPT-CFST(G', Ĝ', K), FPT-CFST(G'', Ĝ'', K'')}
13  end
```

Algorithm 1: Parameterized algorithm for MCFST.

When $|K| = 0$, the resulting conflict graph is in $\mathcal{F}$, and the problem can be solved in polynomial-time by the $\mathcal{F}$-algorithm that we assume exists. Therefore, Algorithm 1 works as a bounded search tree using the $\mathcal{F}$-vertex deletion set K to branch into two branches that distinguish regarding the potential use or not of an element $e \in K$ in the solution. If we use e in the solution, we should delete the elements in $N_{\hat{G}}(e)$ (the neighborhood of e in $\hat{G}$) from both G and $\hat{G}$. In addition, we also remove e from $\hat{G}$ because isolated vertices are irrelevant in conflict graphs. On the other hand, if e is not in the solution, we should remove e from G and $\hat{G}$. Note that e is removed from $\hat{G}$ in both cases. This procedure is applied recursively until $|K| = 0$, resulting in a conflict graph in $\mathcal{F}$ because $\mathcal{F}$ is a hereditary graph class. Note that we don't need to fix a particular edge on a solution; we just need to leave it free to be used if necessary. Finally, when the conflict graph becomes a graph in the class $\mathcal{F}$, we can run the $\mathcal{F}$-algorithm to solve the resulting instance. Since the FPT-MCFST algorithm constructs a bounded search tree with k levels, we have a $2^k \cdot n^{O(1)}$-time algorithm.   □

As a corollary, it follows that MCFST can be solved in FPT time when parameterized by the distance to $2K_2$-free graphs or the distance to P_3-free graphs. We left as an open problem the complexity of the problem when the conflict graph is $(K_2 \cup P_3)$-free or $2P_3$-free.

References

1. Capua, R., Frota, Y., Ochi, L.S., Vidal, T.: A study on exponential-size neighborhoods for the bin packing problem with conflicts. J. Heuristics **24**(4), 667–695 (2018). https://doi.org/10.1007/s10732-018-9372-2
2. Darmann, A., Pferschy, U., Schauer, J.: Determining a minimum spanning tree with disjunctive constraints. In: Rossi, F., Tsoukias, A. (eds.) ADT 2009. LNCS (LNAI), vol. 5783, pp. 414–423. Springer, Heidelberg (2009). https://doi.org/10.1007/978-3-642-04428-1_36
3. Darmann, A., Pferschy, U., Schauer, J., Woeginger, G.J.: Paths, trees and matchings under disjunctive constraints. Discret. Appl. Math. **159**(16), 1726–1735 (2011)
4. Dhanalakshmi, S., Sadagopan, N., Manogna, V.: On 2K2-free graphs. Int. J. Pure Appl. Math. **109**(7), 167–173 (2016)
5. Viana, L., Campêlo, M., Sau, I., Silva, A.: A unifying model for locally constrained spanning tree problems. J. Comb. Optim. **42**(1), 125–150 (2021). https://doi.org/10.1007/s10878-021-00740-2
6. Filho, I.T.F.A.: Characterizing Boolean satisfiability variants. Ph.D. thesis, Massachusetts Institute of Technology (2019)
7. Gendreau, M., Laporte, G., Semet, F.: Heuristics and lower bounds for the bin packing problem with conflicts. Comput. Oper. Res. **31**(3), 347–358 (2004)
8. Graham, R., Hell, P.: On the history of the minimum spanning tree problem. Ann. Hist. Comput. **7**(1), 43–57 (1985)
9. Kruskal, J.B.: On the shortest spanning subtree of a graph and the traveling salesman problem. Proc. Am. Math. Soc. **7**(1), 48–50 (1956)
10. Maňuch, J., Gaur, D.R.: Fitting protein chains to cubic lattice is NP-complete. J. Bioinform. Comput. Biol. **6**(01), 93–106 (2008)
11. Pferschy, U., Schauer, J.: The knapsack problem with conflict graphs. J. Graph Algorithms Appl. **13**(2), 233–249 (2009)
12. Pilz, A.: Planar 3-SAT with a clause/variable cycle. Discrete Math. Theor. Comput. Sci. **21**(3), 18:1–18:20 (2019). https://doi.org/10.23638/DMTCS-21-3-18
13. Prim, R.C.: Shortest connection networks and some generalizations. Bell Syst. Tech. J. **36**(6), 1389–1401 (1957)
14. Tippenhauer, S.: On planar 3-SAT and its variants. Master's thesis, Fachbereich Mathematik und Informatik der Freien Universitat Berlin (2016)
15. Tsur, D.: Faster parameterized algorithm for cluster vertex deletion. Theory Comput. Syst. **65**(2), 323–343 (2021)
16. Zhang, R., Kabadi, S.N., Punnen, A.P.: The minimum spanning tree problem with conflict constraints and its variations. Discrete Optim. **8**(2), 191–205 (2011)

B_0-VPG Representation of AT-free Outerplanar Graphs

Sparsh Jain[1] , Sreejith K. Pallathumadam[2]([⊠]) ,
and Deepak Rajendraprasad[2]

[1] College of Computing, Georgia Institute of Technology, Atlanta, GA, USA
sparsh.jain@gatech.edu
[2] Indian Institute of Technology Palakkad, Palakkad, India
111704002@smail.iitpkd.ac.in, deepak@iitpkd.ac.in

Abstract. B_0-VPG graphs are intersection graphs of axis-parallel line
segments in the plane. We show that all AT-free outerplanar graphs are
B_0-VPG. In the course of the argument, we show that any AT-free outerplanar graph can be identified as an induced subgraph of a 2-connected
outerplanar graph whose weak dual is a path. Our B_0-VPG drawing procedure works for such graphs and has the potential to be extended to
larger classes of outerplanar graphs.

Keywords: Outerplanar · AT-free · B_0-VPG · Grid intersection
graph · Graph drawing

1 Introduction

A *k-bend path* is a simple path in a two-dimensional grid with at most k bends.
Geometrically they are polylines in the plane made of at most $k+1$ axis-parallel
(horizontal or vertical) line segments. *Vertex intersection graphs of Paths on a
Grid (VPG)* (resp., B_k-*VPG*) is the class of graphs which can be represented
as intersection graphs of simple (resp., k-bend) paths in a two-dimensional grid.
The *bend number* of a graph G in VPG is the minimum k for which G is in
B_k-VPG. VPG graphs are equivalent to *string graphs* which are intersection
graphs of curves in the plane. One motivation to study B_k-VPG graphs comes
from VLSI circuit design where the paths correspond to wires in the circuit. A
natural concern in VLSI design is to reduce the number of bends in each path
(wire) in the representation. A second motivation is that certain algorithmic
tasks become easier when restricted to B_1-VPG or B_0-VPG graphs (cf. [21]).

Some of the standard graph theoretic terminology used in the rest of the
introduction are defined in Sect. 1.2. Planar graphs have received the maximum
attention from the perspective of B_k-VPG representations, some of which we
describe in Sect. 1.1. Following up on a series of improvements and conjectures
by various authors, Gonçalves, Isenmann, and Pennarun in 2018 showed that all

---

S. Jain—A part of this work was done while at Indian Institute of Technology Palakkad.

© Springer Nature Switzerland AG 2022
N. Balachandran and R. Inkulu (Eds.): CALDAM 2022, LNCS 13179, pp. 103–114, 2022.
https://doi.org/10.1007/978-3-030-95018-7_9

planar graphs are B_1-VPG [17]. This is tight since many simple planar graphs like 4-wheel, 3-sun, triangular prism, to name a few, are not B_0-VPG. This makes the question of characterizing B_0-VPG planar graphs very appealing. Even though recognizing B_0-VPG graphs is NP-complete in general, we do not know the recognition complexity of the same question restricted to planar graphs. *Segment intersection graphs* are intersection graphs of line segments in the plane. Chalopin and Gonçalves in 2009 showed that every planar graph is a segment intersection graph [7], confirming a conjecture of Scheinerman from 1984 [24]. One way to refine the class of segment intersection graphs is to restrict the number of directions permitted for the segments. If the number of directions is limited to two, we rediscover B_0-VPG. *k-DIR* graphs are intersection graphs of line segments that can lie in at most k directions in the plane. It is known that bipartite planar graphs are 2-DIR [12,13,18] and triangle-free planar graphs are 3-DIR [4]. West conjectures that any planar graph is 4-DIR [25]. This adds to the appeal for characterizing B_0-VPG planar graphs.

Characterizing outerplanar B_0-VPG graphs will be a good step towards the above since some of the structures that forbid a planar graph from having a B_0-VPG representation are also present among outerplanar graphs. Outerplanar graphs were known to be B_1-VPG [5] before the same was shown for planar graphs. In this article we take a small step towards this goal by showing that AT-free outerplanar graphs are B_0-VPG. To do so, we first show that any AT-free outerplanar graph can be identified as an induced subgraph of a 2-connected linear outerplanar graph. We call a 2-connected outerplanar graph *linear* if it's weak dual is a path. This result may be of independent interest. Our B_0-VPG drawing is essentially for 2-connected linear outerplanar graphs and has potential to be extended to larger classes of outerplanar graphs. However, we cannot extend this result to AT-free planar graphs since we have examples of AT-free planar graphs, like 4-wheel and triangular prism, which are not B_0-VPG.

After a brief literature review in Sect. 1.1 and a recall of some standard graph theoretic terminology in Sect. 1.2, we spread the proof of our main result in two sections. In Sect. 2 we introduce a subclass of outerplanar graphs called linear outerplanar graphs and show that every AT-free outerplanar graph is linear. Further, we show that every linear outerplanar graph is an induced subgraph of a 2-connected linear outerplanar graph. In Sect. 3, we show that every 2-connected linear outerplanar graph is B_0-VPG. Since B_0-VPG is easily seen to be a hereditary graph class (closed under induced subgraphs), it follows that all linear outerplanar graphs are B_0-VPG. Furthermore, since we have shown that all AT-free outerplanar graphs are linear the main result of this article follows.

1.1   Literature

The class B_k-VPG was introduced by Asinowski et al. in 2012 [2]. Nevertheless, these graphs were previously studied in various forms. One of them is *grid intersection graphs* (GIG) which are bipartite graphs that can be represented as intersection graphs of horizontal and vertical line segments in the plane where no two parallel line segments intersect [18]. It is easy to see that the class of bipartite B_0-VPG graphs and GIGs are equivalent.

Apart from the celebrated result by Gonçalves et al. that planar graphs are B_1-VPG [17], we can see a chronology of results of B_k-VPG in planar graphs. Hartman et al. proved that bipartite planar graphs are GIG [18] and hence B_0-VPG. In [2], Asinowski et al. showed that planar graphs are B_3-VPG and conjectured that it is tight. Disproving this conjecture, Chaplick and Ueckerdt proved that planar graphs have a B_2-VPG representation [10]. Biedl and Derka further improved the result which proves that planar graphs have 1-string B_2-VPG representation which means that any two paths can intersect at most once [3]. Francis and Lahiri proved that any plane graphs, formed by connecting the leaves of a tree making only simple cycles, have a B_1-VPG representation [15].

The recognition problem for string graphs and hence VPG graphs is NP-complete [19,23]. The recognition problem for 2-DIR graphs and hence B_0-VPG graphs is NP-complete [20]. The recognition of whether a given graph is in B_k-VPG, $k \geq 0$, is NP-complete even if it is guaranteed that the given graph is in B_{k+1}-VPG [9]. The same article also shows that the classes B_k-VPG and B_{k+1}-VPG , $k \geq 0$, are separated. Chaplick et al. left the possibility of such a separation in chordal graphs open [9] which was answered partially by Chakraborty et al. [6]. Cohen et al. showed the existence of a cocomparability graph with bend number k for all $k \geq 0$ (Theorem 3.1 in [11]). Chaplick et al. provided a polynomial time decision algorithm for B_0-VPG chordal graphs in 2011 [8]. B_0-VPG characterizations are known for block graphs [1], split graphs, chordal bull-free graphs, chordal claw-free graphs [16] and cocomparability graphs [22].

1.2   Terminology and Notation

The *closed neighborhood* $N[v]$ of a vertex v in a graph is the set containing v and its neighbors in G. We will refer to vertices of a graph with k neighbors as *k-degree vertices*. A graph G is *H-free* if G does not contain an induced subgraph isomorphic to H. We use C_k to denote the simple cycle on k vertices. A cycle on k vertices $x_0, \ldots, x_{k-1}$ where each x_i is adjacent to x_{i+1} (addition is modulo k) can also be denoted as $x_0, \ldots, x_{k-1}, x_0$. A C_k together with an additional vertex v adjacent to all the vertices of the cycle is called a *wheel graph*, denoted as W_k. A *triangular prism* is the complement of C_6. A triangle v_0, v_1, v_2, v_0 together with an independent set $S = \{u_0, u_1, u_2\}$ is called a *3-sun* if $\forall i \in \{0,1,2\}, N[u_i] = \{u_i, v_i, v_{i+1}\}$ where addition is modulo 3.

A set of three independent vertices is called an *asteroidal triple* or *AT* when there exists a path among each pair of them containing no vertex from the closed neighborhood of the third vertex. An *AT-free* graph is a graph which does not have an AT. A subset of vertices in a graph is called a *separator* if its removal increases the number of components of the graph. A vertex x is a *cutvertex* if $\{x\}$ is a separator. A graph is *k-connected* if it does not have a separator of size smaller than k. A graph is *connected* if it is 1-connected. A *block* of a graph is any maximal 2-connected subgraph of the graph. A *trivial block* is a block containing at most two vertices.

A *plane graph* is an embedding of a planar graph in the plane with no crossing edges. A *face* in a plane graph is any region in the plane bounded by edges in the graph and not containing any other vertex or edge. Among the faces contributed by a plane graph in the plane, exactly one is an *unbounded (outer) face* and the remaining are *bounded (inner) faces*. The *dual graph* H of a plane graph G is a graph that has a vertex for each face of G and an edge between two vertices in H if the corresponding faces of G share an edge. The *weak dual* of a plane graph G is an induced subgraph of its dual whose vertices correspond to the bounded faces of G. A *boundary edge* in a plane graph is an edge that is shared by the unbounded face of the graph. Edges shared only by bounded faces are called *internal* edges. A planar graph is called *outerplanar* if it has a plane embedding in which all the vertices are incident on the outer face. Outerplanar graphs will always be drawn in such a way that the outer (unbounded) face contains all the vertices and hence the above terminology of faces, duals, weak duals, boundary edges and internal edges will be used assuming such a plane drawing.

2   Linear Outerplanar Graphs

In this section, we introduce a subclass of outerplanar graphs called linear outerplanar graphs (Definition 2) and show that every AT-free outerplanar graph is linear. Further, we show that every linear outerplanar graph is an induced subgraph of a 2-connected linear outerplanar graph. Though the definition of general linear outerplanar graphs is technical, 2-connected linear outerplanar graphs turn out to be exactly those 2-connected outerplanar graphs whose weak dual is a path. This simplicity is exploited in the next section to find a B_0-VPG representation for them.

Definition 1. *Let G be an outerplanar graph. A face of G which corresponds to a leaf of the weak dual is called a* leaf *face. A block B of G is called* safe *if B has at most two cutvertices and if B contains more than one face, then to each cutvertex x in B we can associate a different leaf face of B containing x. A block B of G incident to a cutvertex x is called* big *for x if either B has a cutvertex other than x or a face not containing x. A cutvertex x of G is called* safe *if at most two blocks of G incident to x are big.*

Definition 2 (Linear Outerplanar Graph). *An outerplanar graph G is called* linear *if the weak dual of G is a linear forest, and every block and cutvertex of G are safe.*

We start with some observations that will help us prove that AT-free outerplanar graphs are linear.

Observation 1. *If v is a 2-degree vertex in a 2-connected outerplanar graph, then $N[v]$ is not a separator of G.*

Proof. Since degree of v is 2, v and its two neighbors are three consecutive vertices in the outer face of G. Removing three consecutive vertices from the outer cycle of G does not disconnect G.   □

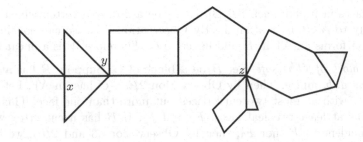

Fig. 1. A linear outerplanar graph with three cutvertices x, y, z and five blocks. Notice that all blocks and cutvertices are safe.

The following list of observations are immediate consequences of applying the above observation to a block of an outerplanar graph and the trivial fact that a single vertex does not separate a 2-connected graph.

Observation 2. *Let B be a block of an outerplanar graph G. A 2-degree vertex in B is a vertex in B whose degree inside B is 2. For every cutvertex x in B, we will denote a neighbor of x outside B by x'.*

  (i) *If a, b, c are three pairwise non-adjacent 2-degree vertices in B, then $\{a, b, c\}$ forms an AT in G.*
 (ii) *If a and b are two non-adjacent 2-degree vertices in B and c is a cutvertex in B non-adjacent to a and b, then $\{a, b, c'\}$ forms an AT in G.*
(iii) *If a and b are two cutvertices in B and c is a 2-degree vertex in B non-adjacent to a and b, then $\{a', b', c\}$ forms an AT in G.*
 (iv) *If a, b, c are three cutvertices in B, then $\{a', b', c'\}$ forms an AT in G.*

Observation 3. *Let B be a block of an outerplanar graph. Every leaf face of B contains a 2-degree vertex in B which is not incident to any other bounded face of B.*

Proof. A face has at least three vertices and at most two vertices of a leaf face can be shared by another face. Hence all the remaining vertices are 2-degree vertices in B not incident to any other bounded face of B.

Theorem 1. *AT-free outerplanar graphs are linear.*

Proof. Let G be an AT-free outerplanar graph. To prove that G is linear, we need to show that G satisfies three conditions.

1. Weak dual of G is a linear forest. Let W be the weak dual of G. Since G is outerplanar, W is a forest [14]. If W is not a linear forest, then, it has a vertex of degree 3 (or more). Let that vertex be f and three of its neighbors be n_1, n_2, and n_3. Let the corresponding faces in G be F for f, and N_i for n_i, $i \in [3]$. Consider the induced subgraph H formed by vertices of F, N_1, N_2, and N_3. It is easy to see that H is a 2-connected outerplanar graph where N_1, N_2, and N_3

are leaf faces in it. For each $i \in [3]$, let a_i be a 2-degree vertex which belongs exclusively to N_i (Observation 3). By Observation 2(i), we can conclude that $\{a_1, a_2, a_3\}$ forms an AT in H and hence in G. Therefore W is a linear forest.

2. *Every block of G is safe.* Let B be a block of G. Suppose B has at least 3 cutvertices a, b, and c. Then by Observation 2(iv), G has an AT. Let B be a block of G with at most two cutvertices but more than one face. This implies that B has at least two leaf faces, F_1 and F_2. If B has a cutvertex which is neither incident to F_1 nor F_2, then by Observations 3 and 2(ii), we have an AT in G. Hence every cutvertex of B is incident to a leaf face of B. If B has two cutvertices x and y and neither of them is incident to one of the leaf faces, then by Observations 3 and 2(iii), we have an AT in G. Hence x and y can be associated with two leaf faces containing them. Hence B is safe.

3. *Every cutvertex of G is safe.* Let x be a cutvertex in G. Suppose x has at least 3 big blocks B_1, B_2, and B_3 incident to it. For each $i \in [3]$, since B_i is big, it either contains a face F_i not containing x or another cutvertex x_i and another block B_i' incident to x_i. In both cases we can find a vertex y_i (in F_i or B_i' respectively) which is at a distance 2 or more from x. One can easily see that $\{y_1, y_2, y_3\}$ forms an AT in G. Hence x is safe.   □

We end this section by showing that every linear outerplanar graph is an induced subgraph of a 2-connected linear outerplanar graph. Figure 2 shows a 2-connected linear outerplanar graph which contains the graph in Fig. 1 as an induced subgraph.

Lemma 1. *Every connected linear outerplanar graph is an induced subgraph of a 2-connected linear outerplanar graph.*

Proof. Let G be a connected linear outerplanar graph. Let B_1 and B_2 be any two blocks of G joined by a cutvertex x. For each $i \in \{1, 2\}$, we choose a neighbor y_i of x from B_i as follows. If B_i is a single edge or a single face L_i, then any neighbor of x in B_i can be chosen as y_i. If B_i has more than one face, then we have a leaf face L_i of B_i associated to x. Pick y_i to be a neighbor of x from L_i such that xy_i is a boundary edge of L_i. Let the supergraph of G obtained by adding a new vertex x' adjacent to y_1 and y_2 be called G' and the new block containing B_1, B_2 and x' be called B'.

Since x is a cutvertex and xy_1, xy_2 are boundary edges of G, the block B' remains outerplanar. Let the weak duals of B_1 and B_2 respectively be (possibly empty) the paths $F_1, \ldots, F_{k-1}$ and $F_{k+1}, \ldots, F_l$ with $F_{k-1} = L_1$ and $F_{k+1} = L_2$ when they are non-empty paths. Then the path $F_1, \ldots, F_{k-1}, F_k, F_{k+1}, \ldots, F_l$ is the weak dual of B' where F_k corresponds to the new face x, y_1, x', y_2, x. This holds true even if B_1 is trivial ($k = 1$) or B_2 is trivial ($k = l$), or both. Hence G' is outerplanar and its weak dual is a linear forest. We argue next that the graph G' is linear outerplanar in two special cases. Since the blocks of G' other than B' and cutvertices of G' not in B' remain safe in G', it suffices to argue that the block B' and the cutvertices in it are safe in G'. Moreover, any cutvertex x' of B' other than x is safe since the number of big blocks incident to x' does not increase due to the merging of B_1 and B_2. This is because if B_i, $i \in \{1,2\}$, has two cutvertices, then B_i is already big for both. Hence it suffices to show that the block B' and the vertex x (if it remains a cutvertex in G') are safe.

Since x is a safe cutvertex in G, at most two blocks of G incident to x are big. Thus we can always find two blocks B_1 and B_2 incident to x such that either B_2 is not big for x or B_1, B_2 are the only two blocks incident to x. Hence the following two cases are exhaustive.

Case 1 (B_2 is not big for x). Since B_2 does not have any cutvertex other than x, B' has at most two cutvertices in G', possibly x and another vertex x_1 from B_1. If B_1 has more than one face, then since x is associated with F_{k-1}, x_1 belongs to F_1. If B_1 is a single edge or a single face, then x_1 still belongs to F_1 in B'. Since B_2 is not big for x, F_l contains x. Hence we can associate x_1 to F_1 and x to F_l in B' and conclude that B' is safe. One can verify that B' is big for x in G' only if B_1 is big for x in G. Hence x is safe in G'.

Case 2 (B_1 and B_2 are the only two blocks incident to x). In this case, x is no more a cutvertex in G'. Hence B' has at most two cutvertices, possibly x_1 from B_1 and x_2 from B_2. One can verify, as in the first case, that x_1 belongs to F_1 whether B_1 is trivial or not and x_2 belongs to F_l whether B_2 is trivial or not. Hence B' is safe in G'.

Repeating the above merging till no cutvertices are left results in a 2-connected linear outerplanar graph which contains G as an induced subgraph. ☐

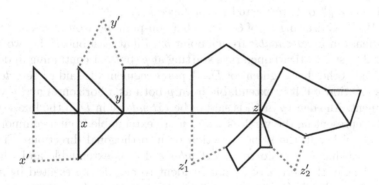

Fig. 2. A 2-connected linear outerplanar graph constructed from the graph in Fig. 1 using the construction procedure employed in the proof of Lemma 1. It can be verified that the addition of new vertices x', y', z_1' and z_2' satisfy the cases 1, 2, 1 and 2 in the proof respectively.

Remark 1. We would also like to point out that not all connected AT-free outerplanar graphs are induced subgraphs of 2-connected AT-free outerplanar graphs. Let G be a C_5 together with a pendant vertex. While G is AT-free outerplanar, any 2-connected outerplanar graph G' containing G as an induced subgraph is not AT-free. To see this, consider the induced subgraph H of G' formed by the C_5 in G and another face F sharing an edge with this C_5. We can pick one 2-degree vertex a from F and two non-adjacent 2-degree vertices b and c from the C_5 which are both non-adjacent to a. From Observation 2(i), it follows that $\{a, b, c\}$ is an AT. This is the obstacle which nudged us to study the larger class of linear outerplanar graphs, which admittedly is rather technical.

3 B_0-VPG Representation of 2-connected Linear Outerplanar Graphs

It's drawing time. In this section, we show that every 2-connected linear outerplanar graph is B_0-VPG (Lemma 2). The proof of Lemma 2 is algorithmic which draws a B_0-VPG representation for any 2-connected linear outerplanar graph (cf. Fig. 3 for example). Since B_0-VPG is easily seen to be a hereditary graph class (closed under induced subgraphs), it follows from Lemma 1 that all linear outerplanar graphs are B_0-VPG. Furthermore, since we have shown that all AT-free outerplanar graphs are linear (Theorem 1), the main result of this article follows.

Lemma 2. *Every 2-connected linear outerplanar graph is B_0-VPG.*

Proof. Let G be a 2-connected linear outerplanar graph with n faces labeled $F_1, \ldots, F_n$ such that the weak dual of G is the path $F_1, \ldots, F_n$. For each $i \in [n-1]$, the edge shared by F_i and F_{i+1} is denoted by e_i. For notational convenience, we set e_n to be any boundary edge of F_n. For each $i \in [n]$, let G_i denote the induced subgraph of G restricted to the faces $F_1, \ldots, F_i$.

In a B_0-VPG drawing D_i of G_i, we call a non-point horizontal (resp., vertical) line segment l in D_i *extendable* from a point $p \in l$ if at least one of the two infinite horizontal (resp., vertical) open rays starting at p (but not containing p) does not intersect any other line segment of D_i. A point segment l is said to be *extendable* from its location p if it is extendable from p both as a horizontal and a vertical line segment. An edge xy in G_i is said to be *extendable* in D_i if the line segments l_x and l_y representing the vertices x and y are extendable from a common point $p \in l_x \cap l_y$ either in the same direction or in orthogonal directions. Finally a B_0-VPG drawing D_i is said to be *extendable* if e_i is extendable and whenever F_i is a triangle, the vertex of F_i not incident to e_{i-1} is represented by a point segment. Note that if F_1 is a triangle, all the vertices of F_1 are represented by point segments in D_1.

If the length of F_1 is 4 or more, then we can represent F_1 as the intersection graph of line segments laid out on the boundary of an axis-parallel rectangle with the endpoints of e_1 being orthogonal (and hence sharing only a corner of the rectangle). This is an extendable B_0-VPG drawing D_1 of G_1. If $F_1 \cong C_3$, then representing all the three vertices as point segments at the same point gives an extendable B_0-VPG drawing D_1 of G_1. Let D_i, $i < n$, be an extendable B_0-VPG drawing of G_i. From D_i, we construct an extendable B_0-VPG drawing D_{i+1} of G_{i+1} as follows.

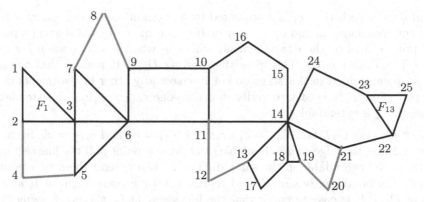

(a) Fig. 2 is redrawn with labels for all the vertices.

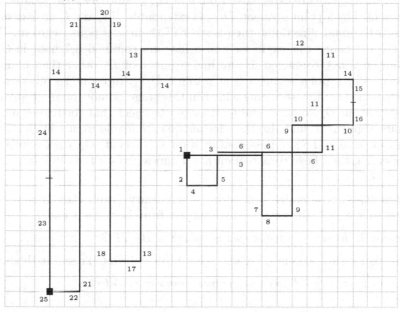

(b) The B_0-VPG drawing of the graph depicted in Fig. 2.

Fig. 3. A 2-connected linear outerplanar graph G and a B_0-VPG representation of it. The collinear overlapping line segments are drawn a little apart for clarity. The point segments (for e.g. vertices 1 and 25) are drawn as black squares.

Case 1 (length of F_{i+1} is 4 or more). Let $F_{i+1} = v_0, \ldots, v_k, v_0$, with $e_i = v_k v_0$ and $e_{i+1} = v_j v_{j+1}$, $j < k$. Since D_i is extendable, the edge $v_k v_0$ is extendable in D_i. Extend the line segments l_k and l_0 (representing v_k and v_0 respectively) in orthogonal directions to two points q_k and q_0 outside of the bounding box of D_i. Let q be the intersection point of the perpendiculars to l_k and l_0 at q_k and q_0 respectively. Represent the path $v_1, \ldots v_{k-1}$ on the two line segments from q_0 to

q and q to q_k such that v_1 is represented by a segment containing q_0, v_{k-1} by a segment containing q_k and v_j, v_{j+1} by orthogonal line segments sharing a point. The point shared by these two segments will be q_0 when $j = 0$, q_k when $j = k-1$ and q in all other cases. This gives the drawing D_{i+1}. It is clear that the new line segments added in this stage do not intersect any other line segments in D_i except l_0 and l_k. It is easy to verify that the edge $e_{i+1} = v_j v_{j+1}$ is extendable. Hence D_{i+1} is extendable.

Case 2 ($F_{i+1} \cong C_3$). Let $F_{i+1} = a, b, c, a$, with $e_i = ca$ and $e_{i+1} = ab$. Since D_i is extendable, the edge ca is extendable in D_i from a point p. If the line segments l_c and l_a are extendable in the same direction, then extend them to a point q outside the bounding box of D_i and represent b by a point segment l_b at q to obtain D_{i+1}. It is easy to check that the line segment l_a, the point segment l_b, and also the edge ab are extendable from q in D_{i+1}. Since ab is extendable and b is represented by a point segment, D_{i+1} is extendable. If l_c and l_a are extendable only in orthogonal directions, then neither of them is a point segment. Hence $F_i \not\cong C_3$ and hence the vertices c and a have no common neighbor in G_i. So the point p is not contained in any line segment of D_i other than l_c and l_a. Represent b by a point segment l_b at p to get D_{i+1}. In both the subcases, it is clear that the new line segments added in this stage do not intersect any other line segments in D_i except l_c and l_a. It is easy to check that the line segment l_a, the point segment l_b, and also the edge ab are extendable from p in D_{i+1}. Since ab is extendable and b is represented by a point segment, D_{i+1} is extendable.

Repeating the above construction $n - 1$ times gives a B_0-VPG drawing D_n of $G_n = G$. □

Since B_0-VPG is easily seen to be a hereditary graph class (closed under induced subgraphs), the next theorem follows from Lemmas 1 and 2.

Theorem 2. *Every linear outerplanar graph is B_0-VPG.*

Together with Theorem 1, the above establishes the main result in this article.

Theorem 3. *Every AT-free outerplanar graph is B_0-VPG.*

4   Concluding Remarks

Even though we showed that all linear outerplanar graphs and in particular, all AT-free outerplanar graphs are B_0-VPG, the characterization of B_0-VPG outerplanar graphs still evades us. It is easy to see that linearity is not necessary for B_0-VPG outerplanar graphs. Planar bipartite graphs, and hence outerplanar bipartite graphs are B_0-VPG [18]. But outerplanar bipartite graphs can be far from being linear, in the sense that their weak duals can be trees with arbitrarily large degrees for internal nodes. More than our result, we think the drawing technique employed in the proof of Lemma 2 might become useful in an attempt to characterize B_0-VPG outerplanar graphs.

Acknowledgments. We thank K. Muralikrishnan for posing the question of characterizing B_0-VPG outerplanar graphs.

References

1. Alcón, L., Bonomo, F., Mazzoleni, M.P.: Vertex intersection graphs of paths on a grid: characterization within block graphs. Graphs and Combinatorics **33**(4), 653–664 (2017)
2. Asinowski, A., Cohen, E., Golumbic, M.C., Limouzy, V., Lipshteyn, M., Stern, M.: Vertex intersection graphs of paths on a grid. J. Graph Algorithms Appl. **16**(2), 129–150 (2012)
3. Biedl, T., Derka, M.: 1-string B_2-VPG representation of planar graphs. J. Comput. Geom. (Old Web Site) **7**(2), 191–215 (2016)
4. de Castro, N., Cobos, F.J., Dana, J.C., Márquez, A., Noy, M.: Triangle-free planar graphs and segment intersection graphs. J. Graph Algorithms Appl. **6**(1), 7–26 (2002)
5. Catanzaro, D., et al.: Max point-tolerance graphs. Discret. Appl. Math. **216**, 84–97 (2017)
6. Chakraborty, D., Das, S., Mukherjee, J., Sahoo, U.K.: Bounds on the bend number of split and cocomparability graphs. Theory Comput. Syst. **63**(6), 1336–1357 (2019). https://doi.org/10.1007/s00224-019-09912-4
7. Chalopin, J., Gonçalves, D.: Every planar graph is the intersection graph of segments in the plane. In: Proceedings of the Forty-First Annual ACM Symposium on Theory of Computing, pp. 631–638 (2009)
8. Chaplick, S., Cohen, E., Stacho, J.: Recognizing some subclasses of vertex intersection graphs of 0-bend paths in a grid. In: Kolman, P., Kratochvíl, J. (eds.) WG 2011. LNCS, vol. 6986, pp. 319–330. Springer, Heidelberg (2011). https://doi.org/10.1007/978-3-642-25870-1_29
9. Chaplick, S., Jelínek, V., Kratochvíl, J., Vyskočil, T.: Bend-bounded path intersection graphs: sausages, noodles, and waffles on a grill. In: Golumbic, M.C., Stern, M., Levy, A., Morgenstern, G. (eds.) WG 2012. LNCS, vol. 7551, pp. 274–285. Springer, Heidelberg (2012). https://doi.org/10.1007/978-3-642-34611-8_28
10. Chaplick, S., Ueckerdt, T.: Planar graphs as VPG-graphs. In: Didimo, W., Patrignani, M. (eds.) GD 2012. LNCS, vol. 7704, pp. 174–186. Springer, Heidelberg (2013). https://doi.org/10.1007/978-3-642-36763-2_16
11. Cohen, E., Golumbic, M.C., Trotter, W.T., Wang, R.: Posets and VPG graphs. Order **33**(1), 39–49 (2016)
12. Czyzowicz, J., Kranakis, E., Urrutia, J.: A simple proof of the representation of bipartite planar graphs as the contact graphs of orthogonal straight line segments. Inf. Process. Lett. **66**(3), 125–126 (1998)
13. De Fraysseix, H., Ossona de Mendez, P., Pach, J.: Representation of planar graphs by segments. Intuitive Geom. **63**, 109–117 (1991)
14. Fleischner, H.J., Geller, D.P., Harary, F.: Outerplanar graphs and weak duals. J. Indian Math. Soc. **38**(1–4), 215–219 (1974)
15. Francis, M.C., Lahiri, A.: VPG and EPG bend-numbers of Halin graphs. Discret. Appl. Math. **215**, 95–105 (2016)
16. Golumbic, M.C., Ries, B.: On the intersection graphs of orthogonal line segments in the plane: characterizations of some subclasses of chordal graphs. Graphs and Combinatorics **29**(3), 499–517 (2013)
17. Gonçalves, D., Isenmann, L., Pennarun, C.: Planar graphs as L-intersection or L-contact graphs. In: Proceedings of the Twenty-Ninth Annual ACMSIAM Symposium on Discrete Algorithms, pp. 172–184. SIAM (2018)

18. Hartman, I.B.-A., Newman, I., Ziv, R.: On grid intersection graphs. Discrete Math. **87**(1), 41–52 (1991)
19. Kratochvíl, J.: String graphs. II. Recognizing string graphs is NP-hard. J. Comb. Theory Ser. B **52**(1), 67–78 (1991)
20. Kratochvíl, J., Matoušek, J.: Intersection graphs of segments. J. Comb. Theory, Series B **62**(2), 289–315 (1994)
21. Mehrabi, S.: Approximation algorithms for independence and domination on B_1-VPG and B_1-EPG graphs. arXiv preprint arXiv:1702.05633 (2017)
22. Pallathumadam, S.K., Rajendraprasad, D.: Characterization and a 2D visualization of B_0-VPG cocomparability graphs. In: GD 2020. LNCS, vol. 12590, pp. 191–204. Springer, Cham (2020). https://doi.org/10.1007/978-3-030-68766-3_16
23. Schaefer, M., Sedgwick, E., Štefankovič, D.: Recognizing string graphs in NP. J. Comput. Syst. Sci. **67**(2), 365–380 (2003)
24. Scheinerman, E.R.: Intersection classes and multiple intersection parameters of graphs. Princeton University (1984)
25. West, D.: Open problems. SIAM J. Discrete Math. Newslett. **2**, 10–12 (1991)

P Versus NPC: Minimum Steiner Trees in Convex Split Graphs

A. Mohanapriya[1](✉), P. Renjith[2], and N. Sadagopan[1]

[1] Design and Manufacturing, Indian Institute of Information Technology,
Kancheepuram, Chennai, India
{coe19d003,sadagopan}@iiitdm.ac.in
[2] Design and Manufacturing, Indian Institute of Information Technology,
Kurnool, India
renjith@iiitk.ac.in

Abstract. We investigate the complexity of finding a minimum Steiner tree in new subclasses of split graphs namely tree-convex split graphs and circular-convex split graphs. It is known that the Steiner tree problem (STREE) is NP-complete on split graphs [1]. To strengthen this result, we introduce convex ordering on one of the partitions (clique or independent set), and prove that STREE is polynomial-time solvable for tree-convex split graphs with convexity on clique (K), whereas STREE is NP-complete on tree-convex split graphs with convexity on independent set (I). Further, we show that STREE is polynomial-time solvable for path (triad)-convex split graphs with convexity on I, and circular-convex split graphs. Finally, we show that STREE can be used as a framework for the dominating set problem in split graphs, and hence the complexity of STREE and the dominating set problem is the same for all these graph classes.

Keywords: Steiner tree · Tree-convex · Path-convex · Triad-convex · Circular-convex · Domination

1 Introduction

The computational complexity of the Steiner tree problem (STREE), Domination and its variants for different classes of graphs has been well studied. Given a graph G with terminal set $R \subseteq V(G)$, STREE asks for a set $S \subseteq V(G) \setminus R$ such that the graph induced on $S \cup R$ is connected. The objective is to minimize the number of vertices in S. STREE is NP-complete for general graphs, chordal bipartite graphs [2], and split graphs [3] whose vertex set can be partitioned into a clique and an independent set. It is polynomial-time solvable in strongly chordal graphs [3], series-parallel graphs [4], and outerplanar graphs [5], and for graphs with fixed treewidth [6]. It is known [1] that STREE is polynomial-time solvable in $K_{1,3}$-free split graphs and $K_{1,4}$-free split graphs, whereas in $K_{1,5}$-free

This work is partially supported by the DST-ECRA Project-ECR/2017/001442.

N. Balachandran and R. Inkulu (Eds.): CALDAM 2022, LNCS 13179, pp. 115–126, 2022.
https://doi.org/10.1007/978-3-030-95018-7_10

split graphs, STREE is NP-complete. In this paper, we focus on new subclasses of split graphs, and study the tractability versus intractability status of STREE in those subclasses of split graphs.

It is important to highlight that many problems that are NP-complete on bipartite graphs become polynomial-time solvable when a linear ordering is imposed on one of the partitions. Such graphs are known as convex bipartite graphs in the literature. For example, the dominating set problem is NP-complete on bipartite graphs, whereas it is polynomial-time solvable in convex bipartite graphs [7]. A bipartite graph $G = (X, Y)$ is said to be tree-convex if there is a tree (imaginary) on X such that the neighborhood of each y in Y is a subtree in X. Apart from linear ordering (path-convex ordering), tree-convex ordering, triad-convex ordering, and circular-convex ordering on bipartite graphs have been considered in the literature [8]. Further, the convex ordering on bipartite graphs yielded many interesting algorithmic results for domination and Hamiltonicity [9]. Similarly, the feedback vertex set problem (FVS) is NP-complete on star-convex bipartite graphs, and comb-convex bipartite graphs, whereas it is polynomial-time solvable in chordal bipartite graphs and convex bipartite graphs [9]. Thus, the convex ordering on bipartite graphs reinforces the borderline separating P-versus-NPC instances of many classical combinatorial problems.

Since the tractability versus intractability status of many combinatorial problems on bipartite graphs can be investigated with the help of convex ordering on bipartite graphs, we wish to extend this line of study to split graphs as well. To the best of our knowledge, this paper makes the first attempt in introducing convex properties on split graphs. There are many classical problems such as Domination, Steiner tree, Hamiltonicity and its variants are NP-complete on split graphs. By imposing convex ordering on one of the partitions (clique or independent set), we wish to investigate P-versus-NPC status of STREE on this new subclass of split graphs. As part of this paper, we consider the following convex properties; star-convex, tree-convex, comb-convex, path-convex, and circular-convex split graphs. Further, we look at split graphs having convexity on I (independent set) and split graphs having convexity on K (clique).

For tree-convex and circular-convex split graphs, the computational complexity of the following graph problems are studied in this paper.

1. The Steiner tree problem (STREE).
   Instance: A graph G, a terminal set $R \subseteq V(G)$, and a positive integer k.
   Question: Does there exist a set $S \subseteq V(G) \setminus R$ such that $|S| \leq k$, and $G[S \cup R]$ is connected ?
2. The Dominating set problem (DS).
   Instance: A graph G, and a positive integer k.
   Question: Does G admit a dominating set of size at most k ?
3. The Connected Dominating set problem (CDS).
   Instance: A graph G, and a positive integer k.
   Question: Does G admit a connected dominating set of size at most k ?

4. The Total Dominating set problem (TDS).

Instance: A graph G, and a positive integer k.

Question: Does G admit a total dominating set of size at most k ?

All these problems are NP-complete for general graphs, bipartite graphs, and split graphs [1,8,10]. The complexity of these problems in tree-convex bipartite graphs have been considered in [9]. In this paper, we analyze the complexity of STREE in tree-convex split graphs and its subclasses (triad-convex, star-convex, comb-convex) with convexity on I (K). An interesting theoretical question is

-What is the boundary between the tractability and intractability of STREE in split graphs ?

In this paper, we answer this question by imposing a convex ordering on clique or independent set. In particular, we show that STREE is polynomial-time solvable for tree-convex split graphs with convexity on K, and is NP-complete for tree-convex split graphs with convexity on I. Further, we investigate path, triad convex properties, and show that STREE is polynomial-time solvable for triad (path)-convex split graphs with convexity on I and circular-convex split graphs.

This paper is structured as follows: In Sect. 2, the complexity results of convex split graphs with convexity on I is shown, and the complexity results of convex split graphs with convexity on K is shown in Sect. 3. Using the results of Sects. 2 and 3, we establish the complexities of DS, CDS, and TDS in convex split graphs.

Graph Preliminaries: In this paper, we consider connected, undirected, unweighted and simple graphs. For a graph G, $V(G)$ denotes the vertex set and $E(G)$ represents the edge set. For a set $S \subseteq V(G)$, $G[S]$ denotes the subgraph of G induced on the vertex set S. The open neighborhood of a vertex v is $N_G(v) = \{u \mid \{u,v\} \in E(G)\}$ and the closed neighborhood of v is $N_G[v] = \{v\} \cup N_G(v)$. The degree of vertex v is $d_G(v) = |N_G(v)|$. A split graph is a graph G in which $V(G)$ can be partitioned into two sets; a clique K and an independent set I. A split graph is written as $G = (K \cup I, E)$, where K is a maximal clique and I is an independent set. In a split graph, for each vertex u in K, $N_G^I(u) = N_G(u) \cap I$, $d_G^I(u) = |N_G^I(u)|$, and for each vertex v in I, $N_G^K(v) = N(v) \cap K$, $d_G^K(v) = |N_G^K(v)|$. For each vertex u in K, $N_G^I[u] = N_G(u) \cap I \cup \{u\}$, and for each vertex v in I, $N_G^K[v] = N(v) \cap K \cup \{v\}$. For a split graph G, $\Delta_G^I = \max\{d_G^I(u)\}, u \in K$ and $\Delta_G^K = \max\{d_G^K(v)\}, v \in I$.

Definition 1. *A split graph $G = (K \cup I, E)$ is called π-convex with convexity on K if there is an associated structure $\pi = (K, F)$ in K such that for each vertex $v \in I$, its neighborhood $N_G(v)$ induces a connected subgraph in π.*

Definition 2. *A split graph $G = (K \cup I, E)$ is called π-convex with convexity on I if there is an associated tree $\pi = (I, F)$ in I such that for each vertex $v \in K$, its neighborhood $N_G^I(v)$ induces a connected subgraph in π.*

In general π can be any arbitrary structure. In this paper, We consider the following structures for π; "tree", "star", "triad", "path", and "cycle".

For STREE, we solve for the case $R = I$, and for all the other cases, we obtain solutions using the following transformations and the algorithm for $R = I$.

Case 1: $R = K$ or $R \subset K$.

For $R = K$ or $R \subset K$, the Steiner set is an empty set.

Case 2: $R \subset I$.

For $R \subset I$, we transform the graph G to G' and the solution to G is obtained using G'. The transformation is defined as follows; $G' = G - I'$, where $I' = I - R$ and $R' = R$.

Case 3: $R \cap K \neq \emptyset$ and $R \cap I \neq \emptyset$.

Similar to Case 2, we obtain the solution for this case using the following transformation. Let $W = R \cap K$ and $G' = G - N_G^I[W] - I'$ with $I' = I - R$. Let $R' = I'$. Then the solution to (G, R) is obtained using (G', R').

2 STREE in Split Graphs with Convexity on I

When we refer to convex split graphs in this section, we refer to convex split graphs with convexity on I. For STREE on split graphs with convexity on I, we establish hardness results for star-convex and comb-convex split graphs, and polynomial-time algorithms for path-convex, triad-convex, and circular-convex split graphs.

2.1 Star-Convex Split Graphs

In this section, we establish a classical hardness of STREE in star-convex split graphs by presenting a polynomial-time reduction from the Exact-3-Cover problem to STREE in star-convex split graphs with convexity on I.

The decision version of Exact-3-Cover problem (X3C) is defined below:

EXACT-3-COVER $(X, \mathcal{C})$

Instance: A finite set X with $|X| = 3q$ and a collection $\mathcal{C} = \{C_1, C_2, \ldots, C_m\}$ of 3-element subsets of X.

Question: Is there a subcollection $\mathcal{C}' \subseteq \mathcal{C}$ such that for every $x \in X$, x belongs to exactly one member of $\mathcal{C}'$ (that is, $\mathcal{C}'$ partitions X) ?

The decision version of Steiner tree problem is defined below:

STREE (G, R, k)

Instance: A graph G, a terminal set $R \subseteq V(G)$, and a positive integer k.

Question: Is there a set $S \subseteq V(G) \setminus R$ such that $|S| \leq k$, and $G[S \cup R]$ is connected ?

Theorem 1. *For star-convex split graphs with convexity on I, STREE is NP-complete.*

Proof. **STREE is in NP** Given a star-convex split graph G with convexity on I and a certificate $S \subseteq V(G)$, we show that there exists a deterministic polynomial-time algorithm for verifying the validity of S. Note that the standard Breadth First Search (BFS) algorithm can be used to check whether $G[S \cup R]$ is connected. It is easy to check whether $|S| \leq k$. The certificate verification can be done in $O(|V(G)| + |E(G)|)$. Thus, we conclude that STREE is in NP.

STREE is NP-Hard It is known [11] that X3C is NP-complete. X3C can be reduced in polynomial time to STREE in star-convex split graphs with convexity on I using the following reduction. We map an instance $(X, \mathcal{C})$ of X3C to the corresponding instance (G, R, k) of STREE as follows: $V(G) = V_1 \cup V_2$, $V_1 = \{c_i \mid 1 \leq i \leq m\}$, $V_2 = \{x_1, x_2, \ldots, x_{3q}, x_{3q+1}\}$, $E(G) = \{\{c_i, x_j\} \mid x_j \in C_i, 1 \leq j \leq 3q, 1 \leq i \leq m\} \cup \{\{x_{3q+1}, c_i\} \mid 1 \leq i \leq m\} \cup \{\{c_i, c_j\} \mid 1 \leq i \leq j \leq m\}$. Let $R = V_2$, $k = q$. Note that G is a split graph with V_1 being a clique and V_2 being an independent set. Now we show that G is a star-convex split graph by defining a star T on V_2 as follows:
Let $V(T) = V_2$ and $E(T) = \{\{x_{3q+1}, x_i\} \mid 1 \leq i \leq 3q\}$. We see that x_{3q+1} is the root of the star T.
An illustration for X3C with $X = \{x_1, x_2, x_3, x_4, x_5, x_6\}$ and $\mathcal{C} = \{C_1 = \{x_1, x_2, x_3\}, C_2 = \{x_2, x_3, x_4\}, C_3 = \{x_1, x_2, x_5\}, C_4 = \{x_2, x_5, x_6\}, C_5 = \{x_1, x_5, x_6\}\}$ and the graph G with $R = I$, $k = 2$ and imaginary star on I rooted at x_7 corresponding to X3C is shown in Fig. 1. For $\mathcal{C}' = \{C_2, C_5\}$ and $k = 2$, we see that $S = \{c_2, c_5\}$ is the desired Steiner set.

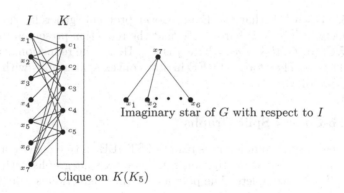

Imaginary star of G with respect to I

Clique on $K(K_5)$

Fig. 1. Reduction: An instance of X3C to STREE in star-convex split graphs with convexity on I

Claim. Exact-3-Cover $(X, \mathcal{C})$ if and only if STREE $(G, R, \frac{|X|}{3})$

Proof. Only if: If there exists $\mathcal{C}' \subseteq \mathcal{C}$ which partitions all elements of X, then the set of vertices $S = \{c_i \in V_1 \mid C_i \in \mathcal{C}'\}$, where c_i is the vertex corresponding to C_i, forms a Steiner set S with $R = V_2$. Also, note that $|S| = q$.

If: Assume that there exists a Steiner tree in G for R. Let $S \subseteq V_1$ be the Steiner set, $|S| = q$. We now construct a $\mathcal{C}' = \{C_i \in C \mid c_i \in S\}$ to X3C. Since $|S| = q$, we have $|\mathcal{C}'| = q$. Further, S is the Steiner set for the terminal set $R = \{x_1, \ldots, x_{3q}, x_{3q+1}\}$. For any $c_i \in S$, we have $|N_G^I(c_i) \setminus \{x_{3q+1}\}| = 3$. Since $|S| = q$, for all $c_i, c_j \in S, i \neq j$, $N_G^I(c_i) \cap N_G^I(c_j) = \{x_{3q+1}\}$. Therefore, $\mathcal{C}'$ is the corresponding solution to X3C. $\qquad\square$

Therefore, STREE is NP-complete on star-convex split graphs with convexity on I. $\qquad\square$

Corollary 1. *STREE in tree-convex split graphs with convexity on I is NP-complete.*

Proof. Since star-convex split graphs are a subclass of tree-convex split graphs, from Theorem 1 this result follows. $\qquad\square$

The study of parameterized complexity is concerned with designing algorithms with complexity $f(k)n^{O(1)}$, where k is the parameter of interest, usually the solution size, and n is the input size. In [12] it is shown that STREE in general graphs is Fixed-parameter Tractable (FPT) if the parameter is the size of the terminal set.

It is known [13] that STREE in general graphs with parameter $|S|$ is W[2]-hard. We now prove a similar result for our graph class.

Theorem 2. *For star-convex split graphs with convexity on I, STREE is W[1]-hard with parameter $|S|$.*

Proof. It is known [14] that the Exact Cover problem (generalization of X3C) with parameter $|\mathcal{C}'|$ is W[1]-hard. Note that the reduction presented in Theorem 1 maps $(X, \mathcal{C})$ to $(G, R, k = q)$, where q is $|\mathcal{C}'|$. Hence the reduction is a parameterized reduction. Therefore, STREE in star-convex split graphs with convexity on I is W[1]-hard. $\qquad\square$

2.2   Comb-Convex Split Graphs

To strengthen the NP-completeness result of STREE on tree-convex split graphs with convexity on I, we show that on comb convex split graphs with convexity on I, STREE is NP-complete. The polynomial-time reduction is from the vertex cover problem on general graphs.

The decision version of Vertex Cover problem (VC) is defined below:

$VC\ (G, k)$
Instance: A graph G, a non-negative integer k.
Question: Does there exist a set $S \subseteq V(G)$ such that for each edge $e = \{u, v\} \in E(G)$, $u \in S$ or $v \in S$ and $|S| \leq k$?

Theorem 3. *For comb-convex split graphs with convexity on I, STREE is NP-complete.*

Proof. **STREE is NP-Hard:** It is known [15] that VC on general graphs is NP-complete and this can be reduced in polynomial time to STREE in comb-convex split graphs using the following reduction. We map an instance (G, k) of VC on general graphs to the corresponding instance $(G^*, R, k' = k)$ of STREE as follows: $V(G^*) = V_1 \cup V_2 \cup V_3$,
$V_1 = \{x_i \mid v_i \in V(G)\}$,
$V_2 = \{y_i \mid e_i \in E(G)\}$,
$V_3 = \{z_i \mid e_i \in E(G)\}$.
We shall now describe the edges of G^*,

$E(G^*) = E_1 \cup E_2 \cup E_3$,
$E_1 = \{\{y_i, x_k\}, \{y_i, x_l\}, \mid e_i = \{v_k, v_l\} \in E(G),\ x_k, x_l \in V_1,\ y_i \in V_2,\ 1 \leq i \leq m,\ 1 \leq k \leq n,\ 1 \leq l \leq n\}$
$E_2 = \{\{x, z_i\}\} \mid x \in V_1,\ z_i \in V_3,\ 1 \leq i \leq m\}$
$E_3 = \{\{x_i, x_j\} \mid 1 \leq i \leq j \leq n\}$.

We define $K = V_1$, $I = V_2 \cup V_3$, and imaginary comb T on I is defined with V_3 as the backbone and V_2 as the pendant vertex set. That is, $V(T) = I$ and $E(T) = \{\{y_1, z_1\}, \{y_2, z_2\}, \dots, \{y_i, z_i\} \mid 1 \leq i \leq m\}$.

An example is illustrated in Fig. 2, the vertex cover instance $G(V, E)$ with $k = 2$ is mapped to STREE instance of comb-convex split graph $G^*(V^*, E^*)$ with $R = \{y_1, y_2, y_3\}$, $k' = 2$.

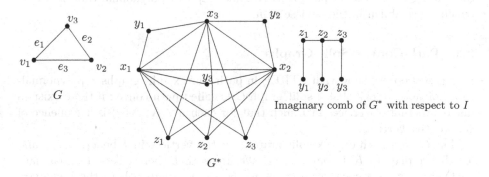

Fig. 2. An example: VC reduces to STREE.

Claim. G^* is a comb-convex split graph with convexity on I.

Proof. For each $x_i \in V_1$, $N_G^I(x_i) = V_3 \cup W$, $W \subseteq V_2$. By construction x_i is adjacent to all of V_3. Therefore, the graph induced on $V_3 \cup W$ is a subtree in G^*. Hence G^* is a comb-convex split graph with convexity on I.   □

Claim. (G, k) has a vertex cover with at most k vertices if and only if $(G^*, R = \{y_i \mid 1 \leq i \leq m\}\}, k' = k)$ has a Steiner tree of size at most $k' = k$ Steiner vertices.

Proof. (Only if) Let $V' = \{v_i \mid 1 \leq i \leq k\}$ is a vertex cover of size k in G. Then we construct the Steiner set S of G^* for $R = \{y_i \mid 1 \leq i \leq m\}$ as follows: $S = \{x_i \mid 1 \leq i \leq k, \ v_i \in V', \ x_i \in V(G^*)\}$. Since V' is a vertex cover, for any edge $e_i = \{v_k, v_l\} \in E(G)$, v_k or v_l is in V'. Hence S contains x_k or x_l. Therefore, for each vertex y_i, there exists a neighbor in S. Since V_1 is a clique by our construction, $G[R \cup S]$ is connected.

(If) For R in G^*, let $S = \{x_i \mid 1 \leq i \leq k'\}$ is a Steiner set of G^* of size k'. Then, we construct the vertex cover V' of size k in G as follows; $V' = \{v_i \mid x_i \in S, \ v_i \in V(G), \ 1 \leq i \leq k'\}$. We now claim that V' is a vertex cover in G. Suppose that there is an edge $e_i = \{v_k, v_l\} \in E(G)$ for which neither v_k nor v_l is in V'. This implies that neither x_k nor x_l is in S. Since R contains y_i, it follows that $N(y_i) \cap S = \emptyset$. Thus S is not a Steiner set. A contradiction. Therefore, V' is a vertex cover of size k in G. $\square$

Therefore, STREE is NP-complete on comb-convex split graphs with convexity on I. $\square$

A closer look at the reduction reveals that the presence of pendant vertices in the comb makes the problem NP-hard. Therefore, we shall investigate the complexity of STREE in a variant of comb-convex split graphs where there are no pendant vertices in the comb which is precisely the class of path-convex split graphs. Interestingly STREE in path-convex split graph is polynomial-time solvable, which we establish in the next section.

2.3   Path-Convex Split Graphs

In this section, we show that STREE in path-convex split graphs is polynomial-time solvable. Recall that a split graph G is called path-convex if there exists a linear ordering of vertices in I such that for each $u \in K$, $N(u)$ is a sequence of consecutive vertices.

Let G be a path-convex split graph. Let the vertices in I be $x_1, \ldots, x_r$ and let the vertices in K be $w_1, \ldots, w_s$. We know that there exists an imaginary path on I and for each vertex $u \in K$, $N_G^I(u)$ is a subpath in the imaginary path. For each $u \in K$, $l(u)$ is the least vertex in $N_G^I(u)$, and $r(u)$ is the greatest vertex in $N_G^I(u)$. For each vertex $x_i \in I$, we define $T(x_i) = \{u \mid u \in N(x_i), \text{ and } r(u) \text{ is maximum}\}$. Let $w(x_i)$ be an arbitrary vertex from $T(x_i)$.

Steiner tree algorithm for path-convex split graphs identifies the vertex $w \in K$ adjacent to x_1 such that $r(w)$ is maximum and we continue this from $r(w)$. This greedy approach is indeed optimum, which we establish in this section.

The following algorithm computes a minimum Steiner set for G.

Algorithm 1. *STREE for path-convex split graphs with convexity on I.*

1: **Input**: A connected path-convex split graph G with convexity on I and $R = I$.
2: Let $a = x_1$, $S = \{\}$.
3: Let $S = S \cup \{w(a)\}$.
4: **if** $x_r \in N(w(a))$ **then**
5:     Output S.
6: **else**
7:     Let x_j be $r(w(a))$.
8:     Let $a = x_{j+1}$ and continue from Step 3.
9: **end if**

By $r(w_i) \preceq r(w_j)$, $w_i, w_j \in K$, we mean that the vertex $r(w_i)$ appears before $r(w_j)$. Let $S = \{u_1, \ldots u_p\}$ be the Steiner vertices chosen by the algorithm. Note that as per our algorithm $r(u_1) \preceq r(u_2) \preceq \ldots \preceq r(u_p)$. Let $S' = \{u_1, \ldots u_p\}$ be the Steiner vertices chosen by any optimal algorithm. Without loss of generality, we arrange S' such that $r(v_1) \preceq r(v_2) \preceq \ldots \preceq r(v_q)$.

Theorem 4. *For $1 \leq k \leq q$, let $x_j = r(u_k)$, and $x_l = r(v_k)$. Then, $j \geq l$.*

Proof. By mathematical induction k, $k \geq 1$.
Base Case: For $k = 1$, Algorithm 1 has chosen u_1. Since by Step 2 of the algorithm, $r(u_1) \succeq r(u_j)$ for any $u_j \in N(u_1)$. Therefore, $r(u_1) \geq r(v_1)$. Thus $j \geq l$ is true for the base case.
Induction Hypothesis: Assume that for $k \geq 2$, $j \geq l$ is true.
Induction Step: We prove that at $k+1$ iteration, $k \geq 2$, $j \geq l$. By our induction hypothesis, we know that up to k^{th} iteration $j \geq l$. We know that $x_j = r(u_k)$, and $x_l = r(v_k)$. The vertex chosen by the algorithm at $(k + 1)^{th}$ iteration be u_{k+1}. Since S' is the optimal solution, let v_{k+1} be the $(k + 1)^{th}$ vertex in S' such that $r(v_{k+1})$ is either less than or equal to x_j or adjacent to x_{j+1}.
Case 1: If $r(v_{k+1})$ is less than or equal to x_j and we know that at $(k + 1)^{th}$ iteration $r(u_{k+1}) \succeq x_j$, then $r(u_{k+1}) \geq r(v_{k+1})$.
Case 2: Consider the case when $r(v_{k+1})$ is adjacent to x_{j+1}. We know that v_{k+1} is in S' such that $S' \cap N(x_{j+1}) \neq \emptyset$. Suppose that $r(u_{k+1}) \preceq r(v_{k+1})$. At $(k+1)^{th}$ iteration, according to Step 3, and Step 2 of the algorithm, $w(x_{j+1})$ is included in S. Hence $u_{k+1} = w(x_{j+1})$. Recall that as per our definition of $w(x_{j+1})$, it is the vertex adjacent to x_{j+1} such that $r(u_k)$ is maximum. Thus $r(u_{k+1}) \preceq r(v_{k+1})$ is a contradiction. Therefore, $r(u_{k+1}) \succeq r(v_{k+1})$, and at $(k + 1)^{th}$ iteration, $j \geq l$. ☐

Theorem 5. *Algorithm 1 outputs a minimum Steiner set, that is $p = q$.*

Proof. By Theorem 4, we know that $r(u_q) \succeq r(v_q)$, and hence $|S| \leq |S'|$. Since S' is an optimal solution, $|S| \geq |S'|$. By Step 3 of the algorithm we know that $x_n \in N(u_k)$. Therefore, $|S| = |S'|$, and $p = q$. ☐

It is easy to see that Algorithm 1 runs in time $O(mn)$.

Now we see that for comb-convex split graph, STREE is NP-complete, whereas for path-convex split graph, STREE is polynomial-time solvable. This clearly brings out P-versus-NPC investigation of STREE in tree-convex split graphs. This is one of the objectives of this research.

It is important to highlight that we can solve STREE on triad-convex and circular-convex split graphs by using the algorithm of path-convex split graphs as a black box.

3   STREE in Split Graphs with Convexity on K

Having analyzed P-versus-NPC status of STREE for convex split graphs with convexity on I, we shall now analyze the same with respect to split graphs having convexity on K.

3.1   Tree-Convex Split Graphs

In this section, we present a polynomial-time algorithm to find a minimum Steiner tree in tree-convex split graphs. We solve for first the case $R = I$, and using which we solve (i) $R \subset I$ and (ii) $R \cap K \neq \emptyset$ and $R \cap I \neq \emptyset$. Let $G(K \cup I, E)$ be a tree-convex split graph with a imaginary tree T.

To construct the Steiner tree for G, we use the following scheme. Initially all vertices in the imaginary tree T is colored gray. The vertex colored gray is changed white or black as per the following rules:

We recolor the gray colored vertex as white or black as per the following rules:

Rule 1:(Gray is colored white) A leaf vertex $u \in T$ recolored white when there does not exist a pendant vertex in $N_G^I(u)$.

Rule 2:(Gray is colored Black) A leaf vertex $u \in T$ recolored white when there exists a pendant vertex in $N_G^I(u)$.

The algorithm that computes Steiner tree for the case $R = I$ works with imaginary tree T.

The Steiner tree algorithm $R = I$ starts from an arbitrary leaf vertex, say $u \in T$. Check with respect to u which rule is applicable. If Rule 1 is applied, then G is modified to $G = G - u$. If suppose Rule 2 is applied, then G is modified to $G = G - N_G^I(u)$. We continue the process for $|K| - 1$ times. In this process the vertices that are colored black are included in the solution and they are precisely the Steiner set, which we prove in Theorem 6.

Theorem 6. *The Steiner set S obtained using the above procedure is a minimum Steiner set.*

Proof. On the contrary, there exists a Steiner set S' for G such that $|S'| < |S|$. Since $|S'| < |S|$, the coloring obtained from the algorithm is not optimal. Hence there exists a coloring such that number of white vertices are more, and the number of black vertices are less compared to the coloring obtained from the algorithm.

A gray colored leaf vertex u is colored black, when there exists a pendant vertex $x \in I$ adjacent to u in that instance. We know that G is a tree-convex split graph with convexity on K, let u be a leaf vertex in T, and let v be a unique vertex adjacent to u in T. Suppose that u is not adjacent to a pendant vertex $x \in I$. Then $N(u) \subseteq N(v)$. Hence including v in the Steiner set instead of u will connect more number of vertices. Suppose that u is adjacent to a pendant vertex $x \in I$. Then u is colored black. This is the invariant followed by the algorithm.

Hence the coloring having less number of black vertices compared to the coloring obtained from our algorithm is not possible. Therefore, $|S'| < |S|$ is a contradiction and S is a minimal Steiner set.

Remarks: Since STREE in tree-convex split graphs with convexity K is polynomial-time solvable, STREE is polynomial-time solvable on well known special structures such as star, path, triad, and comb-convex split graphs with convexity on K. It is important to note that the above approach can be used as a black box for STREE on circular-convex split graphs.

Since Steiner set for convex split graphs $S \subseteq K$, we observe that S is also DS, CDS, and TDS. The P-versus-NPC status of STREE for convex properties discussed in this paper also holds true for DS, CDS, TDS.

Conclusion and Directions for Further Research:
We have shown the complexity of STREE, and domination and its variants in tree-convex and circular-convex split graphs. The results presented in this paper can be used as a framework for Steiner tree variants (Steiner path and cycle) and domination problems (outer connected domination, Roman domination) restricted to split, and bipartite graphs.

References

1. Renjith, P., Sadagopan, N.: The Steiner tree in $K_{1,r}$-free split graphs-a Dichotomy. Discrete Appl. Math. **280**, 246–255 (2020)
2. Müller, H., Brandstädt, A.: The NP-completeness of Steiner tree and dominating set for chordal bipartite graphs. Theoretical Comput. Sci. **53**(2–3), 257–265 (1987)
3. White, K., Farber, M., Pulleyblank, W.: Steiner trees, connected domination and strongly chordal graphs. Networks **15**(1), 109–124 (1985)
4. Joseph, A.W., Charles, J.C.: Steiner trees, partial 2-trees, and minimum ifi networks. Networks **13**(2), 159–167 (1983)
5. Joseph, A.W., Charles, J.C.: Steiner trees in outerplanar graphs. In: Proceedings of 13th Southeastern Conference on Combinatorics, Graph Theory, and Computing, pp. 15–22 (1982)
6. Chimani, M., Mutzel, P., Zey, B.: Improved steiner tree algorithms for bounded treewidth. J. Discr. Algorith. **16**, 67–78 (2012)
7. Mohanapriya, A., Renjith, P., Sadagopan, N., et al.: Steiner tree in k-star caterpillar convex bipartite graphs-a dichotomy. arXiv preprint arXiv:2107.09382, 2021
8. Pandey, A., Panda, B.S.: Domination in some subclasses of bipartite graphs. Discr. Appl. Math. **252**, 51–66 (2019)

9. Chen, H., Lei, Z., Liu, T., Tang, Z., Wang, C., Xu, K.: Complexity of domination, hamiltonicity and treewidth for tree convex bipartite graphs. J. Combin. Optim., 1–16 (2015). https://doi.org/10.1007/s10878-015-9917-3
10. Damaschke, P., Müller, H., Kratsch, D.: Domination in convex and chordal bipartite graphs. Inf. Proc. Lett. **36**(5), 231–236 (1990)
11. Michael, R.G.: Computers and intractability: a guide to the theory of NP-Completeness. WH Freeman and Co., (1979)
12. Stuart, E.D., Robert, A.W.: The Steiner problem in graphs. Networks **1**(3), 195–207 (1971)
13. Dom, M., Lokshtanov, D., Saurabh, S.: Incompressibility through colors and ids. In: International Colloquium on Automata, Languages, and Programming, pp. 378–389. Springer (2009). https://doi.org/10.1007/978-3-642-02927-1_32
14. Ashok, P., Kolay, S., Misra, N., Saurabh, S.: Unique covering problems with geometric Sets. In: Xu, D., Du, D., Du, D. (eds.) Computing and Combinatorics. COCOON 2015. LNCS, vol. 9198. Springer, Cham (2015). https://doi.org/10.1007/978-3-319-21398-9_43
15. Cormen, T.H., Leiserson, C.E., Rivest, R.L., Stein, C.: Introduction to algorithms. MIT Press (2009)

On cd-Coloring of $\{P_5, K_4\}$-free Chordal Graphs

M. A. Shalu[ID] and V. K. Kirubakaran[✉][ID]

Indian Institute of Information Technology, Design and Manufacturing,
Kancheepuram, India
{shalu,mat19d002}@iiitdm.ac.in

Abstract. A k-cd-coloring of a graph G is a partition of the vertex set of G into k independent sets $V_1, \ldots, V_k$, where each V_i is dominated by some vertex of G. The least integer k such that G admits a k-cd-coloring is called the cd-chromatic number, $\chi_{cd}(G)$, of G. We say that $S \subseteq V(G)$ is a subclique in G if $d_G(x, y) \neq 2$ for every $x, y \in S$. The cardinality of a maximum subclique in G is called the subclique number, $\omega_s(G)$, of G. Given a graph G and $k \in \mathbb{N}$, the problem CD-COLORABILITY checks whether $\chi_{cd}(G) \leq k$. The problem CD-COLORABILITY is NP-complete for K_4-free graphs [Merounane et al., 2014], P_5-free graphs, and chordal graphs [Shalu et al., 2020]. In this paper, we show that the problem CD-COLORABILITY is $O(n^2)$-time solvable in the intersection of the above graph classes ($\{P_5, K_4\}$-free chordal graphs). The problem SUBCLIQUE takes a graph G and $k \in \mathbb{N}$ as inputs and checks whether $\omega_s(G) \geq k$. The SUBCLIQUE problem is NP-complete for P_6-free graphs and bipartite gaphs [Shalu et al., 2017]. We prove that the problem SUBCLIQUE is $O(n^3)$-time solvable in the class of P_6-free chordal bipartite graphs (a subclass of P_6-free bipartite graphs). In addition, we show that the cd-chromatic number and the subclique number are equal in these two graph classes.

1   Introduction

A proper (vertex) coloring f of a graph assigns colors to its vertices such that adjacent vertices receive distinct colors. The set of vertices that receive the same color is a color class. Graph coloring is one of the classical problems in the field of combinatorics. The problem aims at reducing the number of colors required for a proper (vertex) coloring. The minimum number of colors required to color a graph G with adjacent vertices receiving different colors is called the chromatic number of G, denoted by $\chi(G)$. Another well-studied problem in this field is the *dominating set problem*. We say that a subset D of the vertex set V of a graph G dominates V if every vertex in $V \setminus D$ is adjacent to some vertex in D. The cardinality of the smallest dominating set in a graph G is called domination number, and is denoted by $\gamma(G)$. The dominator coloring problem [1,2,6,7] and the class domination coloring problem [11,16] are two emerging problems in graph theory involving both

M. A. Shalu—Supported by SERB(DST), MATRICS scheme MTR/2018/000086.

N. Balachandran and R. Inkulu (Eds.): CALDAM 2022, LNCS 13179, pp. 127–139, 2022.
https://doi.org/10.1007/978-3-030-95018-7_11

coloring and domination. In a vertex coloring f of a graph G, if each color class of f is dominated by some vertex in G, then it is called a *class domination coloring (cd-coloring)* of G [11, 16]. The minimum number of colors required for cd-coloring a graph G is called the *cd-chromatic number*, $\chi_{cd}(G)$, of G. If two vertices receive the same color under some cd-coloring of G, then the distance between them in G is two: i.e., any two vertices not at distance two will receive different colors in any cd-coloring of G. This observation gives a lower bound for the cd-chromatic number, called the subclique number. For a graph G, a subset S of the vertex set is a *subclique* if no pair of vertices in S are at distance two from each other in G. The cardinality of a maximum subclique in G is called the *subclique number*, $\omega_s(G)$, of G. Note that the cd-chromatic number of a graph G is at least the subclique number of G, i.e., $\chi_{cd}(G) \geq \omega_s(G)$. The computational version of the problems cd-coloring and subclique are as follows.

CD-COLORABILITY	SUBCLIQUE
Instance : A graph G and $k \in \mathbb{N}$.	**Instance** : A graph G and $k \in \mathbb{N}$.
Question : Is G k-cd-colorable?	**Question** : Is $\omega_s(G) \geq k$?

The problem CD-COLORABILITY is NP-complete for bipartite graphs [11], K_4-free graphs [11], P_5-free graphs [15], and chordal graphs [15]. The problem k-CD-COLORABILITY is NP-complete for $k \geq 4$ [11] and is $O(n^5)$-time solvable for $k \leq 3$ [13]. Kiruthika et al. [8] gave an $O(2^n n^4 \log n)$-time algorithm to find the cd-chromatic number of a graph with n vertices. They also gave FPT algorithms for general graphs and chordal graphs. Chen [3] proved that the cd-chromatic number is hard to approximate within a factor of $\frac{|V|^{(1-\epsilon)}+1}{1.5}$ for every $\epsilon > 0$. Das and Mishra [5] proved (approximation) hardness results of the problem CD-COLORABILITY in chordal graphs and bipartite graphs. Shalu et al. [15] proved that an optimal cd-coloring of split graphs, P_4-free graphs, and claw-free graphs can be found in poly-time. Applications of the cd-coloring problem can be found in [3, 9, 14].

For a graph G, every clique in G is also a subclique in G, and thus $\omega_s(G) \geq \omega(G)$. The SUBCLIQUE problem is NP-complete for chordal graphs, bipartite graphs, P_6-free graphs, and H-free graphs where H is any graph on four or five vertices other than P_4 [14]. The problem is poly-time solvable for split graphs and P_4-free graphs [14]. In our previous work [12], we proved that the cd-chromatic number and the subclique number are equal for trees and co-bipartite graphs.

In this paper, we prove that the cd-chromatic number and the subclique number of a $\{P_5, K_4\}$-free chordal graph can be found in $O(n^2)$ time (see Sect. 3). Also, a maximum subclique of a P_6-free chordal bipartite graph can be found in $O(n^3)$ time (see Sect. 4). In addition, we prove that for any graph G in one of the above two graph classes, $\chi_{cd}(G) = \omega_s(G)$.

2 Preliminaries

We follow West [17] for terminology and notation. The number of vertices and the number of edges in a graph are denoted by n and m, respectively. We denote a path with vertex set $\{x_1, \ldots, x_k\}$ and edge set $\{x_i x_{i+1} : 1 \leq i \leq k-1\}$ by $x_1 \cdots x_k$. A *clique* C of a graph G is a subset of the vertex set such that every pair of vertices in C are adjacent. The *clique number* is the size of a largest clique in G and is denoted by $\omega(G)$. For a graph G, we say that $D \subseteq V(G)$ is a *total dominating set* in G if every vertex of G has a neighbor in D. The cardinality of a smallest total dominating set in G is called the *total domination number* of G, denoted by $\gamma_t(G)$. For $Y \subseteq V(G)$, $G[Y]$ denotes the induced subgraph of G with vertex set Y. We denote the length of a shortest path joining x and y in G by $d_G(x, y)$. For a vertex $u \in V(G)$ and a subset W of the vertex set, $d_G(u, W) = \min\{d_G(u, w) \mid w \in W\}$. Given a graph H, we say that a graph G is H-free if no induced subgraph of G isomorphic to H. For $x \in V(G)$, the neighborhood of x is defined as $N(x) = \{y \in V(G) \mid xy \in E(G)\}$ and $N[x] = \{x\} \cup N(x)$ is the closed neighborhood of x. Let $A(x) = V(G) \setminus N[x]$. For $X, Y \subseteq V(G)$ and $X \cap Y = \emptyset$, we define $[X, Y]$ to be the set of all edges in G with one end vertex in X and the other end vertex in Y. A set $D \subseteq V(G)$ is said to be a *biclique* in G if D can be partitioned into two non-empty independent sets X and Y such that every vertex in X is adjacent to each vertex in Y. We say that a graph H is complete bipartite if $V(H)$ is a biclique.

3 $\{P_5, K_4\}$-free Chordal Graphs

In this section, we show that for a $\{P_5, K_4\}$-free chordal graph G, $\chi_{cd}(G) = \omega_s(G)$. Also, we prove that the cd-chromatic number and the subclique number of G can be found in $O(n^2)$ time.

Observation 1. *If G is a connected P_5-free graph with a maximal clique N_0, then the vertex set can be partitioned as $V(G) = N_0 \cup N_1 \cup N_2$ where $N_i = \{u \in V(G) \mid d(u, N_0) = i\}$ for $i = 1, 2$.*

Proof of Observation 1 is omitted in this paper.

Observation 2. *If G is a connected P_5-free chordal graph with $\omega(G) = 2$, then G is a tree with a dominating edge and hence $\chi_{cd}(G) = \omega_s(G) = 2$. (For an example see Fig. 1)*

Observation 3. *Let G be a connected P_5-free chordal graph with clique number three. Also, let $N_0 = \{x_0, x_1, x_2\}$ be a maximum clique in G and let $V(G)$ be partitioned as in Observation 1. Then the following holds.*

1. *Let $M_i = \{u \in N_1 \mid N(u) \cap N_0 = \{x_i\}\}$ and $L_i = \{u \in N_1 \mid N(u) \cap N_0 = N_0 \setminus \{x_i\}\}$. Then,*
   (i) *$[M_i, M_j] = \emptyset$, for $i \neq j$, else G contains an induced C_4.*
   (ii) *$[L_i, L_j] = \emptyset$, for $i \neq j$, otherwise G contains an induced C_4.*
   (iii) *$[M_i, L_i] = \emptyset$, else G contains an induced C_4.*
   (iv) *For $i = 0, 1, 2$, L_i is an independent set, otherwise G contains a K_4. Thus, $\bigcup_{i=0}^{2} L_i$ is an independent set since $[L_i, L_j] = \emptyset$.*

2. If $u, v \in N_1$ such that uv is an edge in G, then $N(u) \cap N_0 \subseteq N(v) \cap N_0$ or $N(v) \cap N_0 \subseteq N(u) \cap N_0$. Otherwise, G contains an induced C_4.

Lemma 1. Let G be a connected P_5-free chordal graph with $\omega(G) = 3$. Let the vertex set be partitioned as in Observation 1 where N_0 is a maximum clique in G, and $N_2 \neq \emptyset$. Then, $(\bigcup_{x \in N_2} N(x)) \cap N_1$ is a clique of size one or two.

Proof of Lemma 1 is omitted in this paper.

Lemma 2. Let G be a connected P_5-free chordal graph with $\omega(G) = 3$. Let $N_0 = \{x_0, x_1, x_2\}$ be a clique in G, and let the vertex set be partitioned as in Observation 1. Then, G has a minimal dominating clique C such that $C \subseteq N_0 \cup N_1$ and $C \cap N_0 \neq \emptyset$.

Proof.
Case 1: N_2 is empty.
Then, $V(G)$ is dominated by N_0. Thus, a subset of N_0 forms the minimal dominating clique C.
Case 2: N_2 is non-empty.
By Lemma 1, N_2 is dominated by $B = (\bigcup_{x \in N_2} N(x)) \cap N_1$ which induces an edge or a vertex in $G[N_1]$. Let C_1 be a minimal subset of B that dominates N_2.
Case 2.1: $|C_1| = 2$.
Let $C_1 = \{u, v\}$ where $uv \in E(G)$. Then, by Observation 3.2, $N(u) \cap N_0 \subseteq N(v) \cap N_0$ or $N(v) \cap N_0 \subseteq N(u) \cap N_0$. Without loss of generality, assume that $N(u) \cap N_0 \subseteq N(v) \cap N_0$. Since $u \in N_1$, u has a neighbor in N_0. Let $x_0 \in N(u)$. Then, $x_1, x_2 \notin N(u)$ because when $x_1 u$ or $x_2 u$ is an edge in G, $\{x_0, x_1, u, v\}$ or $\{x_0, x_2, u, v\}$ induces a K_4 in G. Note that $x_0 v \in E(G)$ and $\{x_0, u, v\}$ induces a triangle in $V(G)$.

Claim 1: $D = \{x_0, u, v\}$ dominates $V(G)$.
Clearly, the set $N_0 \cup N_2 \cup \{u, v\}$ is dominated by D. Assume that there exists a vertex $y \in N_1$ not dominated by D. Since $y \in N_1$, y has a neighbor in $N_0 \setminus D$. Let $yx_1 \in E(G)$. Then, $\{y, x_1, x_0, u, w\}$ induces a P_5 in G for some $w \in N_2 \cap N(u)$ (since $(\bigcup_{x \in N_2} N(x)) \cap N_1 = \{u, v\}$, $yw \notin E(G)$), a contradiction. Thus, D dominates $V(G)$.

Claim 2: D is a minimal dominating set in G.
On the contrary, suppose that there exists a proper subset D' of D that dominates $V(G)$. Since x_0 does not dominate N_2, u or v belongs to D'. Without loss of generality, assume that $u \in D'$. If $v \notin D'$, then u dominates N_2, a contradiction to $C_1 = \{u, v\}$ being the minimal subset of B that dominates N_2. This implies that $u, v \in D'$. Therefore, $x_0 \notin D'$ because D' is a proper subset of D and $u, v \in D'$. This implies that $D' = \{u, v\}$ dominates N_0. Also, we know that $N(u) \cap N_0 \subseteq N(v) \cap N_0$. Thus, $N_0 \subseteq N(v)$. This implies that $\{x_0, x_1, x_2, v\}$ induces a K_4 in G, a contradiction. Thus, D is a minimal dominating clique. Also, since $x_0 \in D$, $D \cap N_0 \neq \emptyset$.

Case 2.2: $|C_1| = 1$.
Let $C_1 = \{u\}$ and $D_1 = N(u) \cap N_0$. Since $u \in N_1$, $D_1 \neq \emptyset$. Also, since G is K_4-free u has a non-neighbor in N_0. Without loss of generality, assume that $x_0 \in D_1$ and $x_1 \notin D_1$.

Claim 3: $C_2 = D_1 \cup \{u\}$ dominates $V(G)$.
Clearly, $N_0 \cup N_2 \cup \{u\}$ is dominated by C_2. Suppose that C_2 does not dominate $V(G)$. Then, there exists a vertex $y \in N_1$ not adjacent to any vertex in C_2. Since $y \in N_1$, y has a neighbor in $N_0 \setminus C_2$. Let $x_1 \in N(y)$. Then, $\{y, x_1, x_0, u, w\}$ forms an induced P_5 for some $w \in N_2$, a contradiction. Hence, C_2 dominates $V(G)$.

Thus, by Claim 3, there exists a subset C of C_2 that dominates $V(G)$. Since $x_1 \notin N(u)$, in order to dominate x_1, C contains an element of N_0.

From the above cases, it is evident that G has a minimal dominating clique C such that $C \subseteq N_0 \cup N_1$ and $C \cap N_0 \neq \emptyset$. $\qquad\square$

Corollary 1. *For a connected P_5-free chordal graph with clique number three, a minimal dominating clique described in Lemma 2 can be found in $O(n+m)$ time.*

Proof of Corollary 1 runs BFS with respect to some triangle in a connected P_5-free chordal graph to find the sets N_0, N_1 and N_2, and the rest of the proof follows from Lemma 2. A detailed proof is omitted in this paper.

In the following observations and lemmas, we study the subclique number and the cd-chromatic number of a connected P_5-free chordal graph with clique number three based on the size of its minimal dominating clique. These observations and lemmas help us to prove our theorem (Theorem 2) on $\{P_5, K_4\}$-free chordal graphs.

Observation 4.

1. If G is a graph with a universal vertex, then every subclique in G is also a clique and every proper coloring of G is also a cd-coloring. Thus, $\chi_{cd}(G) = \chi(G)$ *[15]* and $\omega_s(G) = \omega(G)$ *[14]*.
2. Let G be a graph. Then, for any subclique S in G, $S \cap N[u]$ (respectively $S \cap N(u)$) is a clique for every vertex $u \in V(G)$ since $S \cap N[u]$ (respectively $S \cap N(u)$) is a subclique in the induced subgraph $G[N[u]]$ (a graph with a universal vertex u).

Theorem 1. *[12] If G is connected graph with a dominating clique D, then for any subclique S in G, the following statements are true.*

(i) If $D \cap S \neq \emptyset$, then $|S| \leq \omega(G)$.
(ii) If $D \cap S = \emptyset$, then $|S| \leq |D|(\omega(G) - 1)$.

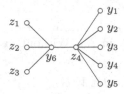

Fig. 1. An edge dominated tree $E_{4,6}$ with a dominating edge $y_6 z_4$.

We denote a tree T with a dominating edge uv as $E_{p,q}$ where $p = |N(u)|$ and $q = |N(v)|$. Clearly, $E_{1,q} \cong K_{1,q}$. Figure 1 shows the edge-dominated tree $E_{4,6}$.

Observation 5. *A tree T is P_5-free if and only if $T \cong E_{p,q}$ for some $p, q \in \mathbb{N}$ or $T \cong K_1$ (see Observation 2).*

Observation 6. *Let G be a P_5-free chordal graph with a universal vertex u and $\omega(G) = 3$. Then, $\chi_{cd}(G) = \chi(G) = 3 = \omega(G) = \omega_s(G)$ by Observation 4. Also, since $G \setminus u$ is a forest, every connected component of $G \setminus u$ is isomorphic to K_1 or $E_{p,q}$ for some $p, q \in \mathbb{N}$ by Observation 5.*

Observation 7. *Let G be a P_5-free chordal graph with $\omega(G) = 3$. Then, $G[N(u)]$ is triangle-free for every vertex $u \in V(G)$. Hence, $G[N(u)]$ is a P_5-free forest.*

Lemma 3. *Let G be a connected P_5-free chordal graph with $\omega(G) = 3$, and let $uv \in E(G)$ such that $D = \{u, v\}$ is a minimal dominating set in G. Then, $\chi_{cd}(G) = \omega_s(G)$.*

Proof. Clearly, $3 = \omega(G) \le \omega_s(G)$. By Theorem 1, $\omega_s(G) \le 4$. So, $\omega_s(G) \in \{3, 4\}$.

Case 1: $\omega_s(G) = 3$.

Claim 1: $N(u) \setminus N(v)$ or $N(v) \setminus N(u)$ is an independent set.
If not, let y_1, y_2 and z_1, z_2 be vertices in $N(u) \setminus N(v)$ and $N(v) \setminus N(u)$ respectively such that $y_1 y_2, z_1 z_2 \in E(G)$. Since $\omega_s(G) = 3$, $\{y_1, y_2, z_1, z_2\}$ is not a subclique, and thus at least two of the four vertices are at distance two from each other. Without loss of generality, assume that $d(y_1, z_1) = 2$. Hence, y_1 and z_1 have a common neighbor w in G. Note that $w \notin \{u, v\}$ because $z_1 u, y_1 v \notin E(G)$. Since $\{y_1, w, z_1, u, v\}$ does not induce a C_5 nor its subsets induce a C_4 in G, $uw, vw \in E(G)$. This implies that $w \notin \{y_1, y_2, z_1, z_2\}$. Also, w is neither adjacent to y_2 nor adjacent to z_2 because $G[N(u)]$ and $G[N(v)]$ are triangle-free by Observation 7. Also, $y_i z_j \notin E(G)$ for all $i, j \in \{1, 2\}$ (otherwise $\{y_i, u, v, z_j\}$ induces a C_4 in G). Thus, $\{y_2, y_1, w, z_1, z_2\}$ induces a P_5 in G, a contradiction. Therefore, $N(u) \setminus N(v)$ or $N(v) \setminus N(u)$ is an independent set.

Let $N(v) \setminus N(u)$ be an independent set. By Observation 7, $G[N(u)]$ is bipartite. Let $X \cup Y$ be a partition of $N(u)$ into independent sets. Then, $V(G) =$

$X \cup Y \cup (N(v) \setminus N(u))$ is a 3-cd-coloring of G where color classes X and Y are dominated by u and $N(v) \setminus N(u)$ is dominated by v. Thus, $3 = \omega_s(G) \leq \chi_{cd}(G) \leq 3$. Hence, $\chi_{cd}(G) = \omega_s(G) = 3$.

Case 2: $\omega_s(G) = 4$.
Clearly, $\chi_{cd}(G) \geq 4$. Then, $N(u) \setminus N(v)$ and $N(v) \setminus N(u)$ aren't independent sets (otherwise G admits a 3-cd-coloring as in Case 1).
Let $X_1 \cup Y_1 = N(u)$ and $X_2 \cup Y_2 = N(v) \setminus N(u)$ be bipartitions of graphs $G[N(u)]$ and $G[N(v) \setminus N(u)]$, respectively. Then, $V(G) = X_1 \cup X_2 \cup Y_1 \cup Y_2$ is a 4-cd-coloring of G where vertices in X_1 and Y_1 are dominated by u and the vertices in X_2 and Y_2 are dominated by v. Thus, $4 = \omega_s(G) \leq \chi_{cd}(G) \leq 4$. This shows that $\chi_{cd}(G) = \omega_s(G) = 4$.   □

Observation 8. *Let G be a connected P_5-free chordal graph with a minimal dominating clique $\{u, v\}$ and $\omega(G) = 3$. Then, finding the subclique number of G is the same as checking whether $N(u) \setminus N(v)$ and $N(v) \setminus N(u)$ are independent sets by Lemma 3 (Case 1 of Lemma 3 shows that if $\omega_s(G) = 3$ then at least one of $N(u) \setminus N(v)$ and $N(v) \setminus N(u)$ is an independent set, and Case 2 of Lemma 3 shows that if $\omega_s(G) = 4$ then neither $N(u) \setminus N(v)$ nor $N(v) \setminus N(u)$ is an independent set). The latter problem can be solved in $O(n^2)$ time. Hence, the subclique number and the cd-chromatic number of G can be found in $O(n^2)$ time.*

Observation 9. *Let $N_0 = \{x_0, x_1, x_2\}$ be a minimal dominating clique in a connected P_5-free chordal graph G with $\omega(G) = 3$. Also, let $N_1 = V(G) \setminus N_0$. Then, for sets M_0 and L_2 defined as in Observation 3.1, $[M_0, L_2] = \emptyset$. In general, $[M_i, L_j] = \emptyset$ for $i, j \in \{0, 1, 2\}$ and $i \neq j$.*

Observation 10. *Let $N_0 = \{x_0, x_1, x_2\}$ be a minimal dominating clique in a connected P_5-free chordal graph G with $\omega(G) = 3$. Let $N_1 = V(G) \setminus N_0$. Then, for $u \in M_i$ and $v \in M_j$, $d(u, v) \neq 2$ where M_i's are as in Observation 3.1 and $i \neq j$.*

Proof of Observations 9 and 10 are omitted.

For a graph G with clique number three and a dominating clique of size three, $\omega_s(G) \in \{3, 4, 5, 6\}$ by Theorem 1. We study the structure of G when $\omega_s(G) = 3$ in Observation 11 and in Observation 12 we discuss the structure of G when $\omega_s(G) = 4, 5, 6$. These Observations help us to prove Lemma 4.

Observation 11. *Let G be a connected P_5-free chordal graph with $\omega(G) = \omega_s(G) = 3$. Let $N_0 = \{x_0, x_1, x_2\}$ be a minimal dominating clique in G. Then, $\chi_{cd}(G) = 3$ and $V(G) \setminus \{x_0, x_1, x_2\}$ is an independent set.*

Proof. Let $N_1 = V(G) \setminus N_0$, and let M_i and L_i be sets defined in Observation 3.1.
Claim 1: $V(G) \setminus \{x_0, x_1, x_2\}$ is an independent set.
  Suppose that there is an edge $u_0 v_0$ in $V(G) \setminus \{x_0, x_1, x_2\}$. Then, by Observations 3.1 and 9, $u_0, v_0 \in M_i$ for some $i = 0, 1, 2$. Without loss of generality, assume that $u_0, v_0 \in M_0$. Since $\{x_0, x_1, x_2\}$ is a minimal dominating set in G, there exist vertices $u_1 \in N(x_1) \cap A(x_0) \cap A(x_2) = M_1$ and

$u_2 \in N(x_2) \cap A(x_0) \cap A(x_1) = M_2$. Then, $\{u_0, u_1, u_2, v_0\}$ forms a subclique of size four (because $d(u_0, v_0) = 1$, $d(u_j, u_k) \neq 2$ and $d(v_0, u_j) \neq 2$ for $j, k \in \{0, 1, 2\}$ by Observation 10). This contradicts the fact that $\omega_s(G) = 3$. Hence, $V(G) \setminus \{x_0, x_1, x_2\}$ is an independent set.

Let $X_i = \{x_{i+1}\} \cup (N(x_i) \cap A(x_{i+1})) = \{x_{i+1}\} \cup M_i \cup L_{i+1}$ for $i = 0, 1, 2$: from here onwards in this section, we take all the subscripts of x_i, M_i and L_i to be mod 3. Each X_i is an independent set since $M_i \cup L_{i+1} \subseteq V(G) \setminus \{x_0, x_1, x_2\}$ is an independent set and x_{i+1} is not adjcent to any vertex in $M_i \cup L_{i+1}$ by definitions of M_i and L_{i+1}.

Claim 2: $V(G) = X_0 \cup X_1 \cup X_2$ is a 3-cd-coloring of G.

Let $x \in V(G) \setminus \{x_0, x_1, x_2\}$. Then, $x \in N(x_i)$ for some $i = 0, 1, 2$. If $x \in A(x_{i+1})$, then $x \in N(x_i) \cap A(x_{i+1}) \subseteq X_i$. Else, $xx_{i+1} \in E(G)$. Since $\{x_i, x_{i+1}, x_{i+2}, x\}$ does not induce a K_4, $xx_{i+2} \notin E(G)$. This implies that $x \in N(x_{i+1}) \cap A(x_{i+2}) \subseteq X_{i+1}$. Thus, every vertex in G belongs to X_i for some $i = 0, 1, 2$. Also, by the definitions of M_i and L_{i+1}, each X_i is dominated by x_i. Hence, $V(G) = X_0 \cup X_1 \cup X_2$ forms a 3-cd-coloring of G. This implies that $3 = \omega_s(G) \leq \chi_{cd}(G) \leq 3$. Thus, $\chi_{cd}(G) = \omega_s(G) = 3$. □

Observation 12. *Let G be a connected P_5-free chordal graph with $\omega(G) = 3$ and $\omega_s(G) = j$ for some $j = 4, 5, 6$. Let $N_0 = \{x_0, x_1, x_2\}$ be a minimal dominating clique in G. Then, there exists $j - 3$ integers $i_1, \ldots, i_{j-3} \in \{0, 1, 2\}$ such that $V(G) \setminus (N(x_{i_1}) \cup \ldots \cup N(x_{i_{j-3}}))$ is an independent set, and every connected component of $G[N(x_{i_1})], \ldots, G[N(x_{i_{j-3}})]$ is isomorphic to K_1 or $E_{p,q}$ for some $p, q \in \mathbb{N}$. Also, $\chi_{cd}(G) = j$.*

Proof of Observations 12 is omitted in this paper.

Lemma 4. *Let G be a connected P_5-free chordal graph with $\omega(G) = 3$, and let $\{x_0, x_1, x_2\}$ be a minimal dominating clique in G. Then, $\chi_{cd}(G) = \omega_s(G)$.*

Proof. Since $\omega(G) \leq \omega_s(G)$, it is clear that $\omega_s(G) \in \{3, 4, 5, 6\}$ by Theorem 1. The rest of the proof follows from Observations 11 and 12. □

Observation 13. *The subclique number and the cd-chromatic number of a connected P_5-free chordal graph with clique number three and a minimal dominating clique $\{x_0, x_1, x_2\}$ can be found in $O(n^2)$ time.*

Observations 11 and 12 will prove Observation 13. A detailed proof of Observation 13 is omitted in this paper.

Theorem 2. *Let G be a $\{P_5, K_4\}$-free chordal graph. Then, $\chi_{cd}(G) = \omega_s(G)$. Also, the cd-chromatic number and the subclique number of G can be found in $O(n^2)$ time.*

Proof. Let G be a connected $\{P_5, K_4\}$-free chordal graph. Then, $\omega(G) \leq 3$. We prove the theorem in the following three cases.

Case 1: $\omega(G) = 1$.
Then, $G \cong K_1$ and $\chi_{cd}(G) = \omega_s(G) = 1$. This can be found in constant time.

Case 2: $\omega(G) = 2$.
Then, by Observation 2, G is a P_5-free tree and $\chi_{cd}(G) = \omega_s(G) = 2$. This can be found in constant time.

Case 3: $\omega(G) = 3$.
By Lemma 2, G admits a minimal dominating clique, say C, and it can be obtained in $O(n^2)$ time by Corollary 1. Then, the following statements are true.

1. If $|C| = 1$, then $\chi_{cd}(G) = \omega_s(G) = 3$ by Observation 6. This result can be obtained in constant time.
2. If $|C| = 2$, then $\chi_{cd}(G) = \omega_s(G)$ by Lemma 3. Also, $\chi_{cd}(G)$ and $\omega_s(G)$ can be found in $O(n^2)$ time by Observation 8.
3. If $|C| = 3$, then $\chi_{cd}(G) = \omega_s(G)$ by Lemma 4. Also, $\chi_{cd}(G)$ and $\omega_s(G)$ can be found in $O(n^2)$ time by Observation 13.

Thus, from the above cases, we can conclude the following.

1. For a connected $\{P_5, K_4\}$-free chordal graph G, $\chi_{cd}(G) = \omega_s(G)$.
2. The cd-chromatic number and the subclique number of a connected $\{P_5, K_4\}$-free chordal graph G can be found in $O(n^2)$ time.

We know that for a disconnected graph H with connected components $G_1, \ldots, G_k$, $\chi_{cd}(H) = \sum_{i=1}^{k} \chi_{cd}(G_i)$ and $\omega_s(H) = \sum_{i=1}^{k} \omega_s(G_i)$. Thus, the above results hold for the class of $\{P_5, K_4\}$-free chordal graphs.   □

4 P_6-free Chordal Bipartite Graphs

A bipartite graph G is said to be *chordal bipartite* if every induced cycle of length at least six has a chord in it. Clearly, a chordal bipartite graph G is C_n-free for every $n \neq 4$. It is known that in the class P_6-free trees, $\omega_s - \omega$ can be arbitrarily large [12]. In this section, we prove that $\chi_{cd}(G) = \omega_s(G)$ when G is a P_6-free chordal bipartite graph. In addition, we show that a maximum subclique of a P_6-free chordal bipartite graph can be found in $O(n^3)$ time.

Theorem 3. *[10] A graph G is $\{P_6, C_6, K_3\}$-free if and only if every connected induced subgraph of G has a dominating set that induces a complete bipartite subgraph.*

Theorem 4. *Let G be a connected P_6-free chordal bipartite graph. Then,* $\chi_{cd}(G) = \omega_s(G)$.

Proof. Clearly, G is a $\{P_6, C_6, K_3\}$-free graph. Then, by Theorem 3, G has a dominating set which induces a complete bipartite subgraph. Let $D = \{x_1, \ldots, x_k, y_1, \ldots, y_l\}$ be such a set of minimum cardinality. Let $D = X' \cup Y'$ be a partition of D into independent sets $X' = \{x_1, \ldots, x_k\}$ and $Y' = \{y_1, \ldots, y_l\}$. Let

$X = \bigcup_{j=1}^{l} N(y_j)$ and $Y = \bigcup_{i=1}^{k} N(x_i)$. Clearly, $X' \subseteq X$ and $Y' \subseteq Y$. Since G is triangle-free, $X \cap Y = \emptyset$. Thus, $X \cup Y$ is a partition of $V(G)$. We show that X is independent set. If not, let $w, z \in X$ such that $wz \in E(G)$. Since $X = \bigcup_{j=1}^{l} N(y_j)$, there exist $p, q \in \{1, \ldots, l\}$ such that $y_p w, y_q z \in E(G)$. Then, either $\{y_p, w, z, y_q, x_1\}$ induces a C_5 or one of its subsets induce a K_3 in G, a contradiction. Thus, X is an independent set. Similarly, we can show that Y is an independent set. Since D is a minimum dominating biclique of G, for every vertex x_i ($1 \le i \le k$), there exists a vertex $u_i \in Y$ such that $x_i u_i \in E(G)$ and $x_\alpha u_i \notin E(G)$ for every $\alpha \in \{1, \ldots, k\} \setminus \{i\}$. Similarly, for every vertex y_j ($1 \le j \le l$), there exists a vertex $v_j \in X$ such that $y_j v_j \in E(G)$ and $y_\beta v_j \notin E(G)$ for every $\beta \in \{1, \ldots, l\} \setminus \{j\}$. Note that u_i's and v_j's are pairwise distinct. Let $S = \{u_1, \ldots, u_k, v_1, \ldots, v_l\}$. Clearly, $|S| = k + l$. The following claims prove that S is a subclique.

Claim 1 : $d(u_1, u_2) \neq 2$.
By the definition of u_i's, $x_1 u_1, x_2 u_2 \in E(G)$ and $x_2 u_1, x_1 u_2 \notin E(G)$. Contrary to Claim 1, assume that $d(u_1, u_2) = 2$. Then, there exists a vertex a in $V(G)$ such that $a u_1$ and $a u_2$ are edges in G. Since $u_1 \in Y$ and Y is an independent set, $a \notin Y$. We claim that $a \notin X'$. If not, let $a = x_i$ for some $i = 1, \ldots, k$. Then, $x_i u_1, x_i u_2 \in E(G)$ for some $i = 1, \ldots, k$. We know that $x_1 u_1 \in E(G)$ and $x_\alpha u_1 \notin E(G)$ for $\alpha \in \{2, \ldots, k\}$. Hence, $a = x_1$. This implies that $a u_2 = x_1 u_2 \in E(G)$, a contradiction to $x_1 u_2 \notin E(G)$. Thus, $a \notin X'$. Hence, $a \in X \setminus X'$.

Next, we show that the sets $N(x_1)$ and $N(x_2)$ are dominated by a. On the contrary, suppose that there exists a vertex $u \in N(x_1) \setminus \{u_1\}$ such that $au \notin E(G)$. Then, $\{u, x_1, u_1, a, u_2, x_2\}$ induces P_6 or C_6 in G (because $uu_1, ua, uu_2, x_1 a, x_1 u_2, x_1 x_2, u_1 u_2, u_1 x_2, a x_2$ are not edges in G). This implies that every vertex in $N(x_1)$ is adjacent to a. Similarly, we can show that every vertex in $N(x_2)$ is adjacent to a. Thus, a dominates the set $N(x_1) \cup N(x_2) \supseteq Y'$. This implies that $(D \setminus \{x_1, x_2\}) \cup \{a\}$ is a biclique of size $k + l - 1$ dominating $V(G)$, a contradiction to the fact that D is a minimum dominating biclique. Therefore, $d(u_1, u_2) \neq 2$.

Similarly, we can prove that for $1 \le i_1 < i_2 \le k$, $d(u_{i_1}, u_{i_2}) \neq 2$, and for $1 \le j_1 < j_2 \le l$, $d(v_{j_1}, v_{j_2}) \neq 2$.

Claim 2 : $d(u_1, v_1) \neq 2$.
Clearly, $u_1 \in Y$ and $v_1 \in X$. Contrary to Claim 2, assume that $d(u_1, v_1) = 2$. Thus, $u_1 v_1 \notin E(G)$. In addition, there exists a vertex a in $V(G)$ such that au_1, av_1 are edges in G. Then, $a \notin X \cup Y = V(G)$ because X and Y are independent sets, a contradiction. This proves Claim 2.

Similarly, we can prove that $d(u_i, v_j) \neq 2$ for $1 \le i \le k$ and $1 \le j \le l$.

By Claims 1 and 2, it is evident that $S = \{u_1, \ldots, u_k, v_1, \ldots, v_l\}$ is a subclique of size $k + l$ in G. Therefore, $\omega_s(G) \ge k + l$.

Next, we produce a $(k+l)$-cd-coloring of G. Let $U_1 = N(x_1)$, and $U_i = N(x_i) \setminus \bigcup_{\alpha=1}^{i-1} N(x_\alpha)$ for $2 \le i \le k$. Clearly, $u_i \in U_i$ and each U_i is dominated by the vertex x_i for $1 \le i \le k$. Similarly, define $V_1 = N(y_1)$, and $V_j = N(y_j) \setminus \bigcup_{\beta=1}^{j-1} N(y_\beta)$ for $2 \le j \le l$. Clearly, $v_j \in V_j$ and each V_j is dominated by the vertex y_j for $1 \le j \le l$. Also, since D dominates $V(G)$, each vertex in Y belongs to U_i for some i and each vertex in X belongs to V_j for some j. Thus, $V(G) = (\bigcup_{i=1}^{k} U_i) \bigcup (\bigcup_{j=1}^{l} V_j)$ is a $(k+l)$-cd-coloring of G. This implies that $\chi_{cd}(G) \le k+l$. Thus, we have, $k+l \le \omega_s(G) \le \chi_{cd}(G) \le k+l$. Hence, $\chi_{cd}(G) = \omega_s(G) = k+l$. $\qquad\square$

Remark 1.

1. A cycle on five vertices (C_5) is a $\{P_6, C_6, K_3\}$-free graph with $\omega_s(C_5) = 2$ and $\chi_{cd}(C_5) = 3$. Since $\chi_{cd}(C_5) \neq \omega_s(C_5)$, the above theorem (Theorem 4) on P_6-free chordal bipartite graphs cannot be extended to $\{P_6, C_6, K_3\}$-free graphs.
2. Theorem 4 on P_6-free chordal bipartite graphs cannot be extended to P_7-free chordal bipartite graphs. A P_7-free chordal bipartite graph G with $\chi_{cd}(G) = 4$ and $\omega_s(G) = 2$ is shown in Fig. 2.

Lemma 5. *[4] The total dominating set problem in chordal bipartite graphs can be solved in $O(n^2)$ time.*

Theorem 5. *[11] The cd-chromatic number of a triangle-free graph is equal to its total domination number.*

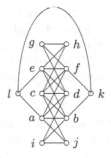

Fig. 2. A P_7-free chordal bipatite graph G where $\{k, l, a, d, g, h\}$ induces a P_6. Here, $\chi_{cd}(G) = \gamma_t(G) = 4$ (see Theorem 5). Also, $\omega_s(G) = 2$: choose any two vertices $x, y \in V(G)$ such that $d(x, y) \neq 2$, then for any other vertex $z \in V(G)$, either $d(x, z) = 2$ or $d(y, z) = 2$.

Remark 2. If D is a total dominating set of size $\gamma_t(G)$ in a triangle-free graph G, then Merouane et al.'s algorithm [11] produces a $\gamma_t(G)$-cd-coloring in $O(n+m)$ time. Thus, by Lemma 5 and Theorem 5, an optimal cd-coloring of a chordal bipartite graph can be found in $O(n^2)$ time using Merouane et al.'s algorithm.

Lemma 6. *Let G be a connected P_6-free chordal bipartite graph. Then, every total dominating set in G of size $\gamma_t(G)$ is a biclique.*

Proof of Lemma 6 is omitted in this paper.

Corollary 2. *A maximum subclique of a P_6-free chordal bipartite graph can be found in $O(n^3)$ time.*

Proof. Let G be a connected P_6-free chordal bipartite graph. Then, by Lemmas 5 and 6, a biclique $D = \{x_1, \ldots, x_k, y_1, \ldots, y_l\}$ of size $\gamma_t(G)$ dominating $V(G)$ can be found in $O(n^2)$ time (here, $\gamma_t(G) = k + l$). Since D is a minimum dominating biclique of G, for every vertex x_i ($1 \le i \le k$), there exists a vertex $u_i \in Y$ such that $x_i u_i \in E(G)$ and $x_\alpha u_i \notin E(G)$ for every $\alpha \in \{1, \ldots, k\} \setminus \{i\}$. Similarly, for every vertex y_j ($1 \le j \le l$), there exists a vertex $v_j \in X$ such that $y_j v_j \in E(G)$ and $y_\beta v_j \notin E(G)$ for every $\beta \in \{1, \ldots, l\} \setminus \{j\}$. It is proved in Theorem 4 that the set $S = \{u_1, \ldots, u_k, v_1, \ldots, v_l\}$ is a maximum subclique in G. The set S can be found by choosing exactly one vertex from each set $X_i' = N(x_i) \setminus \bigcup_{\substack{1 \le \alpha \le k \\ \alpha \ne i}} N(x_\alpha)$

for $i \in \{1, \ldots, k\}$ and $Y_j' = N(y_j) \setminus \bigcup_{\substack{1 \le \beta \le l \\ \beta \ne j}} N(x_\beta)$ for $j \in \{1, \ldots, l\}$. A set X_i'

(respectively Y_j') can be found in $O(n^2)$ time and at most n such sets are found. Thus, a maximum subclique of G can be found in $O(n^3)$ time.

From the above result, we can conclude that a maximum subclique in a P_6-free chordal bipartite graph can be found in $O(n^3)$ time because $\omega_s(H) = \sum_{i=1}^{k} \omega_s(G_i)$ for a graph H with connected components $G_1, \ldots, G_k$. $\qquad \square$

5   Conclusion

In this paper, we proved that the cd-chromatic number and the subclique number are equal for the class of $\{P_5, K_4\}$-free chordal graphs and the class of P_6-free chordal bipartite graphs. In addition, we proved that the cd-chromatic number and the subclique number of a $\{P_5, K_4\}$-free chordal graph can be found in $O(n^2)$ time. Also, a maximum subclique in a P_6-free chordal bipartite graph can be found in $O(n^3)$ time.

Acknowledgement. We thank Cyriac Antony for proof-reading this paper. We thank the reviewers of CALDAM 2022 whose suggestions improved the paper.

References

1. Arumugam, S., Bagga, J., Chandrasekar, K.R.: On dominator colorings in graphs. Proc. Math. Sci. **122**, 561–571 (2012). https://doi.org/10.1007/s12044-012-0092-5
2. Arumugam, S., Chandrasekar, K.R., Misra, N., Philip, G., Saurabh, S.: Algorithmic aspects of dominator colorings in graphs. In: Iliopoulos, C.S., Smyth, W.F. (eds.) Combinatorial Algorithms. IWOCA 2011. LNCS, vol. 7056. Springer, Heidelberg (2011). https://doi.org/10.1007/978-3-642-25011-8_2
3. Chen, Y.H.: The dominated coloring problem and its application. In: Murgante, B., et al. (eds.) Computational Science and Its Applications. LNCS, vol. 8584. Springer, Cham (2014). https://doi.org/10.1007/978-3-319-09153-2_10

4. Damaschke, P., Muller, H., Kratsch, D.: Domination in convex and chordal bipartite graphs. Inf. Proc. Lett. **36**, 231–236 (1990). https://doi.org/10.1016/0020-0190(90)90147-P
5. Das, S., Mishra, S.: Lower bounds on approximating some variations of vertex coloring problem over restricted graph classes. Discrete Math. Algorith. Appl. **12** (2020). https://doi.org/10.1142/S179383092050086X
6. Gera, R., Horton, S., Rasmussen, C.: Dominator colorings and safe clique partitions. Congressus Numerantium **181**, 19–32 (2006)
7. Gera, R.M.: On dominator colorings in graphs. Graph Theory Notes of New York LII, pp. 25–30 (2007)
8. Kiruthika, R., Rai, A., Saurabh, S., Tale, P.: Parametrized and exact algorithms for class domination coloring. In: Proceedings of the 43rd Conference on Current Trends in Theory and Practice of Computer Science, LNCS vol. 10139, pp. 336–349 (2017). https://doi.org/10.1007/978-3-319-51963-0_26
9. Klavžar, S., Tavakoli, M.: Dominated and dominator colorings over (edge) corona and hierarchical products. Appl. Math. Comput. **390** (2021). https://doi.org/10.1016/j.amc.2020.125647
10. Liu, J., Zhou, H.: Dominating subgraphs in graphs with some forbidden structures. Discrete Math. **135**, 163–168 (1994). https://doi.org/10.1016/0012-365X(93)E0111-G
11. Merouane, H.B., Haddad, M., Chellali, M., Kheddouci, H.: Dominated colorings of graphs. Graphs Combin. **31**(3), 713–727 (2015). https://doi.org/10.1007/s00373-014-1407-3
12. Shalu, M.A., Kirubakaran, V.K.: On cd-coloring of trees and co-bipartite graphs. In: Mudgal A., Subramanian C.R. (eds.) Algorithms and Discrete Applied Mathematics. LNCS, vol. 12601, pp. 209–221 (2021). https://doi.org/10.1007/978-3-030-67899-9_16
13. Shalu, M.A., Sandhya, T.P.: The cd-Coloring of graphs. In: Govindarajan, S., Maheshwari, A. (eds.) Algorithms and Discrete Applied Mathematics. LNCS, vol. 9602. Springer, Cham (2016). https://doi.org/10.1007/978-3-319-29221-2_29
14. Shalu, M.A., Vijayakumar, S., Sandhya, T.P.: A lower bound of the cd-Chromatic number and its complexity. In: Gaur, D., Narayanaswamy, N. (eds.) Algorithms and Discrete Applied Mathematics. LNCS, vol. 10156. Springer, Cham (2017). https://doi.org/10.1007/978-3-319-53007-9_30
15. Shalu, M.A., Vijayakumar, S., Sandhya, T.P.: On complexity of cd-coloring of graphs. Discr. Appl. Math. **280**, 171–185 (2020). https://doi.org/10.1016/j.dam.2018.03.004
16. Swaminathan, V., Sundareswaran, R.: Color class domination in graphs. Narosa Publishing House, Mathematical and Experimental Physics (2010)
17. West, D.B.: Introduction to Graph Theory. Second Edition, Pearson (2018)

An Output-Sensitive Algorithm for All-Pairs Shortest Paths in Directed Acyclic Graphs

Andrzej Lingas[1], Mia Persson[2(⊠)], and Dzmitry Sledneu[3]

[1] Department of Computer Science, Lund University, 22100 Lund, Sweden
Andrzej.Lingas@cs.lth.se
[2] Department of Computer Science and Media Technology, Malmö University,
20506 Malmö, Sweden
mia.persson@mau.se
[3] Malmö, Sweden

Abstract. First, we present a new algorithm for the single-source shortest paths problem (SSSP) in edge-weighted directed graphs, with n vertices, m edges, and both positive and negative real edge weights. Given a positive integer parameter t, in $O(tm)$ time the algorithm finds for each vertex v a path distance from the source to v not exceeding that yielded by the shortest path from the source to v among the so called $t+$*light paths*. A directed path between two vertices is $t+$light if it contains at most t more edges than the minimum edge-cardinality directed path between these vertices. For $t = O(n)$, our algorithm yields an $O(nm)$-time solution to SSSP in directed graphs with real edge weights matching that of Bellman and Ford.

Our main contribution is a new, output-sensitive algorithm for the all-pairs shortest paths problem (APSP) in directed acyclic graphs (DAGs) with positive and negative real edge weights. The running time of the algorithm depends on such parameters as the number of leaves in (lexicographically first) shortest-paths trees, and the in-degrees in the input graph. If the trees are sufficiently thin on the average, the algorithm is substantially faster than the best known algorithm.

Finally, we discuss an extension of hypothetical improved upper time-bounds for APSP in non-negatively edge-weighted DAGs to include directed graphs with a polynomial number of large directed cycles.

1 Introduction

The *length* of a path in an *edge-weighted* graph is the sum of the weights of edges on the path. A *shortest* path between two vertices in a graph has minimal length among all paths between these vertices. The *distance* between vertices v and u is the length of a shortest path from v to u. If the graph is directed, the paths are supposed to be also directed.

Shortest path problems, in particular the single-source shortest paths problem (SSSP) and the all-pairs shortest paths problem (APSP), belong to the most basic and important problems in graph algorithms [5, 17]. There are several variants of SSSP and APSP depending among other things on the restrictions on edge weights and the input graphs. The input to these problems is a directed or an undirected edge-weighted graph.

© Springer Nature Switzerland AG 2022
N. Balachandran and R. Inkulu (Eds.): CALDAM 2022, LNCS 13179, pp. 140–151, 2022.
https://doi.org/10.1007/978-3-030-95018-7_12

The output is a representation of shortest paths between the source and all other vertices or between all pairs of vertices in the graph, respectively.

In the general case of directed graphs (without negative cycles), when both positive and negative real edge weights are allowed, the difference between the best known asymptotic upper time-bounds for SSSP and APSP respectively is surprisingly small. Namely, if the input directed graph has n vertices and m edges with real weights, then the best known SSSP algorithm due to Bellman [3], Ford [7], and Moore [12] runs in $O(nm)$ while the APSP can be solved already in $O(nm + n^2 \log n)$ time [11,17]. The APSP solution uses Johnson's $O(nm)$-time reduction of the general edge weight case to the non-negative edge case and then it runs Dijkstra's algorithm [6] n times [11,17]. The latter upper time-bound for APSP with arbitrary real edge-weights has been more recently improved to $O(nm + n^2 \log \log n)$ by Pettie in [14]. Note that the aforementioned best asymptotic upper time bounds for SSSP and APSP are different only for sparse graphs with $o(n \log \log n)$ edges. Interestingly, when edge weights are integers, the best known upper time-bound for APSP just in terms of n is $n^3 / 2^{\Omega(\sqrt{\log n})}$ [4].

The situation alters dramatically when the input directed graph is acyclic, i.e., when it does not contain directed cycles. Then, a simple dynamic programming algorithm processing vertices in a topologically sorted order solves the SSSP problem in $O(n+m)$ time [5], an $O(n(n + m))$-time solution to the APSP problem in this case follows.

In fact, Yen could use the aforementioned method for SSSP in DAGs iteratively in order to improve the time complexity of Bellman-Ford algorithm for directed graphs by a constant factor [15]. Bellman-Ford algorithm runs in $n-1$ iterations. In each iteration, for each edge e, the current distance (from the source) at the head of e is compared to the sum of the current distance at the tail of e and the weight if e. If the sum is smaller the distance at the head of e is updated. To achieve the improvement, Yen imposes a linear order on the vertices of the input directed graph which yields a decomposition of the graph into two DAGs. Next, the SSSP method for DAGs is run on each of the two DAGs instead of an iteration of Bellman-Ford algorithm [15]. Bannister and Eppstein obtained a further improvement of the time complexity of Bellman-Ford algorithm by a constant factor using a random linear order [2].

A pair of vertices in an edge weighted undirected or directed graph can be connected by several paths, in particular several shortest paths. Beside the length of a path, the number of edges forming it can be an important characteristic. For example, Zwick provided several exact and approximation algorithms for all pairs *lightest* (i.e., having minimal number of edges) shortest paths in directed graphs with restricted edge weights in [18].

In this paper, first we consider $t+light$ *paths*, i.e., directed paths that have at most t more edges than the paths with the same endpoints having the minimal number of edges. In part following [15], we iterate $O(t)$ times the SSSP method for DAGs on two implicit DAGs yielded by an extension of the BFS partial order to a linear order. The iterations alternatively process the vertices in a breadth-first sorted order and the reverse order. In result, we obtain path distances from the source to all other vertices that are not greater than the corresponding shortest-path distances for $t+$light paths. It takes $O(tm)$ time totally. For $t = n - 2$, our method matches that of Bellman-Ford for SSSP in directed graphs with real edge weights.

A vertex v is an *ancestor* (a *direct ancestor*, respectively) of a vertex u in a DAG if there is a directed path (edge, respectively) from v to u in the DAG.

Our main result is a new, output-sensitive algorithm for the APSP problem in DAGs. It runs in time $O(\min\{n^\omega, nm + n^2 \log n\} + \sum_{v \in V} \mathrm{indeg}(v)|\mathrm{leaf}(T_v)|)$, where n is the number of vertices, m is the number of edges, ω is the exponent of fast $n \times n$ matrix multiplication[1], $\mathrm{indeg}(v)$ stands for the indegree of v, T_v is a tree of lexicographically-first shortest directed paths from all ancestors of v to v, $\mathrm{leaf}(T_v)$ is the set of leaves in T_v, and for a set X, $|X|$ stands for its size. Note that if T_v is a path the term $O(\mathrm{indeg}(v)|\mathrm{leaf}(T_v)|)$ equals $O(\mathrm{indeg}(v))$ while when T_v is a star with v as a sink the term becomes $O(\mathrm{indeg}(v)|T_v|)$. Thus, the running time of the APSP algorithm can be so low as $O(n^\omega)$ and so high as $O(n^\omega + nm)$. It follows also that if α is defined by $\max_{v \in V} |\mathrm{leaf}(T_v)| = O(n^\alpha)$ then the algorithm runs in $O(n^\omega + mn^\alpha)$ time. Similarly, if β is defined by $\frac{\sum_{v \in V} |\mathrm{leaf}(T_v)|}{n} = O(n^\beta)$ then the algorithm runs in $O(n^\omega + n^{2+\beta})$ time.

Finally, we provide an extension of hypothetical, improved upper time-bounds for APSP in DAGs with non-negative edge weights to include directed graphs with a polynomial number of large directed cycles.

In the full version [9], we additionally present experimental comparisons of our SSSP algorithm with the Bellman-Ford one. They show that our SSSP algorithm converges to the true shortest-path distances on dense edge-weighted pseudorandom graphs faster than the Bellman-Ford algorithm does.

1.1 Paper Organization

In the next section, we provide our solution to the SSSP problem in directed graphs with real edge weights based on the SSSP method for DAGs and the BFS partial order in terms of $t+$light paths. Section 3 is devoted to our output-sensitive algorithm for the APSP problem in DAGs with real edge weights and its analysis. In Sect. 4, we discuss the extension of hypothetical, improved bounds for APSP in DAGs with non-negatively weighted edges to directed graphs with a polynomial number of large directed cycles. We conclude with Final remarks. A description of our experimental results can be found in the full version [9].

2 An Application of DAG SSSP Method to Arbitrary Digraphs

The SSSP problem for directed acyclic graphs can be solved by topologically sorting the DAG vertices and applying straightforward dynamic programming. For consecutive vertices v in the sorted order, the distance $dist(v)$ of v from the source is set to the minimum of $dist(u) + weight(u, v)$ over all direct ancestors u of v, where $weight(u, v)$ stands for the weight of the edge (u, v). It takes linear (in the size of the DAG) time. Yen used the dynamic programming method iteratively to improve the time complexity of Bellman-Ford algorithm for directed graphs by a constant factor in [15]. Interestingly, we can similarly apply this method iteratively to determine shortest-path

---

[1] ω is not greater than 2.3729 [1].

distances among paths using almost the minimal number of edges. To formulate our algorithm (Algorithm 1), we need the following definition and two procedures.

Definition 1. *A directed path from a vertex u to a vertex v in a directed graph is lightest if it consists of the smallest possible number of edges. A path from u to v is $t+$light if it includes at most t more edges than a lightest path from u to v.*

procedure $SSSPDAG(G, D)$
Input: A directed graph (V, E) with real edge weights, linearly ordered vertices $v_1, ..., v_n$, and a 1-dimensional table D of size n with upper bounds on the distances from v_1 to all vertices in V.
Output: Improved upper bounds on the shortest-path distances from v_1 to all vertices in V in the table D.
for $j = 2, ..., n$ **do**
For each edge (v_i, v_j) where $i < j$
$D(v_j) \leftarrow \min\{D(v_j), D(v_i) + weight(v_i, v_j)\}$
procedure $reverseSSSPDAG(G, D)$
Input and output: the same as in $SSSPDAG(G, D)$
for $j = n - 1, ..., 1$ **do**
For each edge (v_i, v_j) where $i > j$
$D(v_j) \leftarrow \min\{D(v_j), D(v_i) + weight(v_i, v_j)\}$

Algorithm 1
Input: A directed graph (V, E) with n vertices, real edge weights and a distinguished source vertex s, and a positive integer t.
Output: Upper bounds on the shortest-path distances from s to all other vertices in V not exceeding the corresponding shortest-path distances constrained to $t+$light paths.

1. Run BFS from the source s.
2. Order the vertices of G extending the BFS partial order according to the levels of the BFS tree, i.e., s comes first, then the vertices reachable by direct edges from s, then the vertices reachable by paths composed of two edges and so on. We may assume w.l.o.g. that all vertices are reachable from s or alternatively extend the aforementioned order with the non-reachable vertices arbitrarily.
3. Initialize a 1-dimensional table D of size n, setting $D(v_1) \leftarrow 0$ and $D(v_j) \leftarrow \infty$ for $1 < j \leq n$
4. $SSSPDAG(G, D)$
5. **for** $k = 1, ..., t$ **do**
   (a) $reverseSSSPDAG(G, D)$
   (b) $SSSPDAG(G, D)$

Theorem 1. *Let G be a directed graph with n vertices, m real-weighted edges, and a distinguished source vertex s. For all vertices v of G different from s, an upper bound on their distance from the source vertex s, not exceeding the length of a shortest path among $t+$light paths from s to v, can be computed in $O((t + 1)(m + n))$ total time.*

Proof. Consider Algorithm 1 and in particular the ordering of the vertices specified in its second step. We shall refer to an edge (v_i, v_j) as forward if $i < j$ otherwise we shall call it backward. Note that the vertices at the same level of the BFS tree can be connected both by forward as well as backward edges. See also Fig. 1. Let ℓ be the number of (forward) edges in a lightest path from s to a given vertex v. It follows that any path from s to v, in particular a shortest t+light one, has to have at least ℓ forward edges.

To see this, consider the BFS tree from the source s. Define the level of a vertex in the tree as the number of edges on the path from s to the vertex in the tree. Thus, in particular, $level(s) = 0$ while $level(v) = \ell$. Recall that the linear order extending the partial BFS order used in Algorithm 1 is non-decreasing with respect of the levels of vertices. Also, if (u, w) is a forward edge then $level(u) \leq level(w) \leq level(u) + 1$ and if (u, w) is a backward edge then $level(u) \geq level(w)$. Hence, any path from s to v has to have at least ℓ forward edges, each increasing the level by one.

Consequently, a shortest t+light path from s to v can have at most t backward edges. Thus, it can be decomposed into at most $2t+1$ maximal fragments of consecutive edges of the same type (i.e., forward or backward, respectively), where the even numbered fragments consist of backward edges. Thus, the at most $2t + 1$ calls of the procedures $SSSPDAG(G, s, D)$, $reverseSSSPDAG(G, s, D)$ in the algorithm are sufficient to detect a distance from s to v not exceeding the length of a shortest path among t+light paths from s to v. The asymptotic running time of the algorithm is dominated by the aforementioned procedure calls. Hence, it is $O((t + 1)(m + n))$. $\qquad\square$

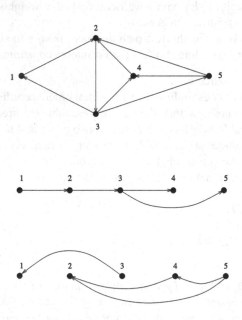

Fig. 1. An example of a graph with a BFS vertex numbering and the two DAGs implied by forward and backward edges, respectively.

We can obtain a representation of directed paths achieving the upper bounds on the distances from the source provided in Theorem 1 in a form of a tree of paths emanating from the source by backtracking. By setting $t = n - 2$ in this theorem, we can match the best known SSSP algorithm for directed graphs with positive and negative real edge weights, i.e., the Bellman-Ford algorithm and its constant factor improvements [11, 17], running in $O(nm)$ time. Similarly as in the case of Bellman-Ford algorithm, by calling additionally $reverseSSSPDAG(G, D)$ and $SSSPDAG(G, D)$ after the last iteration in Algorithm 1, we can detect the existence of negative cycles.

Comparing our algorithm with the Bellman-Ford one, note that if the lightest path from the source to a vertex v has ℓ edges then $\ell + t$ iterations in the Bellman-Ford algorithm may be needed to obtain an upper bound on the distance of v from the source comparable to that obtained after $O(t)$ iterations in Algorithm 1.

3   An Output-Sensitive APSP Algorithm for DAGs

The APSP problem in DAGs with both positive and negative real edge weights can be solved in $O(n(n + m))$ time by running n times the SSSP algorithm for DAGs. It is an intriguing open problem if there exist substantially more efficient algorithms for APSP in edge-weighted DAGs. In this section, we make a progress on this question by providing an output-sensitive algorithm for this problem. Its running time depends on the structure of shortest path trees. Although in the worst-case it does not break the $O(nm)$ barrier it seems to be substantially more efficient in the majority of cases.

The standard algorithm for APSP for DAGs just runs the SSSP algorithm for DAGs for each vertex of the DAG as a source separately. Our APSP algorithm (Algorithm 2) does everything in one sweep along the topologically sorted order. Its main idea is for each vertex v to compute the tree of lexicographically-first shortest paths from the ancestors u of the currently processed vertex v to v, in the topologically sorted order. For each ancestor u of v, Algorithm 2 proceeds as follows. In case the tree of lexicographically-first shortest paths from the already considered ancestors of v includes u (as some intermediate vertex) then the algorithm is done as for u. Otherwise, Algorithm 2 finds the direct ancestor of v on the lexicographically-first shortest path P from u to v and adds an initial fragment of P to the tree. By the topologically sorted order in which the ancestors u of v are considered, the latter situation can happen only when u is a leaf of the (final) tree of lexicographically-first shortest paths from the ancestors of v to v. Algorithm 2 finds the direct ancestor of v on P by comparing the lengths of shortest paths from u to v with different direct ancestors of v as the next to the last vertex on the paths in time proportional to the indegree of v. It also finds the initial fragment of P to add by using the link to the lexicographically-first shortest path from u to the direct ancestor of v that is on P. The correctness of the algorithm is immediate. The issues are an implementation of these steps and an estimation of the running time.

To specify our output-sensitive algorithm (Algorithm 2) more exactly, we need the following definition.

Definition 2. *Assume a numbering of vertices in an edge-weighted DAG extending the topological partial order. A shortest (directed) path P from v_k to v_i in the DAG is* first in a lexicographic order *if the direct ancestor v_j of v_i on P has the lowest number j among all direct ancestors of v_i on shortest paths from v_k to v_i and the subpath of P from v_k to v_j is the lexicographically-first shortest path from v_k to v_j. For a vertex v_i in the DAG, the tree T_{v_i} of (lexicographically-first) shortest paths is the union of lexicographically-first paths from all ancestors of v_i to v_i. Note that the vertex v_i is a sink of T_{v_i}. It is assumed to be the root of T_{v_i} and leaf(T_{v_i}) stands for the set of leaves of T_{v_i}.*

Algorithm 2

Input: A DAG (V, E) with real edge weights.
Output: For each vertex $v \in V$, the tree T_v of lexicographically-first shortest paths from all ancestors of v to v given by the table $NEXT_v$, where for each ancestor u of v, $NEXT_v(u)$ is the direct successor of u in the tree T_v, (i.e., the head of the unique directed edge having u as the tail in the tree).

1. Determine the source vertices, topologically sort the remaining vertices in V, and number the vertices in V accordingly, assigning to the sources the lowest numbers.
2. Set n to $|V|$ and r to the number of sources in G.
3. Initialize an $n \times n$ table *dist* by setting $dist(u, u) = 0$ and $dist(u, v) = \infty$ for $u, v \in V$, $u \neq v$.
4. **for** $i = r + 1, ..., n$ **do**
   (a) Compute the set $A(v_i)$ of ancestors of v_i.
   (b) Initialize a 1-dimensional table $NEXT_{v_i}$ of size $|A(v_i)|$, setting $NEXTv_i(v_j)$ to 0 for $v_j \in A(v_i)$.
   (c) **for** $v_k \in A(v_i)$ in increasing order of the index k **do**
      i. **if** $NEXT_{v_i}(v_k) \neq 0$ **then** proceed to the next iteration of the interior for block.
      ii. Determine a direct ancestor v_j of v_i that minimizes the value of $dist(v_k, v_j) + weight(v_j, v_i)$. In case of ties the vertex v_j with the smallest index j is chosen among those yielding the minimum.
      iii. $v_{current} \leftarrow v_k$
      iv. **while** $v_{current} \neq v_j \wedge NEXT(v_{current}, v_i) = 0$ **do**
         $dist(v_{current}, v_i) \leftarrow dist(v_{current}, v_j) + weight(v_j, v_i)$
         $NEXT_{v_i}(v_{current}) \leftarrow NEXT_{v_j}(v_{current})$
         $v_{current} \leftarrow NEXT_{v_i}(v_{current})$
      v. **if** $NEXT_{v_i}(v_j) = 0$ **then** $dist(v_j, v_i) \leftarrow weigh(v_j, v_i) \wedge NEXT_{v_i}(v_j) \leftarrow v_i$

Lemma 1. *Steps 4.c.iii-v add the missing fragments of a lexicographically shortest path from v_k to v_i and set the distances from vertices in the fragments to v_i in time proportional to the number of vertices added to T_{v_i}.*

Proof. Follow the path from v_k to v_j in T_{v_j} extended by (v_j, v_i) until a vertex $v_q \in T_{v_i}$ is encountered. This is done in Steps 4.c.iii-v. The membership of $v_{current}$ in T_{v_i} is verified by checking whether or not $NEXT_{v_i}(v_{current}) = 0$. Also, if $v_{current}$ is not yet in T_{v_i} then its distance to v_i is set by $dist(v_{current}, v_i) \leftarrow dist(v_{current}, v_j) + weight(v_j, v_i)$ and it is added to T_{v_i} by $NEXT_{v_i}(v_{current}) \leftarrow NEXT_{v_j}(v_{current})$ in

Step 4.c.iv. By the inclusion of v_q in T_{v_i}, a whole shortest path Q from v_q to v_i is already included in T_{v_i} by induction on the number of steps performed by the algorithm. We claim that Q exactly overlaps with the final fragment of the extended path starting from v_q. To see this encode Q and the aforementioned fragment of the extended path by the indices of their vertices in the reverse order. By our rule of resolving ties in Step 4.c.ii both encodings should be first in the lexicographic order so we have an exact overlap. For this reason, it is sufficient to add the initial fragment of the extended path ending at v_q to T_{v_i} and if necessary also the edge (v_j, v_i) to T_{v_i}, and to update the distances from vertices in the added fragment to v_i, i.e., to perform Steps 4.c.iii-v. □

Theorem 2. *The APSP algorithm for a DAG (V, E) with n vertices, m edges and real edge weights (Algorithm 2) runs in time $O(\min\{n^\omega, nm + n^2 \log n\} + \sum_{v \in V} indeg(v)|leaf(T_v)|)$.*

Proof. The sets of ancestors can be determined in Step 4.a by computing the transitive closure of the input DAG in $O(\min\{n^\omega, nm\})$ time by using fast matrix multiplication [13] or BFS [5], first. In fact, to implement the loop in Step4.c, we need the sets of ancestors to be ordered according to the numbering of vertices provided in Step 1. If the transitive closure matrix is computed such an ordered set of ancestors can be easily retrieved in $O(n)$ time. Otherwise, additional preprocessing sorting the unordered sets of ancestors is needed. The total cost of the additional preprocessing is $O(n^2 \log n)$.

All the remaining steps, excluding Steps 4.c.ii-v for vertices v_k not yet in T_{v_i}, can be done in total (i.e., over all iterations) time $O(\sum_{v \in V}(1 + |A(v)|)) = O(n^2)$, where $A(v)$ stands for the set of ancestors of v in the DAG. The time taken by Step 4.c.ii, when v_k is not yet in the current T_{v_i}, is $O(indeg(v_i))$. Suppose that v_k is not a leaf of the final tree T_{v_i}. Then, there must exist some leaf v_p of the final tree such that there is path from v_p via v_k to v_i in this tree. By the numbering of vertices extending the partial topological order, we have $p < k$. We infer that the aforementioned path is already present in the current T_{v_i}. Thus, in particular the vertex v_k is in the current tree. Hence, the total time taken by Step 4.c.ii is $O(\sum_{v \in V} indeg(v)|leaf(T_v)|)$. Finally, the total time taken by Steps 4.c.iii-v is $O(\sum_{v \in V}(1 + |A(v)|))$ by Lemma 1. □

Note that the following inequalities hold:

$$\sum_{v \in V} indeg(v)|leaf(T_v)| \leq m \max_{v \in V} |leaf(T_v)|,$$

$$\sum_{v \in V} indeg(v)|leaf(T_v)| \leq n^2 \frac{\sum_{v \in V} |leaf(T_v)|}{n}.$$

They immediately yield the following corollary from Theorem 2.

Corollary 1. *Let $G = (V, E)$ be an n-vertex DAG with n vertices and m edges with real edge weights. Suppose $\max_{v \in V} |leaf(T_v)| = O(n^\alpha)$ and $\frac{\sum_{v \in V} |leaf(T_v)|}{n} = O(n^\beta)$. The APSP problem for G is solved by Algorithm 2 in time $O(\min\{n^\omega, nm + n^2 \log n\} + \min\{mn^\alpha, n^{2+\beta}\})$.*

Observe that $|\text{leaf}(T_v)|$ is equal to the minimum number of directed paths covering the tree T_v. Hence, $\alpha < 1$ if the maximum of the minimum number of paths covering T_v over v is substantially sublinear. Similarly, $\beta < 1$ if the average of the minimum number of paths covering T_v over v is substantially sublinear.

To illustrate the superiority of Algorithm 2 over the standard $O(n(n+m))$-time method for APSP in DAGs, consider the following simple, extreme example. Suppose M is a positive integer. Let D be a DAG with vertices $v_1, v_2, ..., v_n$, and edges (v_i, v_j), where $i < j$, such that the weight of (v_i, v_j) is -1 if $j = i+1$ and M otherwise. It is easy to see the tree T_{v_i} is just the path $v_1, v_2, ..., v_i$ and hence $|\text{leaf}(T_{v_i})| = 1$. Consequently, Algorithm 2 on the DAG D runs in $O(n^\omega)$ time while the standard method requires $O(n^3)$ time. If $M = 1$, one could also run Zwick's APSP algorithm for directed graphs with edge weights in $\{-1, 0, 1\}$ on this example in $O(n^{2.575})$ time [16].

For a refinement of Theorem 2, see the full version [9].

4 A Potential Extension to Digraphs with Large Cycles

As we have already noted the APSP problem in DAGs with both positive and negative real edge weights can be solved in $O(n(n+m))$ time. It is also an interesting open problem if one can derive substantially more efficient algorithms for APSP in DAGs than the $O(n(n+m))$-time method in case of restricted edge weights, e.g., non-negative edge weights etc. In this section, under the assumption of the existence of such substantially more efficient algorithms for DAGs with non-negative edge weights, we show that they could be extended to include directed graphs having a polynomial number of large cycles.

The idea of the extension is fairly simple, see Fig. 2. We pick uniformly at random a sample of vertices of the input directed graph that hits all the directed cycles with high probability (cf. [16]). Here, we use the assumption on the minimum size of the cycles and on the polynomially bounded number of the cycles. Next, we remove the vertices belonging to the sample and run the hypothetical fast algorithm for APSP in DAGs on the resulting subgraph of the input graph which is acyclic with high probability. In order to take into account shortest path connections using the removed vertices, we run the Dijkstra's SSSP algorithm from each vertex in the sample on the original input graph two times. In the second run we reverse the directions of the edges in the input graph. Finally, we update the shortest path distances appropriately.

Algorithm 3
Input: A directed graph (V, E) with n vertices, m non-negatively weighted edges and a polynomial number of directed cycles, each with at least d vertices.
Output: The shortest-path distances for all ordered pairs of vertices in V.

1. Initialize an $n \times n$ array D by setting all its entries outside the main diagonal to $+\infty$ and those on the diagonal to zero.
2. Uniformly at random pick a sample S of $O(n \ln n/d)$ vertices from V.
3. Run the hypothetical APSP algorithm for DAGs on the graph $(V \setminus S, E \cap \{(u,v)|u, v \in V \setminus S\})$ and for each pair $u, v \in V \setminus S$, set $D(u, v)$ to the distance determined by the algorithm.

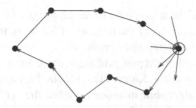

Fig. 2. An example of a directed cycle that can be broken by removing the encircled vertex belonging to the sample. To find shortest-path connections passing through this vertex two SSSP from it are performed, in the original and the reversed edge directions, respectively.

4. For each $s \in S$, run the Dijkstra's SSSP algorithm with s as the source in (V, E) and for all $v \in V \setminus \{s\}$ update the $D(s, v)$ entries respectively.
5. For each $s \in S$, run the Dijkstra's SSSP algorithm with s as the source on the directed graph resulting from reversing the directions of the edges in (V, E), and for all $v \in V \setminus \{s\}$ update the $D(v, s)$ entries respectively.
6. For all pairs u, v of distinct vertices in $V \setminus S$, and for all vertices $s \in S$, set $D(u, v) = \min\{D(u, v), D(u, s) + D(s, v)\}$.

Theorem 3. *Let $t(n, m)$ be the time required by APSP in DAGs with n vertices and m non-negatively weighted edges. Algorithm 3 solves the APSP problem for a directed graph with n vertices, m non-negatively weighted edges and a polynomial number of directed cycles, each with at least d vertices, in $O(t(n, m) + n^3 \ln n / d)$ time with high probability.*

Proof. Suppose that the number of directed cycles in the input graph (V, E) is $O(n^c)$. By picking enough large constant for the expression $n \ln n / d$ specifying the size of the sample S, the probability that a given directed cycle in G is not hit by S can be made smaller than n^{-c-1}. Hence, the probability that the graph resulting from removing the vertices in S is not acyclic becomes smaller than n^{-1}. It follows that Algorithm 3 is correct with high probability. It remains to estimate its running time. Steps 1, 2 can be easily implemented in $O(n^2)$ time. Step 3 takes $t(n, m)$ time. Steps 4, 5 can be implemented in $O((n \ln n / d) \times m + n^2 \ln^2 n / d)$ time [5]. Finally, Step 6 takes $O(n^3 \ln n / d)$ time.   □

Note that because of the term $n^3 \ln n / d$ in the upper time-bound given by Theorem 3, the upper bound can be substantially subcubic only when $d = \Omega(n^\delta)$ for some $\delta > 0$.

5   Final Remarks

In the absence of substantial asymptotic improvements to the time complexity of basic shortest-path algorithms, often formulated at the end of 50s, like the Bellman-Ford algorithm and Dijkstra's algorithm, the results presented in this paper should be of interest. Our output-sensitive algorithm for the general APSP problem in DAGs possibly could lead to an improvement of the asymptotic time complexity of this problem in the average case. A probabilistic analysis of the number of leaves in the lexicographically-first shortest-path trees is an interesting open problem.

In the vast literature on shortest path problems, there are several examples of output-sensitive algorithms. For instance, Karger et al. [8] and McGeoch [10] could orchestrate the n runs of Dijkstra's algorithm in order to solve the APSP problem for directed graphs with non-negative edge weights in $O(m^*n + n\log n)$ time, where m^* is the number of (essential) edges that participate in shortest paths.

Finally, note that DAGs have several important scientific and computational applications in among other things scheduling, data processing networks, biology (phylogenetic networks, epidemiology), sociology (citation networks), and data compression. For these reasons, efficient algorithms for shortest paths in DAGs are of not only theoretical interest.

References

1. Alman, J., Vassilevska Williams, V.: A refined laser method and faster matrix multiplication. In: Proceeding of SODA (2021)
2. Bannister, M.J., Eppstein, D.: Randomized speedup of the Bellman-Ford algorithm. In: Proceedings of ANALCO (2011)
3. Bellman, R.: On a routing problem. Quarter. Appl. Math. **16**(1), 87–90 (1958)
4. Chan, T.M., Williams†, R.: Deterministic APSP, orthogonal vectors, and more: quickly derandomizing Razborov-Smolensky. In: Proceedings of 27th ACM-SIAM Symposium on Discrete Algorithms, pp. 1246–1255 (2016)
5. Cormen, T., Leiserson, C.E., Rivest, R.L., Stein, C.: Introduction to algorithms, 3rd edn. The MIT Press (2009)
6. Dijkstra, E.W.: A note on two problems in connexion with graphs. Numerische Mathematik, pp. 269–271 (1959)
7. Ford, L.R.: Network flow theory. The Rand Corporation, p. 923 (1956)
8. Karger, D.R., Koller, D., Phillips, S.J.: Finding the hidden path: time bounds for all-pairs shortest paths. SIAM J. Comput. **22**, 1199–1217 (1993)
9. Lingas, A., Persson, M., Sledneu, D.: An output-sensitive algorithm for all-pairs shortest paths in directed acyclic graphs. CoRR abs/2108.03455 (2021)
10. McGeoch, C.C.: All-pairs shortest paths and the essential subgraph. Algorithmica **13**, 426–461 (1995)
11. Madkour, A., Aref1, W.G., Rehman, F.U., Rahman, M.A., Basalamah, S.: Shortest-path algorithms. CoRR abs/1705.02044 (2017)
12. Moore, E.F.: The shortest path through a maze. In: Proceedings of an International Symposium on the Theory of Switching, 1957, Part II, pp. 285–292 (1959)
13. Munro, I.: Efficient determination of the transitive closure of a directed graph. Inf. Proc. Lett. **1**(2), 56–58 (1971)
14. Pettie, S.: A new approach to all-pairs shortest paths on real-weighted graphs. Theoretical Comput. Sci. **312**(1), 47–74 (2004)
15. Yen, J.Y.: An algorithm for finding shortest routes from all source nodes to a given destination in general networks. Quarter. Appl. Math. **27**(4), 526–530 (1970)
16. Zwick, U.: All Pairs Shortest Paths using bridging sets and rectangular matrix multiplication. J. ACM **49**, 289–317 (2002)

17. Zwick, U.: Exact and approximate distances in graphs-a survey. In: Proceedings of 9th ESA, pp. 33–48 (2001)
18. Zwick, U.: All pairs lightest shortest paths. In: Proceedings of the STOC, pp. 61–69 (1999)

Covering a Graph with Densest Subgraphs

Riccardo Dondi[1]([✉]) and Alexandru Popa[2]

[1] Università degli Studi di Bergamo, Bergamo, Italy
riccardo.dondi@unibg.it
[2] University of Bucharest, Bucharest, Romania
alexandru.popa@fmi.unibuc.ro

Abstract. Finding densest subgraphs is a fundamental problem in graph mining, with several applications in different fields. In this paper, we consider two variants of the problem of covering a graph with k densest subgraphs, where $k \geq 2$. The first variant aims to find a collection of k subgraphs of maximum density, the second variant asks for a set of k subgraphs such that they maximize an objective function that includes the sum of the subgraphs densities and a distance function, in order to differentiate the computed subgraphs. We show that the first variant of the problem is solvable in polynomial time, for any $k \geq 2$. For the second variant, which is NP-hard for $k \geq 3$, we present an approximation algorithm that achieves a factor of $\frac{2}{5}$.

1 Introduction

Identifying cohesive subgraphs is fundamental in graph mining and graph theory. In several fields, from social networks analysis [18] to computational biology [9] cohesive groups of elements are often related to functionalities of a complex system and it is indeed a fundamental task to identify such cohesive subgraphs.

Several models of cohesive subraphs have been considered in the literature. The first model to be studied has been clique [20], that is a complete graph. However several alternative definitions of cohesive subgraphs have been introduced and studied in the literature (see for example the review in [16]), including densest subgraph. This latter model asks for a subgraph that maximizes the ratio between the number of edges and the number of vertices. Compared to other models, densest subgraphs have the advantage that they can be found in polynomial time [11,12,14] and also approximable in linear time within factor of $\frac{1}{2}$ [2,5,15,17]. The computational tractability and the natural definition of density have lead to a prominent position in graph mining [1,3,4,6,10,21,23,24,27].

The first goal of research in graph mining has been the identification of a single cohesive subgraph, with few exceptions, for example the problem of covering a graph with cliques [13]. In the last years, the interest has moved to the identification of more than a dense subgraph inside a network [4,8,10,26], rather than a single subgraph. This interest is motivated by new approaches to network analysis that require in several cases the identification of the main cohesive

© Springer Nature Switzerland AG 2022
N. Balachandran and R. Inkulu (Eds.): CALDAM 2022, LNCS 13179, pp. 152–163, 2022.
https://doi.org/10.1007/978-3-030-95018-7_13

groups of a network, rather than a single subgraph. Indeed, many real-world networks contain several cohesive groups that may also share common elements, like hubs, that usually belong to many communities [10,19]. Considering the densest subgraph model, the proposed approaches in this direction ask for a collection of dense subgraphs that may share vertices [4,10,21,25]. While the approaches ask for a collection of $k > 1$ densest subgraphs, they differ in the way the overlapping of the subgraphs is handled. Balalau et al. [4] define a hard constraint for the overlapping, based on the Jaccard's index, allowing only a fraction of the vertices to be shared by two subgraphs. Galbrun et al. [10] do not define a hard constraint on the overlapping of two subgraphs that can be identical or very similar, but include a distance function in the objective function to be maximized. Both versions of the problem are NP-hard [4,7], the second is also known to be approximable within factor $\frac{1}{2}$ [7,10].

In this paper, following an approach proposed by Rozenshtein et al. [22] related to temporal graphs, we consider the problem of finding a collection of k dense subgraphs that cover the input graph. Notice that when $k = 1$ the problem is trivial, as the solution must be the input graph. Hence in the paper we focus on the case $k \geq 2$ and we consider two variants of the problem. In the first variant, the objective function is the sum of the densities of the k subgraphs, without any constraints except that they must cover the input graph. For the second variant, similar to [7,10], the objective function includes both the sum of the densities of the k subgraphs and a distance function, in order to differentiate the subgraphs included in a solution. However, notice that the approximation algorithms presented in [7,10] cannot be directly be applied to this variant of the problem, as they may produce solutions that do not cover the input graph.

The paper is organized as follows. Next, in Sect. 2, we introduce the main concepts and we give the formal definitions of the problems we are interested into. In Sect. 3, we show that the first variant of the problem is solvable in polynomial time, for any $k \geq 2$, by showing that there exists an optimal solution that consists of $k - 1$ densest subgraphs and the input graph. For the second variant, which as we observe in Sect. 2 is NP-hard for $k \geq 3$, we present an approximation algorithm that achieves an approximation factor of $\frac{2}{5}$ in Sect. 4. In Sect. 5 we present conclusions and open problems.

Some of the proofs are omitted due to page constraint.

2   Definitions

We consider undirected and unweighted graphs. Given a graph $G = (V, E)$, the density of G, denoted by $dens(G)$, is equal to $dens(G) = \frac{|E|}{|V|}$.

Given a graph $G = (V, E)$ and a subset $V' \subseteq V$, we denote by $G[V'] = (V', E')$ the subgraph of G induced by V'. Two subgraphs $G_1 = (V_1, E_1)$, $G_2 = (V_2, E_2)$ of $G = (V, E)$ are distinct if $V_1 \neq V_2$.

We are now able to consider the first problem we are interested into, called k-Densest Cover Subgraphs.

Problem 1. k-Densest Cover Subgraphs
Input: A graph $G = (V, E)$, an integer k, with $2 \leq k \leq |V|$.
Output: A collection $\mathcal{S} = \{G_1 = (V_1, E_1), \ldots, G_k = (V_k, E_k)\}$ of k subgraphs of G such that $\cup_{i=1}^{k} V_i = V$ and the profit $p(\mathcal{S})$ of $\mathcal{S}$,

$$p(\mathcal{S}) = \sum_{i=1}^{k} dens(G_i)$$

is maximimxed.

Notice that some subgraphs in $\mathcal{S}$ can be identical.

The second problem we consider asks for a set of k subgraphs that optimize an objective function that includes a distance between subgraphs. We start by defining the distance we consider. The distance function we define is inspired by that proposed in [10], except that it has value in the range $[0, 1]$ and not in range $[0, 2]$ (actually the distance function defined in [10] has value in $[1, 2]$ when two subgraphs are different).

Definition 1. *Given a graph $G = (V, E)$ and two subgraphs $G[A]$, $G[B]$, with $A, B \subseteq V$, the distance function $d : 2^V \times 2^V \to \mathbb{R}_+$ between two sets $A, B \subseteq V$ that induce subgraph $G[A]$ and $G[B]$, respectively, is defined as follows:*

$$d(A, B) = 1 - \frac{|A \cap B|^2}{|A||B|}$$

Notice that $0 \leq d(A, B) \leq 1$. Indeed, if $G[A]$ and $G[B]$ are disjoint, that is if $A \cap B = \emptyset$, then $d(A, B) = 1$, while $d(A, B) = 0$ if $G[A]$ and $G[B]$ are the same subgraph, that is $A = B$.

Now, we are able to define the second problem we are interested into, called Top k-Cover-Densest Subgraphs.

Problem 2. Top k-Cover-Densest Subgraphs
Input: A graph $G = (V, E)$, a parameter $\lambda > 0$, an integer k, with $2 \leq k \leq |V|$.
Output: A set $\mathcal{S} = \{G_1 = (V_1, E_1), \ldots, G_k = (V_k, E_k)\}$ of k pairwise distinct subgraphs, with $V_i \subseteq V$, $1 \leq i \leq k$ and $\bigcup_{i=1}^{k} V_i = V$, that maximizes the following value

$$p(\mathcal{S}) = \sum_{i=1}^{k} dens(G_i) + \lambda \sum_{i=1}^{k-1} \sum_{j=i+1}^{k} d(S_i, S_j).$$

Notice that in Top k-Cover-Densest Subgraphs, since $\mathcal{S}$ is a set, the subgraphs in $\mathcal{S}$ must be distinct, unlike in k-Densest Cover Subgraphs. The first term of $p(\mathcal{S})$ is called the *density profit*, while the second them of $p(\mathcal{S})$ is called the *distance profit*.

In [7] a similar problem, termed Top k-Densest Subgraphs, was proven to be NP-hard. Using a similar reduction we can prove the NP-hardness of Top k-Cover-Densest Subgraphs problem is NP-hard.

Corollary 1 (of Theorem 5 [7]). Top k-Cover-Densest Subgraphs *problem is NP-hard for $k \geq 3$.*

2.1   Algorithms for Densest Subgraph

The algorithms we will present are based on computing a densest subgraph of a graph, that is the Densest Subgraph problem. Given a graph $G = (V, E)$, Densest Subgraph can be solved in polynomial time [11, 12] via a reduction to Minimum Cut, with a complexity of $O(|V||E|\log|E|)$ or $O(|V|^3)$ [14].

We also consider the problem of computing a constrained densest subgraph, that is a subgraph that is forced to contain a given set V' of vertices. The problem can be solved in polynomial time, with essentially the same time complexity of Goldberg's Algorithm [27]. Given a graph $G = (V, E)$ and a set $U \subseteq V$, we denote by Densest(G, U) a densest subgraph of G that is forced to contain U. If U consists of a single vertex u, we abuse the notation and write Densest(G, u). If $\mathcal{S}$ is a set of subgraphs such that $G[U]$ is not in $\mathcal{S}$, then Densest$(G, U, \mathcal{S})$ is Densest(G, U), if this subgraph is not in $\mathcal{S}$, else it is $G[U]$.

3   A Polynomial Time Algorithm for the k-Densest Cover Subgraphs

In this section we show a polynomial time algorithm that provides an optimal solution to the k-Densest Cover Subgraphs problem. Our algorithm is very simple: the optimal solution $\mathcal{S} = \{G_1 = (V_1, E_1), \ldots, G_k = (V_k, E_k)\}$ has $G_1 = G_2 = \cdots = G_{k-1}$ equal to the densest subgraph, while $G_k = G$, that is the entire input graph. In what follows we prove the correctness of the algorithm.

We first prove an auxiliary lemma regarding the densest subgraph of a graph.

Lemma 1. *Let $G_d = (V_d, E_d)$ be a densest subgraph of a graph $G = (V, E)$. Let $X \subseteq V_d$, let $G_x = (V_d - X, E_x)$ be the subgraph induced by $V_d - X$ and let $Y = E_d - E_x$. That is, Y is the set of edges that are removed from the densest subgraph after removing the set of vertices X, or, in other words, the edges between the vertices of X and the edges with one endpoint in X and another endpoint in the set $V_d - X$. It holds that:*

$$\frac{|Y|}{|X|} \geq \frac{|E_d|}{|V_d|}$$

Proof. We know that:

$$\frac{|E_d|}{|V_d|} \geq dens(G_x) = \frac{|E_d| - |Y|}{|V_d| - |X|}$$

otherwise we can simply remove the set X of vertices and obtain a subgraph denser than G_d. Then, we get that:

$$|E_d|(|V_d| - |X|) \geq |V_d|(|E_d| - |Y|)$$

$$-|E_d||X| \geq -|V_d||Y|$$

$$|E_d||X| \leq |V_d||Y|$$

$$\frac{|E_d|}{|V_d|} \leq \frac{|Y|}{|X|}$$

□

We know that $G_1, G_2, \ldots, G_k$ must cover all the vertices of G. Now, we show that there exists an optimal solution for the k-Densest Cover Subgraphs such that all the graphs in the solution include a densest subgraph.

Lemma 2. *There exists an optimal solution* $S = \{G_1 = (V_1, E_1), G_2 = (V_2, E_2), \ldots, G_k = (V_k, E_k)\}$ *to the* k-Densest Cover Subgraphs *problem that has the property that all the graphs* $G_i \in S$, $1 \leq i \leq k$, *include a densest graph.*

Proof. Assume by contradiction that there exists a densest subgraph $G_d = (V_d, E_d)$ and a graph $G_i = (V_i, E_i) \in S$ such that $V_d - V_i \neq \emptyset$. Let $X = V_d - V_i$ and let $Y = E_d - E_i$. According to Lemma 1:

$$\frac{|Y|}{|X|} \geq \frac{|E_d|}{|V_d|}$$

By construction, $G_i \cup G_d$ has at least $|Y| + |E_i|$ edges and precisely $|X| + |V_i|$ vertices. We show that:

$$\frac{|Y| + |E_i|}{|X| + |V_i|} \geq \frac{|E_i|}{|V_i|}$$

Indeed, from Lemma 1 and since G_d is a densest subgraph, it holds that

$$\frac{|Y|}{|X|} \geq \frac{|E_d|}{|V_d|} \geq \frac{|E_i|}{|V_i|}$$

thus

$$\frac{|Y| + |E_i|}{|X| + |V_i|} \geq \frac{\frac{|E_i|}{|V_i|}|X| + |E_i|}{|X| + |V_i|} = \frac{|E_i|}{|V_i|}\frac{|X| + |V_i|}{|X| + |V_i|} = \frac{|E_i|}{|V_i|}$$

□

An immediate consequence of Lemma 2 is the fact that the maximal densest subgraph (a densest subgraph of maximum size) is unique. We use this property later in the proof of Theorem 1.

Lemma 3. *Given a graph* $G = (V, E)$, *the maximal densest subgraph* $G_d = (V_d, E_d)$ *in* G *is unique.*

Proof. Assume by contradiction that there exist two distinct maximal densest subgraphs $G_d = (V_d, E_d)$ and $G'_d = (V'_d, E'_d)$. Since G_d and G'_d are distinct, then $V_d - V'_d \neq \emptyset$ and $V'_d - V_d \neq \emptyset$. Thus, we can apply the same argument as in Lemma 2 and by taking the union of V_d, V'_d we obtain another densest subgraph that strictly includes both G_d and G'_d. Thus, we contradict the fact that G_d and G'_d are maximal and the lemma follows.   □

Lemma 3 shows that there exists a unique maximal densest subgraph. Next, we discuss how to compute it in polynomial time.

1. Compute a densest subgraph $G_d = (V_d, E_d)$, using the Goldberg's algorithm of [12]
2. For each $v \in V - V_d$, by applying the algorithm presented in [27] (see Sect. 2.1), compute in polynomial time $G'_d(v) \leftarrow \mathsf{Densest}(G, V_d \cup \{v\})$, a densest subgraph in G that includes $V_d \cup \{v\}$.
3. If no subgraph $G'_d(v)$ is as dense as G_d, rteturn G_d
4. Else, for a subgraph $G'_d(v)$ that is as dense as G_d, define $G_d \leftarrow G'_d(v)$ and iterate the algorithm from point 2.

From Lemma 2, we can conclude that there exists an optimal solution of k-Densest Cover Subgraphs problem where all the subgraphs include the maximal densest subgraph of G.

Corollary 2. *There exists an optimal solution* $S = \{G_1 = (V_1, E_1), G_2 = (V_2, E_2), \ldots, G_k = (V_k, E_k)\}$ *to the* k-Densest Cover Subgraphs *problem that has the property that all the graphs* $G_i \in S$, $1 \leq i \leq k$, *include the maximal densest subgraph of* G.

Next, we show a property of two subgraphs in an optimal solution of k-Densest Cover Subgraphs, that will be useful us to prove the main result of this section.

Lemma 4. *Let* $G_1 = (V_1, E_1)$ *and* $G_2 = (V_2, E_2)$ *be two graphs in an optimal solution of* k-Densest Cover Subgraphs *problem that satisfies Corollary 2 and let* $G_u = G[V_1 \cup V_2]$ *and* G_d *be the maximal densest subgraph. Then:*

$$dens(G_d) + dens(G_u) \geq dens(G_1) + dens(G_2).$$

We now prove the main theorem of this section.

Theorem 1. *There exists an optimal solution* $S = \{G_1, G_2, \ldots, G_k\}$ *to the* k-Densest Cover Subgraphs *problem such that* $G_1 = G_2 = \cdots = G_{k-1} = G_d$, *where* G_d *is the maximal densest subgraph and* $G_k = G$.

Proof. Let $S' = \{G'_1, G'_2, \ldots, G'_k\}$ be an optimal solution to the k-Densest Cover Subgraphs problem such that G'_k is not a densest subgraph. We show that we can modify S' so that the resulting solution $S = \{G_1, G_2, \ldots, G_k\}$ has the following properties: (1) $G_1 = G_2 = \cdots = G_{k-1} = G_d$, where G_d is the maximal densest subgraph, and $G_k = G$, (2) $dens(S) \geq dens(S')$.

First, observe that both solutions S and S' cover all the vertices of G. Now, assume that we have two graphs G'_i and G'_j in S' that are not identical to the maximal densest subgraph G_d. Then, according to Lemma 4, we can replace G'_i and G'_j with the maximal densest subgraph G_d and $G'_i \cup G'_j$.

Thus, we can assume that there exists only one graph $G'_\ell \in S'$ such that G'_ℓ is not the maximal densest subgraph of G. Assume without loss of generality that $\ell \neq 1$. $G'_1 \cup G'_\ell = G$, as the input graph must be covered. Thus G'_ℓ includes $G - G_d$ and, according to Corollary 2, G'_ℓ includes G_d and therefore $G'_\ell = G$ and the theorem follows.

4   An Approximation Algorithm for Top k-Cover-Densest Subgraphs

In this section, we present an approximation algorithm for Top k-Cover-Densest Subgraphs. The approximation algorithm outputs the set of k subgraphs having largest profit between the sets of k subgraphs computed by two algorithms, called Approx-Dens and Dist. Both Approx-Dens and Dist computes a set of k subgraphs that cover G, but Approx-Dens aims at maximizing the density of the subgraphs, while Dist aims at maximizing the distance between the subgraphs. We start by presenting the Approx-Dens algorithm.

Approx-Dens

Approx-Dens computes a solution S of Top k-Cover-Densest Subgraphs. We assume that $k \geq 3$. Indeed, for $k = 2$, we can compute two distinct subgraphs that cover G and that have maximum density by applying the algorithm of Sect. 3, as described in the following. If we obtain two distinct subgraphs, we know that they are two subgraphs of maximum density that cover G. If the algorithm of Sect. 3 returns two identical subgraphs G_1, G_2, then they must be identical to the input graph G. Then we can compute in polynomial time a densest subgraph of G distinct from G_1 by applying the modification of Goldberg's Algorithm described in [7].

Hence assume $k \geq 3$. Approx-Dens first adds to S two densest distinct subgraphs of G, denoted by $G_1 = (V_1, E_1)$ and $G_2 = (V_2, E_2)$. G_1 and G_2 are computed in polynomial time as follows:

- $G_1 = (V_1, E_1)$ is computed by applying Goldberg's Algorithm [12] on G
- G_2 is computed by applying the modification of Goldberg's Algorithm described in [7], with input G and V_1.

We assume that $|V_1| \geq |V_2|$ and $|V_2| \geq 2$. Notice that if this latter condition does not hold, $|V_2| = 1$ and $dens(G_2) = 0$, hence we can compute a solution of Top k-Cover-Densest Subgraphs by computing any subgraph of G, since it contains at least $k - 1$ subgraphs of density 0.

Starting with $S = \{G_1, G_2\}$, Approx-Dens computes the remaining subgraphs with two phases (described later). The first phase is applied when there exists at least one vertex of G that is not covered by S. In this phase, Approx-Dens iteratively adds a subgraph G_i to $S = \{G[V_1], \ldots G[V_{i-1}]\}$, for $3 \leq i \leq k - 1$. In what follows, $C_{i-1} = \bigcup_{j=1}^{i-1} V_j$, that is C_{i-1} is the set of vertices covered by the subgraphs already in S. In the second phase, when $C_{i-1} = V$, that is the input graph is already covered, Approx-Dens adds $k - i + 1$ subgraphs, which are computed depending on the size of G_1. If Phase 2 is never executed, then Approx-Dens adds subgraph $G_k = G$ to S.

Phase 1. Let $S = \{G[V_1], \ldots G[V_{i-1}]\}$. If there exists $u \in V - C_{i-1}$, $G_i \leftarrow$ Densest(G, u).

Phase 2. If $C_{i-1} = V$, then:

- If $|V_1| \leq 2\log_2 |V|$, $G_i,\dots,G_k$ are $k - i$ distinct subgraphs among Densest$(G, V_1 \cup \{u\}, S)$ and Densest$(G, V_2 \cup \{u\}, S)$, with $u \in V - V_1 - V_2$. Notice that the algorithm computes, and possibly adds to S, these subgraphs based on some ordering of the vertices in $V - V_1 - V_2$: it starts with the first vertex of $V - V_1 - V_2$, then the second one, and so on until S contains k subgraphs.
- If $2\log_2 |V| \leq |V_1| \leq |V| - \log_2 |V|$, $G_i,\dots,G_k$ are $k - i$ densest distinct subgraphs among $G[V_1 \cup U]$, with U any subset of $V - V_1$ of size at most $\log_2 |V|$.
- If $|V_1| > |V| - \log_2 |V|$, $G_i,\dots,G_k$ are $k - i$ densest distinct subgraphs among $G[V_1 - U]$, with U a subset of V_{MIN}, where $V_{MIN} \subseteq V_1$ consists of the $\log_2 |V|$ vertices of V_1 having smallest degree.

We recall that if Phase 2 is never executed, $G_k = G$. We start by showing that Approx-Dens returns a feasible solution, that is a set of k subgraphs that covers G.

Lemma 5. Approx-Dens *returns a set of k subgraphs that cover G.*

Next we show that Approx-Dens achieves an approximation factor of $\frac{2}{3}$ for the density profit.

Lemma 6. *Each subgraph G_i, with $3 \leq i \leq k - 1$, computed by Phase 1 of* Approx-Dens *has density at least $\frac{2}{3}dens(G_1)$.*

Proof. Consider the subgraph $G' = G[V_1 \cup \{u\}]$. By construction $dens(G_i) \geq dens(G')$. Now, consider the density of G', it holds that

$$dens(G') \geq \frac{|E_1|}{|V_1| + 1} = \frac{|E_1|}{|V_1|} \frac{|V_1|}{|V_1| + 1}.$$

Since $|V_1| \geq 2$, $\frac{|V_1|}{|V_1|+1} \geq \frac{2}{3}$, thus

$$dens(G') \geq \frac{|E_1|}{|V_1|} \frac{|V_1|}{|V_1| + 1} \geq \frac{2}{3}dens(G_1).$$

It follows that $dens(G_i) \geq dens(G') \geq \frac{2}{3}dens(G_i)$, thus concluding the proof. $\square$

Lemma 7. *Each G_i, with $3 \leq i \leq k$, computed by Phase 2 of* Approx-Dens *has density at least $\frac{2}{3}dens(G_x)$, with $x \in \{1, 2\}$.*

Proof. We consider the three cases of Phase 2.

Case 1. $|V_1| \leq 2\log_2 |V|$ Since $|V_x| \geq 2$, with $x \in \{1, 2\}$, it follows that a subgraph $G_i = (V_i, E_i)$ has the following density:

$$\frac{|E_i|}{|V_i|} \geq \frac{|E_x|}{|V_x| + 1} = \frac{|E_x|}{|V_x|} \frac{|V_x|}{|V_x| + 1} \geq \frac{2}{3} \frac{|E_x|}{|V_x|}$$

Case 2. $2\log_2|V| \le |V_1| \le |V| - \log_2|V|$. Consider a set U_i of vertices added to V_1. Since $|U_i| \le \log_2|V|$ and $|V_1| \ge 2\log_2|V|$, it holds that

$$\frac{|E_i|}{|V_i|} \ge \frac{|E_1|}{|V_1| + \log_2|V|} = \frac{|E_1|}{|V_1|}\frac{|V_1|}{|V_1| + \log_2|V|} \ge \frac{2}{3}\frac{|E_1|}{|V_1|}.$$

Case 3. $|V_1| > |V| - \log_2|V|$. Consider a set $V_{MIN} \subseteq V_1$, with $|V_{MIN}| = \log_2|V|$, a set of vertices having smallest degree in G_1. Let U_i, $3 \le i \le k$, be a subset of V_{MIN} and $G_i = G[V_1 - U_i]$.

Let d be the average degree of vertices in G_1, then by removing U_i from G_1, at most $d|U_i|$ edges are removed from E_1. Thus:

$$\frac{|E_i|}{|V_i|} \ge \frac{|E_1| - d|U_i|}{|V_1|} = \frac{|E_1|}{|V_1|} - \frac{d|U_i|}{|V_1|} \ge \frac{|E_1|}{|V_1|} - d\left(\frac{\log_2|V|}{|V| - \log_2|V|}\right)$$

For $|V|$ larger than a constant ($|V| \ge 37$; notice that if $|V|$ is a constant we can solve the problem by brute force), since $\frac{|E_1|}{|V_1|} \ge \frac{1}{2}d$, it holds that

$$d\left(\frac{\log_2|V|}{|V| - \log_2|V|}\right) \le \frac{1}{6}d \le \frac{1}{3}\frac{|E_1|}{|V_1|}.$$

It follows that

$$\frac{|E_i|}{|V_i|} \ge \frac{2}{3}\frac{|E_1|}{|V_1|}$$

thus concluding the proof. □

Next, we show that Approx-Dens approximates the optimal density within a factor of $\frac{2}{3}$.

Lemma 8. *The subgraphs $G_1, \ldots, G_k$ computed by Approx-Dens have density at least $\frac{2}{3}OPT(Dens)$.*

Proof. Consider an optimal solution consisting of subgrpahs $G_1^*, \ldots, G_k^*$. Since G_1, G_2 are two densest subgraphs of G, it follows from Lemma 6 and Lemma 7 that for i with $3 \le i \le k - 1$, $dens(G_i) \ge \frac{2}{3}dens(G_i^*)$.

Now, consider the subgraphs G_1, G_2, G_k. Since G_1, G_2 are two densest subgraphs of G, it follows that

$$dens(G_1) + dens(G_2) \ge \frac{2}{3}\left(dens(G_1^*) + dens(G_2^*) + dens(G_k^*)\right)$$

thus

$$dens(G_1) + dens(G_2) + dens(G_k) \ge \frac{2}{3}\left(dens(G_1^*) + dens(G_2^*) + dens(G_k^*)\right)$$

and the lemma holds. □

Dist

Dist starts with $z = |V|$ subgraphs, each one containing a vertex of V and, while $k < z$, merges any two of these subgraphs. Let $D_1, \ldots, D_k$ be the subgraphs returned by Dist.

Lemma 9. $D_1, \ldots, D_k$ *have maximum profit distance.*

Proof. The lemma follows from the fact that $D_1, \ldots, D_k$ are disjoint.   □

Now, we show that we achieve an approximation factor of $\frac{2}{5}$ for Top k-Cover-Densest Subgraphs. We denote by OPT the profit of an optimal solution $\mathcal{S}^*$ of Top k-Cover-Densest Subgraphs, by $OPT(Dens)$ (by $OPT(Dist)$, respectively) the density profit (distance profit, respectively) of $\mathcal{S}^*$.

Theorem 2. *Let* $\mathcal{S} = \{G_1, \ldots G_k\}$ *be the solution returned by* Approx-Dens *and let* $\mathcal{D} = \{D_1, \ldots D_k\}$ *be the solution returned by* Dist. *Then* $\max(p(\mathcal{S}), p(\mathcal{D})) \geq \frac{2}{5}OPT$.

Proof. Recall that by Lemma 8, it holds that $p(\mathcal{S}) \geq \frac{2}{3}OPT(Dens)$, and by Lemma 9 it holds that $p(\mathcal{D}) \geq \lambda OPT(Dist)$.
First, assume that $\lambda OPT(Dist) \geq \frac{2}{3} OPT(Dens)$. Then

$$p(\mathcal{D}) \geq \lambda OPT(Dist) \geq \frac{2}{5}\lambda OPT(Dist) + \frac{3}{5}\lambda OPT(Dist) \geq$$

$$\frac{2}{5}\lambda OPT(Dist) + \frac{2}{5}OPT(Dens),$$

thus in this case Dist returns a solution having approximation factor $\frac{2}{5}$.
Assume that $\lambda OPT(Dist) < \frac{2}{3} OPT(Dens)$. Then

$$p(\mathcal{S}) \geq \frac{2}{3}OPT(Dens) \geq \frac{2}{5}OPT(Dens) + \frac{4}{15}OPT(Dens) \geq$$

$$\frac{2}{5}OPT(Dens) + \frac{2}{5}\lambda OPT(Dist).$$

thus in this case Approx-Dens returns a solution having approximation factor $\frac{2}{5}$, concluding the proof.   □

5   Conclusions and Open Problems

We have considered two variants of the problem of covering a graph with k densest subgraphs, where $k \geq 2$. For the first variant, we have shown that it is solvable in polynomial time, for any $k \geq 2$. For the second variant, which is NP-hard for $k \geq 3$, we have presented an approximation algorithm that achieves a factor of $\frac{2}{5}$.

There are some interesting open problems related to k-Densest Cover Subgraphs and Top k-Cover-Densest Subgraphs. It would be nice to study whether the version k-Densest Cover Subgraphs that asks for k densest distinct subgraphs that cover G is polynomial time solvable. A positive answer will help also to improve the approximation of Top k-Cover-Densest Subgraphs.

References

1. Andersen, R., Chellapilla, K.: Finding dense subgraphs with size bounds. In: Avrachenkov, K., Donato, D., Litvak, N. (eds.) WAW 2009. LNCS, vol. 5427, pp. 25–37. Springer, Heidelberg (2009). https://doi.org/10.1007/978-3-540-95995-3_3
2. Asahiro, Y., Hassin, R., Iwama, K.: Complexity of finding dense subgraphs. Discret. Appl. Math. **121**(1–3), 15–26 (2002)
3. Bahmani, B., Kumar, R., Vassilvitskii, S.: Densest subgraph in streaming and mapreduce. PVLDB **5**(5), 454–465 (2012)
4. Balalau, O.D., Bonchi, F., Chan, T.H., Gullo, F., Sozio, M.: Finding subgraphs with maximum total density and limited overlap. In: Cheng, X., Li, H., Gabrilovich, E., Tang, J. (eds.) Proceedings of the Eighth ACM International Conference on Web Search and Data Mining, WSDM 2015, pp. 379–388. ACM (2015)
5. Charikar, M.: Greedy approximation algorithms for finding dense components in a graph. In: Jansen, K., Khuller, S. (eds.) APPROX 2000. LNCS, vol. 1913, pp. 84–95. Springer, Heidelberg (2000). https://doi.org/10.1007/3-540-44436-X_10
6. Dondi, R., Hosseinzadeh, M.M., Guzzi, P.H.: A novel algorithm for finding top-k weighted overlapping densest connected subgraphs in dual networks. Appl. Netw. Sci. **6**(1), 1–17 (2021). https://doi.org/10.1007/s41109-021-00381-8
7. Dondi, R., Hosseinzadeh, M.M., Mauri, G., Zoppis, I.: Top-k overlapping densest subgraphs: approximation algorithms and computational complexity. J. Comb. Optim. **41**(1), 80–104 (2021)
8. Dondi, R., Mauri, G., Sikora, F., Zoppis, I.: Covering a graph with clubs. J. Graph Algorithms Appl. **23**(2), 271–292 (2019)
9. Fratkin, E., Naughton, B.T., Brutlag, D.L., Batzoglou, S.: Motifcut: regulatory motifs finding with maximum density subgraphs. Bioinformatics **22**(14), 156–157 (2006)
10. Galbrun, E., Gionis, A., Tatti, N.: Top-k overlapping densest subgraphs. Data Min. Knowl. Discov. **30**(5), 1134–1165 (2016)
11. Gallo, G., Grigoriadis, M.D., Tarjan, R.E.: A fast parametric maximum flow algorithm and applications. SIAM J. Comput. **18**(1), 30–55 (1989)
12. Goldberg, A.V.: Finding a maximum density subgraph. Technical report, Berkeley, CA, USA (1984)
13. Karp, R.M.: Reducibility among combinatorial problems. In: Miller, R.E., Thatcher, J.W. (eds.) Proceedings of a Symposium on the Complexity of Computer Computations, The IBM Research Symposia Series, pp. 85–103. Plenum Press, New York (1972)
14. Kawase, Y., Miyauchi, A.: The densest subgraph problem with a convex/concave size function. Algorithmica **80**(12), 3461–3480 (2018)
15. Khuller, S., Saha, B.: On finding dense subgraphs. In: Albers, S., Marchetti-Spaccamela, A., Matias, Y., Nikoletseas, S., Thomas, W. (eds.) ICALP 2009, Part I. LNCS, vol. 5555, pp. 597–608. Springer, Heidelberg (2009). https://doi.org/10.1007/978-3-642-02927-1_50
16. Komusiewicz, C.: Multivariate algorithmics for finding cohesive subnetworks. Algorithms **9**(1), 21 (2016)
17. Kortsarz, G., Peleg, D.: Generating sparse 2-spanners. J. Algorithms **17**(2), 222–236 (1994)
18. Kumar, R., Raghavan, P., Rajagopalan, S., Tomkins, A.: Trawling the web for emerging cyber-communities. Comput. Netw. **31**(11–16), 1481–1493 (1999)

19. Leskovec, J., Lang, K.J., Dasgupta, A., Mahoney, M.W.: Community structure in large networks: natural cluster sizes and the absence of large well-defined clusters. Internet Math. **6**(1), 29–123 (2009)
20. Luce, R.D., Perry, A.D.: A method of matrix analysis of group structure. Psychometrika **14**(2), 95–116 (1949)
21. Nasir, M.A.U., Gionis, A., Morales, G.D.F., Girdzijauskas, S.: Fully dynamic algorithm for top-k densest subgraphs. In: Lim, E., (eds.) Proceedings of the 2017 ACM on Conference on Information and Knowledge Management, CIKM 2017, pp. 1817–1826. ACM (2017)
22. Rozenshtein, P., Bonchi, F., Gionis, A., Sozio, M., Tatti, N.: Finding events in temporal networks: segmentation meets densest subgraph discovery. Knowl. Inf. Syst. **62**(4), 1611–1639 (2019). https://doi.org/10.1007/s10115-019-01403-9
23. Sozio, M., Gionis, A.: The community-search problem and how to plan a successful cocktail party. In: Rao, B., Krishnapuram, B., Tomkins, A., Yang, Q. (eds.) Proceedings of the 16th ACM SIGKDD International Conference on Knowledge Discovery and Data Mining, Washington, DC, USA, 25–28 July 2010, pp. 939–948. ACM (2010)
24. Tatti, N., Gionis, A.: Density-friendly graph decomposition. In: Gangemi, A., Leonardi, S., Panconesi, A. (eds.) Proceedings of the 24th International Conference on World Wide Web, WWW 2015, Florence, Italy, 18–22 May 2015, pp. 1089–1099. ACM (2015)
25. Valari, E., Kontaki, M., Papadopoulos, A.N.: Discovery of top-k dense subgraphs in dynamic graph collections. In: Ailamaki, A., Bowers, S. (eds.) SSDBM 2012. LNCS, vol. 7338, pp. 213–230. Springer, Heidelberg (2012). https://doi.org/10.1007/978-3-642-31235-9_14
26. Zou, P., Li, H., Wang, W., Xin, C., Zhu, B.: Finding disjoint dense clubs in a social network. Theor. Comput. Sci. **734**, 15–23 (2018)
27. Zou, Z.: Polynomial-time algorithm for finding densest subgraphs in uncertain graphs. In: Proceedings of Internation Workshop on Mining and Learning with Graphs (2013)

Computational Geometry

Coresets for (k, ℓ)-Median Clustering Under the Fréchet Distance

Maike Buchin [ID] and Dennis Rohde[✉][ID]

Faculty of Informatics, Ruhr University Bochum, Bochum, Germany
{maike.buchin,dennis.rohde-t1b}@rub.de

Abstract. We present an algorithm for computing ε-coresets for (k, ℓ)-median clustering of polygonal curves in $\mathbb{R}^d$ under the Fréchet distance. This type of clustering is an adaption of Euclidean k-median clustering: we are given a set of n polygonal curves in $\mathbb{R}^d$, each of complexity (number of vertices) at most m, and want to compute k median curves such that the sum of distances from the given curves to their closest median curve is minimal. Additionally, we restrict the complexity of the median curves to be at most ℓ each, to suppress overfitting, a problem specific for sequential data. Our algorithm has running time linear in n, sub-quartic in m and quadratic in ε^{-1}. With high probability it returns ε-coresets of size quadratic in ε^{-1} and logarithmic in n and m. We achieve this result by applying the improved ε-coreset framework by Langberg and Feldman to a generalized k-median problem over an arbitrary metric space. Later we combine this result with the recent result by Driemel et al. on the VC dimension of metric balls under the Fréchet distance. Furthermore, our framework yields ε-coresets for any generalized k-median problem where the range space induced by the open metric balls of the underlying space has bounded VC dimension, which is of independent interest. Finally, we show that our ε-coresets can be used to improve the running time of an existing approximation algorithm for $(1, \ell)$-median clustering.

Keywords: Clustering · Coresets · Median · Polygonal curves

1 Introduction

At the present time even efficient approximation algorithms are often incapable of handling massive data sets, which have become common. Here, we need efficient methods to reduce data while (approximately) maintaining the core properties of the data. A popular approach to this topic are ε-coresets; see for example [14,27] for comprehensive surveys. An ε-coreset is a small (weighted) set that aggregates certain properties of a given (massive) data set up to some small error. ε-coresets are very popular in the field of clustering, cf. [11,19,20,22] and they are becoming a topic in other fields, too, cf. [16,28]. The technique for computing an ε-coreset for a given data set highly depends on the application at hand, but mostly ε-coresets are computed by filtering the given data set.

© Springer Nature Switzerland AG 2022
N. Balachandran and R. Inkulu (Eds.): CALDAM 2022, LNCS 13179, pp. 167–180, 2022.
https://doi.org/10.1007/978-3-030-95018-7_14

While ε-coresets can be computed efficiently for k-clustering of points in the Euclidean space, less is known for clustering of curves. In particular, only little effort (cf. [9]) has been made in designing methods for computing ε-coresets for (k, ℓ)-clustering of polygonal curves in $\mathbb{R}^d$ under the Fréchet distance. This type of clustering, which has recently drawn increasing popularity due to a growing number of applications, see [4,5,8,12], is an adaption of the Euclidean k-clustering: we are given a set of polygonal curves and seek to compute k center curves that minimize either the maximum Fréchet distance (center objective), or the sum of Fréchet distances (median objective), among the given curves and their closest center curve. In addition, we restrict the complexity—the number of vertices—of each center curve to be at most ℓ to suppress overfitting, a problem specific for sequential data. This means that input curves and center curves are in general of different complexities, which is the reason why we need a specialized algorithm for computing ε-coresets and can not apply ε-coreset algorithms for discrete metric spaces (cf. [15]) on the input.

The Fréchet distance is a natural dissimilarity measure for curves that is a pseudo-metric and can be computed efficiently [1]. Unlike other measures for curves, like the dynamic time warping distance (or the discrete version of the Fréchet distance), it takes the whole course of the curves into account, not only the pairwise distances among their vertices. This can be particularly useful, e.g. when the input consists of irregularly sampled trajectories, cf. [12]. Unfortunately, since the Fréchet distance is a bottleneck distance measure, i.e., it boils down to a single distance between two points on the curves, it is sensitive to outliers, which may negatively affect its applications. In clustering, we can counteract by choosing an appropriate clustering objective and indeed, the (k, ℓ)-median objective is a good choice, because the median is a robust measure of central tendency. However, the state of the art (k, ℓ)-median clustering algorithms (cf. [8,12]) have exponential running time dependencies and cannot be used in practice, while the practical algorithms for (k, ℓ)-clustering (cf. [4,5]) rely on the (k, ℓ)-center objective, which is not robust and therefore amplifies the sensitivity on outliers.

In this work, we present an algorithm for computing ε-coresets for (k, ℓ)-median clustering under the Fréchet distance and improve an $(1, \ell)$-median clustering algorithm by Buchin et al. [7], using ε-coresets and rendering it much more practical.

1.1   Related Work

(k, ℓ)-clustering of polygonal curves was introduced by Driemel et al. [12]. They developed the first approximation schemes for (k, ℓ)-center and (k, ℓ)-median clustering of polygonal curves in $\mathbb{R}$, which run in near-linear time. They proved that both problems are NP-hard, when k is part of the input. Further, they showed that the doubling dimension of the space of polygonal curves under the Fréchet distance is unbounded, even when the curves are of bounded complexity. Subsequently, Buchin et al. [4] presented a constant factor approximation algorithm for (k, ℓ)-center clustering of polygonal curves in $\mathbb{R}^d$, with running time linear in the number of given curves and polynomial in their maximum complexity. Also they showed that (k, ℓ)-center clustering is NP-hard and NP-hard to approximate within a factor of $(1.5 - \varepsilon)$ for curves in $\mathbb{R}$, respectively $(2.25 - \varepsilon)$ for curves in $\mathbb{R}^d$ with $d \geq 2$, even for $k = 1$. Buchin et al. [5] provided practical algorithms for (k, ℓ) clustering under the Fréchet distance and thereby introduce a new technique, the so called Fréchet centering, for computing better cluster centers. Also, Meintrup et al. [26] provided a practical $(1 + \varepsilon)$-approximation algorithm for discrete (k, ℓ)-median clustering under the presence of a certain number of outliers. Buchin et al. [6] proved that (k, ℓ)-median clustering is also NP-hard, even for $k = 1$. Furthermore, they presented polynomial-time approximation schemes for (k, ℓ)-center and (k, ℓ)-median clustering of polygonal curves under the discrete Fréchet distance. Nath and Taylor [29] gave a near-linear time approximation scheme for (k, ℓ)-median clustering of polygonal curves in $\mathbb{R}^d$ under the discrete Fréchet distance and a polynomial-time approximation scheme for k-median clustering of sets of points from $\mathbb{R}^d$ under the Hausdorff distance. Furthermore, they showed that k-median clustering of point sets under the Hausdorff distance is NP-hard (for constant k). Recently, Buchin et al. [8] developed an approximation scheme for (k, ℓ)-median clustering under the Fréchet distance with running time linear in the number of curves and polynomial in their complexity, where the computed centers have complexity up to $2\ell - 2$.

Langberg and Schulman [25] developed a framework for computing relative error approximations of integrals over any function from a given family of unbounded and non-negative real functions. In particular, this framework can be used to compute ε-coresets for k-clustering of points in $\mathbb{R}^d$ with objective functions based on sums of distances among the points and their closest center. The idea of their framework is to sub-sample the input with respect to a certain non-uniform probability distribution, which is computed using an approximate solution to the problem. More precisely, the approximate solution is used to compute an upper bound on the sensitivity of each data element. The sensitivity is the maximum fraction of cost that the element may cause for any possible solution. It is a notion of the data elements *importance* for the problem and the probability distribution is set up such that each element has probability proportional to its importance. A sample of a certain size drawn from this distribution and properly weighted, is an ε-coreset for the underlying clustering problem with high probability. Feldman and Langberg [15] developed a unified framework for approximate clustering, which is largely based on ε-coresets. They combine the techniques by Langberg and Schulman [25] with ε-approximations, which stem from the framework of range spaces and VC dimension developed in statistical learning theory. As a result, they address a spectrum of clustering problems, such as k-median clustering, k-line median clustering, projective clustering and also other problems like subspace approximation. Braverman et al. [2] improved the aforementioned framework by switching to (ε, η)-approximations, which leads to substantially smaller sample sizes in many cases. Also, they simplified and further generalized the framework and applied it to k-means clustering of points in $\mathbb{R}^d$.

Following, Feldman et al. [16] improved this framework by switching to another range space, thereby obtaining smaller coresets for k-means, k-line means and affine subspace clustering.

1.2 Our Contributions

In this work we develop an algorithm for computing ε-coresets for (k,ℓ)-median clustering of polygonal curves in $\mathbb{R}^d$ under the Fréchet distance (where in the following we assume k,ℓ and d to be constant):

Theorem 1. *There exists an algorithm that, given a set of n polygonal curves in $\mathbb{R}^d$ of complexity at most m each and a parameter $\varepsilon \in (0,1)$, returns with constant positive probability an ε-coreset for (k,ℓ)-median clustering under the Fréchet distance of size $O(k^2 \log^2(k)\varepsilon^{-2} \log(m) \log(kn))$, in time*

$$O(nm \log(m) + nm^3 \log(m) + \varepsilon^{-2} \log(m) \log(n))$$

for $k > 1$ and

$$O(nm \log(m) + m^3 \log(m) + \varepsilon^{-2} \log(m) \log(n))$$

for $k = 1$.

Also we show that ε-coresets can be used to improve the running time of an existing $(1,\ell)$-median $(5 + \varepsilon)$-approximation algorithm [7], thereby facilitating its application in practice.

We start by defining *generalized k-median clustering*, where input and centers come from a subset (not necessarily the same) of an underlying metric space, each, and then derive our ε-coreset result in this setting. This notion captures (k,ℓ)-median clustering under the Fréchet distance in particular, but the analysis holds for any metric space. In doing so, we first give a universal bound on the so called sensitivity of the elements of the given data set and their total sensitivity, i.e., the sum of their sensitivities. The sensitivities are a measure of the data elements importance, i.e., the maximum fraction of the cost an element might cause for any center set, and later they determine the sample probabilities. Our analysis is based on the analysis of Langberg and Schulman [25].

Next, we apply the improved ε-coreset framework by Feldman and Langberg [15]. Here, our analysis is based on the analysis of Feldman et al. [16], but our sample size depends on the VC-dimension of the range space induced by the open metric balls. The open metric balls form a basis of the metric topology, hence it is more natural to study the VC dimension of their associated range space in a geometric setting. Indeed, for the ℓ_p^d spaces these range spaces have already been studied [17, Theorem 2.2] and recently, results for the (continuous and discrete) Fréchet, weak Fréchet and Hausdorff distance were obtained [13], enabling our main result. Finally, we show how an existing $(1,\ell)$-median $(5+\varepsilon)$-approximation algorithm [7] can be improved by means of our ε-coresets.

Theorem 2. *There exists an algorithm that, given a set T of n polygonal curves of complexity at most m each, and a parameter $\varepsilon \in (0, 1/2]$, computes a polygonal curve c of complexity $2\ell - 2$, such that with constant positive probability, it holds that*

$$\text{cost}\,(T, \{c\}) = \sum_{\tau \in T} d_F(\tau, c) \le (5 + \varepsilon) \sum_{\tau \in T} d_F(\tau, c^*) = (5 + \varepsilon)\text{cost}\,(T, \{c^*\}),$$

where c^* is an optimal $(1, \ell)$-median for T under the Fréchet distance. The algorithm has running time

$$O\left(nm \log(m) + m^2 \log(m) + m^{2\ell-1}\varepsilon^{-2\ell d + 2d - 2} \log^2(m) \log(n)\right).$$

Theorems 1 and 2 will follow from Theorems 7 and 9, respectively. Note that although we do not present algorithms for computing ε-coresets for the weak and discrete Fréchet and Hausdorff distance, our results also imply the existence of ε-coresets of similar size for these metrics.

1.3   Organization

In Sect. 2 we give the results for general metric spaces: we derive a universal bound on the sensitivities in Sect. 2.1 and the ε-coreset result in Sect. 2.2. In Sect. 3 we present the algorithm for computing ε-coresets for (k, ℓ)-median clustering. Finally, in Sect. 4 we demonstrate the use of ε-coresets in an existing $(5 + \varepsilon)$-approximation algorithm for $(1, \ell)$-median clustering. All proofs can be found in the full paper [10].

2   Coresets for Generalized k-Median Clustering in Metric Spaces

In this section, we first derive general results for ε-coreset based on the sensitivity sampling framework [15,25]. In the following $d \in \mathbb{N}$ is an arbitrary constant. By $\|\cdot\|$ we denote the Euclidean norm and for $n \in \mathbb{N}$, we define $[n] = \{1, \ldots, n\}$. For a closed logical formula Ψ we define by $\mathbb{1}(\Psi)$ the function that is 1 if Ψ is true and 0 otherwise.

Let $\mathcal{X} = (X, \rho)$ be an arbitrary metric space, where X is any non-empty set and $\rho \colon X \times X \to \mathbb{R}_{\ge 0}$ is a distance function. We introduce a generalized definition of k-median clustering, where the input is restricted to come from a predefined subset $Y \subseteq X$ and the medians are restricted to come from a predefined subset $Z \subseteq X$.

Definition 1. *The generalized k-median clustering problem is defined as follows, where $k \in \mathbb{N}$ is a fixed (constant) parameter of the problem: given a finite and non-empty set $T = \{\tau_1, \ldots, \tau_n\} \subseteq Y$, compute a set C of k elements from Z, such that $\text{cost}\,(T, C) = \sum_{\tau \in T} \min_{c \in C} \rho(\tau, c)$ is minimal.*

We analyze the problem in terms of functions. This allows us to apply the improved ε-coreset framework by Feldman and Langberg [15]. Therefore, given a set $T = \{\tau_1, \ldots, \tau_n\} \subseteq Y$ we define $F = \{f_1, \ldots, f_n\}$ to be a set of functions with $f_i \colon 2^Z \setminus \{\emptyset\} \to \mathbb{R}_{\geq 0}$, $C \mapsto \min_{c \in C} \rho(c, \tau_i)$. For each $C \in 2^Z \setminus \{\emptyset\}$ we now have cost $(T, C) = \sum_{i=1}^{n} f_i(C)$.

In the following, we bound the sensitivity of each $\tau \in T$. That is the maximum fraction of cost (T, C) that is caused by τ, for all C. To comply with the k-median problem we only take into account the k-subsets $C \subseteq Z$.

2.1   Sensitivity Bound

First, we formally define the sensitivities of the inputs $\tau \in T$ in terms of the respective functions.

Definition 2 ([15]). *Let F be a finite and non-empty set of functions $f \colon 2^Z \setminus \{\emptyset\} \to \mathbb{R}_{\geq 0}$. For $f \in F$ we define the sensitivity with respect to F:*

$$\mathfrak{s}(f, F) = \sup_{\substack{C = \{c_1, \ldots, c_k\} \subseteq Z \\ \sum_{g \in F} g(C) > 0}} \frac{f(C)}{\sum_{g \in F} g(C)}.$$

We define the total sensitivity of F as $\mathfrak{S}(F) = \sum_{f \in F} \mathfrak{s}(f, F)$.

We now prove a bound on the sensitivity of all $f \in F$, which then yields a bound on the total sensitivity of F. Later, our coreset will be a weighted sample from a distribution whose probabilities are determined by the derived bounds. To compute the bounds, any (bi-criteria) approximate solution to the generalized k-median problem can be used. Our analysis is an adaption of the analysis of the sensitivities for sum-based k-clustering of points in $\mathbb{R}^d$ by Langberg and Schulman [25]. We note that similar bounds have already been derived in the literature, see e.g., [30].

Lemma 1. *Let $k' \in \mathbb{N}$, $C^* = \{c_1^*, \ldots, c_k^*\} \subseteq Z$ with $\Delta^* = \sum_{i=1}^{n} f_i(C^*)$ minimal and $\hat{C} = \{\hat{c}_1, \ldots, \hat{c}_{k'}\} \subseteq X$ with $\hat{\Delta} = \sum_{i=1}^{n} f_i(\hat{C}) \leq \alpha \cdot \Delta^*$ for an $\alpha \in [1, \infty)$. Breaking ties arbitrarily, we assume that every $\tau \in T$ has a unique nearest neighbor in $\hat{C}$ and for $i \in [k']$, we define $\hat{V}_i = \{\tau \in T \mid \forall j \in [k'] : \rho(\tau, \hat{c}_i) \leq \rho(\tau, \hat{c}_j)\}$ to be the Voronoi cell of $\hat{c}_i$ and $\hat{\Delta}_i = \sum_{\tau \in \hat{V}_i} \rho(\tau, \hat{c}_i)$ to be its cost. For each $i \in [k']$ and $\tau_j \in \hat{V}_i$ it holds that*

$$\gamma(f_j) = \left(1 + \sqrt{\frac{2k'}{3\alpha}}\right)\left(\frac{\alpha \rho(\tau_j, \hat{c}_i)}{\hat{\Delta}} + \frac{2\alpha \hat{\Delta}_i}{\hat{\Delta}|\hat{V}_i|}\right) + \left(1 + \sqrt{\frac{3\alpha}{2k'}}\right)\frac{2}{|\hat{V}_i|} \geq \mathfrak{s}(f_j, F)$$

and $\Gamma = \sum_{f \in F} \gamma(f) = 2k' + 2\sqrt{6\alpha k'} + 3\alpha \geq \mathfrak{S}(F)$.

2.2   Coresets by Sensitivity Sampling

We apply the framework of Feldman and Langberg [15]. First, we formally define ε-coresets for generalized k-median clustering.

Definition 3. *Given $\varepsilon \in (0, 1)$ and a finite non-empty set $T \subseteq Y$, a (multi-)set $S \subseteq X$ together with a weight function $w \colon S \to \mathbb{R}_{>0}$ is a weighted ε-coreset for k-median clustering of T, if for all $C \subseteq Z$ with $|C| = k$ it holds that*

$$(1 - \varepsilon)\mathrm{cost}\,(T, C) \leq \mathrm{cost}_w\,(S, C) \leq (1 + \varepsilon)\mathrm{cost}\,(T, C),$$

where $\mathrm{cost}_w\,(S, C) = \sum_{s \in S} w(s) \cdot \min_{c \in C} \rho(s, c)$.

We define range spaces and the associated concepts.

Definition 4. *A range space is a pair $(X, \mathcal{R})$, where X is a set, called ground set and $\mathcal{R}$ is a set of subsets $R \subseteq X$, called ranges.*

The projection of a range space $(X, \mathcal{R})$ onto a subset $Y \subseteq X$ is the range space $(Y, \{Y \cap R \mid R \in \mathcal{R}\})$. Furthermore, for each range space there exists a complementary range space.

Definition 5. *Let $F = (X, \mathcal{R})$ be a range space. We call $\overline{F} = (X, \overline{\mathcal{R}})$, the range space over $\overline{\mathcal{R}} = \{X \setminus R \mid R \in \mathcal{R}\}$, the complementary range space of F.*

A measure of the combinatorial complexity of a range space is the VC dimension.

Definition 6. *The VC dimension of a range space $(X, \mathcal{R})$ is the cardinality of a maximum cardinality subset $Y \subseteq X$, such that $|\{Y \cap R \mid R \in \mathcal{R}\}| = 2^{|Y|}$.*

Note that F and $\overline{F}$ have equal VC dimension and for any $Y \subseteq X$, the projection of F onto Y has VC dimension at most the VC dimension of F, see for example [18]. We define (ε, η)-approximations of range spaces.

Definition 7 ([21, **Definition 2.3**]). *Let $\varepsilon, \eta \in (0, 1)$ and $(X, \mathcal{R})$ be a range space with finite non-empty ground set. An (η, ε)-approximation of $(X, \mathcal{R})$ is a set $S \subseteq X$, such that for all $R \in \mathcal{R}$*

$$\left| \frac{|R \cap X|}{|X|} - \frac{|R \cap S|}{|S|} \right| \leq \begin{cases} \varepsilon \cdot \frac{|R \cap X|}{|X|}, & \text{if } |R \cap X| \geq \eta \cdot |X| \\ \varepsilon \cdot \eta, & \text{else.} \end{cases}$$

The following theorem is useful for obtaining (ε, η)-approximations.

Theorem 3 ([21, **Theorem 2.11**]). *Let $(X, \mathcal{R})$ be a range space with finite non-empty ground set and VC dimension $\mathcal{D}$. Also, let $\varepsilon, \delta, \eta \in (0, 1)$. There is an absolute constant $c \in \mathbb{R}_{>0}$ such that a sample of*

$$\frac{c}{\eta \cdot \varepsilon^2} \cdot \left(\mathcal{D} \log \left(\frac{1}{\eta} \right) + \log \left(\frac{1}{\delta} \right) \right)$$

elements drawn independently and uniformly at random with replacement from X is a (η, ε)-approximation for $(X, \mathcal{R})$ with probability at least $1 - \delta$.

We define open metric balls, which are the ranges used to derive our result.

Definition 8. *For $r \in \mathbb{R}_{\geq 0}$, $z \in Z$ and $Y \subseteq X$ we denote by $\mathrm{B}(z, r, Y) = \{y \in Y \mid \rho(y, z) < r\}$ the open metric ball with center z and radius r. We denote the set of all open metric balls by $\mathbb{B}(Y, Z) = \{\mathrm{B}(z, r, Y) \mid z \in Z, r \in \mathbb{R}_{\geq 0}\}$.*

Now, we are ready to analyze the computation of the actual ε-coresets. We use the reduction to uniform sampling, introduced by Feldman and Langberg [15] and improved by Braverman et al. [2] (using Theorem 3). Preferably we would apply Theorem 31 by Feldman et al. [16], which however is not possible since it depends on a range space where each function $f \in F$ may be assigned a distinct scaling factor. This is incompatible with the range space induced by the open metric balls we use to obtain our result. However, by adapting and modifying the proof of their theorem we can derive the desired and more versatile result. To handle necessary scaling factors still involved in the analysis, we incorporate results by Munteanu et al. [28] for bounding the VC dimension. The proof can be found in the full paper [10].

Theorem 4. *For $f \in F$ we let $\lambda(f) = \lceil |F| \cdot 2^{\lceil \log_2(\gamma(f)) \rceil} \rceil / |F|$, $\Lambda = \sum_{f \in F} \lambda(f)$, $\psi(f) = \frac{\lambda(f)}{\Lambda}$ and $\mathcal{D}$ be the VC dimension of the range space $(Y, \mathbb{B}(Y, Z))$. Let $\delta, \varepsilon \in (0, 1)$. A sample S of $\Theta\left(\varepsilon^{-2} \alpha k'(\mathcal{D}k \log(k) \log(\alpha k' n) \log(\alpha k') + \log(1/\delta))\right)$ elements τ_i from T, drawn independently with replacement with probability $\psi(f_i)$ and weighted by $w(f_i) = \frac{\Lambda}{|S|\lambda(f_i)}$ is an ε-coreset with probability at least $1 - \delta$.*

3   Coresets for (k, ℓ)-Median Clustering Under the Fréchet Distance

Now we present an algorithm for computing ε-coresets for (k, ℓ)-median clustering of polygonal curves under the Fréchet distance. We start by defining polygonal curves.

Definition 9. *A (parameterized) curve is a continuous mapping $\tau \colon [0, 1] \to \mathbb{R}^d$. A curve τ is polygonal, iff there exist $v_1, \ldots, v_m \in \mathbb{R}^d$, no three consecutive on a line, called τ's vertices, and $t_1, \ldots, t_m \in [0, 1]$ with $t_1 < \cdots < t_m$, $t_1 = 0$ and $t_m = 1$, called τ's instants, such that τ connects every two consecutive vertices $v_i = \tau(t_i), v_{i+1} = \tau(t_{i+1})$ by a line segment.*

We call the segments $\overline{v_1 v_2}, \ldots, \overline{v_{m-1} v_m}$ edges of τ and m the complexity of τ, denoted by $|\tau|$.

Definition 10. *Let $\mathcal{H}$ denote the set of all continuous bijections $h \colon [0, 1] \to [0, 1]$ with $h(0) = 0$ and $h(1) = 1$, which we call reparameterizations. The Fréchet distance between curves σ and τ is $d_F(\sigma, \tau) = \inf_{h \in \mathcal{H}} \max_{t \in [0,1]} \|\sigma(t) - \tau(h(t))\|$.*

Now we introduce the classes of curves we are interested in.

Definition 11. *For $d \in \mathbb{N}$, we define by $\mathbb{X}^d$ the set of equivalence classes of polygonal curves (where two curves are equivalent, iff they can be made identical by a reparameterization) in ambient space $\mathbb{R}^d$. For $m \in \mathbb{N}$ we define by $\mathbb{X}_m^d$ the subclass of polygonal curves of complexity at most m.*

Finally, we define the (k, ℓ)-median clustering problem for polygonal curves.

Definition 12. *The (k, ℓ)-median clustering problem is defined as follows, where $k, \ell \in \mathbb{N}$ are fixed (constant) parameters of the problem: given a set $T \subset \mathbb{X}_m^d$ of n polygonal curves, compute a set of k curves $C^* \subset \mathbb{X}_\ell^d$, such that $\mathrm{cost}\,(T, C^*) = \sum_{\tau \in T} \min_{c^* \in C^*} d_F(\tau, c^*)$ is minimal.*

We bound the VC dimension of metric balls under the Fréchet distance by showing that a result of Driemel et al. [13] holds also in our setting.

Theorem 5. *The VC dimension of $(\mathbb{X}_m^d, \mathbb{B}(\mathbb{X}_m^d, \mathbb{X}_\ell^d))$ is $O\left(\ell^2 \log(\ell m)\right)$.*

Proof. We argue that the claim follows from Theorem 18 by Driemel et al. [13]. First, in their paper polygonal curves do not need to adhere the restriction that no three consecutive vertices may be collinear and they define $\mathbb{X}_m^d$ to be the polygonal curves of exactly m vertices. However, our definitions match by simulating the addition of collinear vertices to those curves in $\mathbb{X}_m^d$ with less than m vertices.

Now, looking into their proof, we can slightly modify the geometric primitives by letting $B_r(p) = \{x \in \mathbb{R}^d \mid \|x - p\| < r\}$, $D_r(\overline{st}) = \{x \in \mathbb{R}^d \mid \exists p \in \overline{ul} : \|p - x\| < r\}$, $C_r(\overline{st}) = \{x \in \mathbb{R}^d \mid \exists p \in \ell(\overline{st}) : \|p - x\| < r\}$ and $R_r(\overline{st}) = \{p + u \mid p \in \overline{st}, u \in \mathbb{R}^d, \langle t - s, u \rangle = 0, \|u\| < r\}$, which does not affect the remainder of the proof and thus yields the same bound on the VC dimension. $\square$

To compute ε-coresets for (k, ℓ)-median clustering under the Fréchet distance, we first need to compute the sensitivities and to do so, we utilize constant factor approximation algorithms. We use [8, Algorithm 1], which only works for $k = 1$ but is very efficient in this case. For $k > 1$ we use Algorithm 1, a modification of [12, Algorithm 3], which we now present. This algorithm uses (approximate) minimum-error ℓ-simplifications, which we now define.

Definition 13. *An α-approximate minimum-error ℓ-simplification of a polygonal curve $\tau \in \mathbb{X}^d$ is a curve $\sigma \in \mathbb{X}_\ell^d$ with $d_F(\tau, \sigma) \leq \alpha \cdot d_F(\tau, \sigma')$ for all $\sigma' \in \mathbb{X}_\ell^d$.*

The following lemma is useful to obtain simplifications.

Lemma 2 ([4, **Lemma 7.1**]). *Given a curve $\sigma \in \mathbb{X}_m^d$, a 4-approximate minimum-error ℓ-simplification can be computed in $O(m^3 \log m)$ time, by combining the algorithms by Alt and Godau [1] and Imai and Iri [23].*

We now present the constant factor approximation algorithm.

Algorithm 1. Constant Factor Approximation for (k, ℓ)-Median Clustering

---
1: **procedure** (k, ℓ)-MEDIAN-96-APPROXIMATION$(T = \{\tau_1, \ldots, \tau_n\})$
2:     **for** $i = 1, \ldots, n$ **do**
3:         $\hat{\tau}_i \leftarrow$ approximate minimum-error ℓ-simplification of τ_i
4:     $C \leftarrow$ Chen's algorithm with $\varepsilon = 0.\dot{5}, \lambda = \delta$ on $\{\hat{\tau}_1, \ldots, \hat{\tau}_n\}$ [11, Theorem 6.2]
5:     **return** C

We prove the correctness and analyze the running time of Algorithm 1.

Theorem 6. *Given* $\delta \in (0, 1)$ *and* $T = \{\tau_1, \ldots, \tau_n\} \subset \mathbb{X}_m^d$, *Algorithm 1 returns with probability at least* $1 - \delta$ *a 109-approximate* (k, ℓ)-*median solution for* T *in time* $O(nm \log(1/\delta) \log(m) + nm^3 \log(m))$.

The proof can be found in the full paper [10]. We now present the algorithm for computing weighted ε-coresets for (k, ℓ)-median clustering.

Algorithm 2. Coresets for (k, ℓ)-Median Clustering

1: **procedure** (k, ℓ)-MEDIAN-CORESET$(T = \{\tau_1, \ldots, \tau_n\}, \delta, \varepsilon)$
2:     **if** $k = 1$ **then**
3:         $\hat{c} \leftarrow \ell$-Median-34-Approximation$(T, \delta/2)$ [8, Algorithm 1]
4:         $\hat{C} = \{\hat{c}\}$
5:     **else**
6:         $\hat{C} = \{\hat{c}_1, \ldots, \hat{c}_k\} \leftarrow$ Algorithm 1$(T, \delta/2)$
7:     compute $\hat{V}_1, \ldots, \hat{V}_k$, $\hat{\Delta}_1, \ldots, \hat{\Delta}_k$ and γ w.r.t. $\hat{C}$ (cf. Lemma 1)
8:     compute λ, Λ w.r.t. γ and ψ w.r.t. λ (cf. Theorem 4)
9:     $S \leftarrow$ sample $\Theta(k\varepsilon^{-2}(d^2\ell^2 k \log(d\ell m) \log(kn) \log^2(k) + \log(1/(2\delta))))$
              elements from T independently with replacement with respect to ψ
10:    compute w w.r.t. λ, Λ and S (cf. Theorem 4)
11:    **return** S and w

We prove the correctness and analyze the running time of Algorithm 2. Also, we analyze the size of the resulting ε-coreset.

Theorem 7. *Given a set* $T = \{\tau_1, \ldots, \tau_n\} \subset \mathbb{X}_m^d$ *and* $\delta, \varepsilon \in (0, 1)$, *Algorithm 2 computes a weighted ε-coreset of size* $O(\varepsilon^{-2}(\log(m) \log(n) + \log(1/\delta)))$ *for* (k, ℓ)-*median clustering with probability at least* $1 - \delta$, *in time*

$$O(nm \log(m) \log(1/\delta) + nm^3 \log(m) + \varepsilon^{-2}(\log(m) \log(n) + \log(1/\delta)))$$

for $k > 1$ *and*

$$O(nm \log(m) + m^2 \log(m) \log^2(1/\delta) + m^3 \log(m) + \varepsilon^{-2}(\log(m) \log(n) + \log(1/\delta)))$$

for $k = 1$.

We prove this theorem (in the full paper [10]) by combining Lemma 1, Theorem 4, and Theorem 6, respectively [8, Corollary 3.1].

4   Towards Practical $(1, \ell)$-Median Approximation Algorithms

In this section, we present a modification of Algorithm 3 from [7]. Our modification uses ε-coresets to improve the running time of the algorithm, rendering it more tractable in a big data setting. We start by giving some definitions. For $p \in \mathbb{R}^d$ and $r \in \mathbb{R}_{\geq 0}$ we denote by $B(p, r) = \{q \in \mathbb{R}^d \mid \|p - q\| \leq r\}$ the closed Euclidean ball of radius r with center p. We give a standard definition of grids.

Definition 14 (grid). *Given a number $r \in \mathbb{R}_{>0}$, for $(p_1, \ldots, p_d) \in \mathbb{R}^d$ we define by $G(p, r) = (\lfloor p_1/r \rfloor \cdot r, \ldots, \lfloor p_d/r \rfloor \cdot r)$ the r-grid-point of p. Let $P \subseteq \mathbb{R}^d$ be a subset of $\mathbb{R}^d$. The grid of cell width r that covers P is the set $\mathbb{G}(P, r) = \{G(p, r) \mid p \in P\}$.*

Such a grid partitions the set P into cubic regions and for each $r \in \mathbb{R}_{>0}$ and $p \in P$ we have that $\|p - G(p, r)\| \leq \sqrt{d}r$. The following theorem by Indyk [24] is useful for evaluating the cost of a curve at hand.

Theorem 8 ([24, Theorem 31]). *Let $\varepsilon \in (0, 1]$ and $T \subset \mathbb{X}^d$ be a set of polygonal curves. Further let W be a non-empty sample, drawn uniformly and independently at random from T, with replacement. For $\tau, \sigma \in T$ with $\mathrm{cost}\,(T, \tau) > (1 + \varepsilon)\mathrm{cost}\,(T, \sigma)$ it holds that $\Pr[\mathrm{cost}\,(W, \tau) \leq \mathrm{cost}\,(W, \sigma)] < \exp\left(-\varepsilon^2 |W|/64\right)$.*

The following theorem, which we combine with fine-tuned grids, allows us to obtain low-complexity center curves.

Lemma 3 ([7, Lemma 4.1]). *Let $\sigma, \tau \in \mathbb{X}^d$ be polygonal curves. Let $v_1^\tau, \ldots, v_{|\tau|}^\tau$ be the vertices of τ and let $r = d_F(\sigma, \tau)$. There exists a polygonal curve $\sigma' \in \mathbb{X}^d$ with every vertex contained in at least one of $B(v_1^\tau, r), \ldots, B(v_{|\tau|}^\tau, r)$, $d_F(\sigma', \tau) \leq d_F(\sigma, \tau)$ and $|\sigma'| \leq 2|\sigma| - 2$.*

Finally, we present our improved modification of Algorithm 3 from [7]. This algorithm uses ε-coresets every time it has to evaluate the cost of a center set. The dramatic effect of this small modification is that we nearly lose the original linear running time dependency on n in the most time consuming part of the algorithm, rendering it practical in the setting where we have a lot of curves of much smaller complexity than number $(\ell < m \ll n)$.

Algorithm 3. $(1, \ell)$-Median by Simple Shortcutting and ε-Coreset

---
1: **procedure** $(1, \ell)$-MEDIAN-$(5 + \varepsilon)$-APPROXIMATION$(T = \{\tau_1, \ldots, \tau_n\}, \delta, \varepsilon)$
2:     $\hat{c} \leftarrow (1, \ell)$-Median-34-Approximation$(T, \delta/4)$ [8, Algorithm 1]
3:     $\varepsilon' \leftarrow \varepsilon/67, \ P \leftarrow \emptyset$
4:     $(T', w) \leftarrow (1, 2\ell - 2)$-Median-Coreset$(T, \delta/4, \varepsilon')$
5:     $\Delta \leftarrow \text{cost}_w(T', \{\hat{c}\}), \ \Delta_u \leftarrow \Delta/(1 - \varepsilon'), \ \Delta_l \leftarrow \Delta/((1 + \varepsilon')34)$
6:     $S \leftarrow$ sample $\lceil -2(\varepsilon')^{-1}(\ln(\delta) - \ln(4)) \rceil$ curves from T uniformly
        and independently with replacement
7:     $W \leftarrow$ sample $\lceil -64(\varepsilon')^{-2}(\ln(\delta) - \ln(\lceil -8(\varepsilon')^{-1}(\ln(\delta) - \ln(4)) \rceil)) \rceil$ curves
        from T uniformly and independently with replacement
8:     $c \leftarrow$ arbitrary element from $\arg\min_{s \in S} \text{cost}(W, s)$
9:     **for** $i = 1, \ldots, |c|$ **do**
10:        $P \leftarrow P \cup \mathbb{G}(B(v_i^c, (3 + 4\varepsilon')\Delta_u/n), \varepsilon'\Delta_l/(n\sqrt{d}))$ $(v_i^c: i^{\text{th}}$ vertex of $c)$
11:     $C \leftarrow$ set of all polygonal curves with $2\ell - 2$ vertices from P
12:     **return** $\arg\min_{c' \in C} \text{cost}_w(T', \{c'\})$

We show the correctness and analyze the running time of Algorithm 3.

Theorem 9. *Given two parameters* $\delta \in (0, 1)$, $\varepsilon \in (0, 1/2]$ *and a set* $T = \{\tau_1, \ldots, \tau_n\} \subset \mathbb{X}_m^d$ *of polygonal curves, with probability at least* $1 - \delta$ *Algorithm 3 returns a* $(5 + \varepsilon)$-*approximate* $(1, \ell)$-*median for* T *with* $2\ell - 2$ *vertices, in time*

$$O\left(nm\log(m) + m^2\log(m)\log^2(1/\delta) + m^{2\ell-1}\frac{\log(m)\log(n) + \log(1/\delta))\log(m)}{\varepsilon^{2\ell d - 2d + 2}}\right).$$

We prove Theorem 9 (in the full paper [10]) by modifying the proof of [7, Theorem 5.1].

5   Conclusion

We presented an algorithm for computing ε-coresets for (k, ℓ)-median clustering of polygonal curves under the Fréchet distance and used these to improve the running time of an existing approximation algorithm for $(1, \ell)$-median clustering. Unfortunately, it was not possible to improve the existing (k, ℓ)-median approximation algorithms in [8,12] by means of ε-coresets. This is due to the recursive approximation scheme used in these works, where the candidate center sets are not necessarily evaluated against the input, but against subsets of the input. Thus, we would need an ε-coreset for any subset of the input, which is not practical. We note that, to the best of our knowledge, no (k, ℓ)-median clustering algorithm exists that do not employ this approximation scheme.

It is still an interesting open problem whether there exist sublinear size ε-coresets for weighted sets of polygonal curves. To derive such a result one may need a sublinear bound on the VC dimension of the range space of metric balls under scaled Fréchet distances, which is not evident at the moment. We note that such a result would enable the use of the iterative size reduction technique recently introduced by Braverman et al. [3].

References

1. Alt, H., Godau, M.: Computing the Fréchet distance between two polygonal curves. Int. J. Comput. Geom. Appl. **5**, 75–91 (1995)
2. Braverman, V., Feldman, D., Lang, H.: New Frameworks for Offline and Streaming Coreset Constructions. CoRR abs/1612.00889 (2016)
3. Braverman, V., Jiang, S.H., Krauthgamer, R., Wu, X.: Coresets for clustering in excluded-minor graphs and beyond. In: Proceedings of the 2021 ACM-SIAM Symposium on Discrete Algorithms, pp. 2679–2696. SIAM (2021)
4. Buchin, K., et al.: Approximating (k, l)-center clustering for curves. In: Proceedings of the 2019 ACM-SIAM Symposium on Discrete Algorithms, pp. 2922–2938 (2019)
5. Buchin, K., Driemel, A., van de L'Isle, N., Nusser, A.: klcluster: center-based clustering of trajectories. In: Proceedings of the 27th ACM SIGSPATIAL International Conference on Advances in Geographic Information Systems, pp. 496–499 (2019)
6. Buchin, K., Driemel, A., Struijs, M.: On the hardness of computing an average curve. In: 17th Scandinavian Symposium and Workshops on Algorithm Theory. LIPIcs, vol. 162, pp. 19:1–19:19. Schloss Dagstuhl - Leibniz-Zentrum für Informatik (2020)
7. Buchin, M., Driemel, A., Rohde, D.: Approximating (k, ℓ)-Median Clustering for Polygonal Curves. arXiv e-prints arXiv:2009.01488 (2020)
8. Buchin, M., Driemel, A., Rohde, D.: Approximating (k, ℓ)-median clustering for polygonal curves. In: Proceedings of the 2021 ACM-SIAM Symposium on Discrete Algorithms, pp. 2697–2717. SIAM (2021)
9. Buchin, M., Rohde, D.: Coresets for (k, l)-clustering under the Fréchet distance. In: Proceedings of the 35th European Workshop on Computational Geometry (2019)
10. Buchin, M., Rohde, D.: Coresets for (k, ℓ)-Median Clustering under the Fréchet Distance. CoRR abs/2104.09392 (2021)
11. Chen, K.: On coresets for k-median and k-means clustering in metric and euclidean spaces and their applications. SIAM J. Comput. **39**(3), 923–947 (2009)
12. Driemel, A., Krivosija, A., Sohler, C.: Clustering time series under the Fréchet distance. In: Proceedings of the Twenty-Seventh Annual ACM-SIAM Symposium on Discrete Algorithms, pp. 766–785 (2016)
13. Driemel, A., Phillips, J.M., Psarros, I.: The VC dimension of metric balls under Fréchet and Hausdorff distances. In: 35th International Symposium on Computational Geometry, pp. 28:1–28:16 (2019)
14. Feldman, D.: Introduction to Core-sets: an Updated Survey. CoRR abs/2011.09384 (2020)
15. Feldman, D., Langberg, M.: A unified framework for approximating and clustering data. In: Proceedings of the 43rd ACM Symposium on Theory of Computing, pp. 569–578. ACM (2011)
16. Feldman, D., Schmidt, M., Sohler, C.: Turning big data into tiny data: constant-size coresets for k-means, PCA, and projective clustering. SIAM J. Comput. **49**(3), 601–657 (2020)
17. Goldberg, P.W., Jerrum, M.: Bounding the Vapnik-Chervonenkis dimension of concept classes parameterized by real numbers. Mach. Learn. **18**(2–3), 131–148 (1995)
18. Har-Peled, S.: Geometric Approximation Algorithms. American Mathematical Society (2011)
19. Har-Peled, S., Kushal, A.: Smaller coresets for k-median and k-means clustering. Discrete Comput. Geom. **37**(1), 3–19 (2006). https://doi.org/10.1007/s00454-006-1271-x

20. Har-Peled, S., Mazumdar, S.: On coresets for k-means and k-median clustering. In: Proceedings of the 2004 ACM Symposium on Theory of Computing, pp. 291–300 (2004)
21. Har-Peled, S., Sharir, M.: Relative (p,ε)-approximations in geometry. Discrete Comput. Geom. **45**(3), 462–496 (2011)
22. Huang, L., Jiang, S.H., Li, J., Wu, X.: Epsilon-coresets for clustering (with outliers) in doubling metrics. In: 59th IEEE Annual Symposium on Foundations of Computer Science, pp. 814–825. IEEE Computer Society (2018)
23. Imai, H., Iri, M.: Polygonal approximations of a curve - formulations and algorithms. Mach. Intelligence Pattern Recogn. **6**, 71–86 (1988)
24. Indyk, P.: High-dimensional Computational Geometry. Ph.D. thesis, Stanford University, CA, USA (2000)
25. Langberg, M., Schulman, L.J.: Universal epsilon-approximators for Integrals. In: Proceedings of the Twenty-First Annual ACM-SIAM Symposium on Discrete Algorithms, pp. 598–607 (2010)
26. Meintrup, S., Munteanu, A., Rohde, D.: Random projections and sampling algorithms for clustering of high-dimensional polygonal curves. In: Advances in Neural Information Processing Systems 32, pp. 12807–12817 (2019)
27. Munteanu, A., Schwiegelshohn, C.: Coresets-methods and history: a theoreticians design pattern for approximation and streaming algorithms. KI - Künstliche Intell. **32**(1), 37–53 (2017). https://doi.org/10.1007/s13218-017-0519-3
28. Munteanu, A., Schwiegelshohn, C., Sohler, C., Woodruff, D.P.: On coresets for logistic regression. In: Advances in Neural Information Processing Systems 31: Annual Conference on Neural Information Processing Systems 2018, pp. 6562–6571 (2018)
29. Nath, A., Taylor, E.: k-median clustering under discrete Fréchet and Hausdorff distances. In: 36th International Symposium on Computational Geometry. LIPIcs, vol. 164, pp. 58:1–58:15. Schloss Dagstuhl - Leibniz-Zentrum für Informatik (2020)
30. Varadarajan, K.R., Xiao, X.: On the sensitivity of shape fitting problems. In: IARCS Annual Conference on Foundations of Software Technology and Theoretical Computer Science. LIPIcs, vol. 18, pp. 486–497 (2012)

Bounds and Algorithms for Geodetic Hulls

Sabine Storandt[✉]

University of Konstanz, 78464 Konstanz, Germany
sabine.storandt@uni-konstanz.de

Abstract. The paper is devoted to the study of geodetic convex hulls
in graphs from a theoretical and practical perspective. The notion of
convexity can be transferred from continuous geometry to discrete graph
structures by defining a node subset to be (geodetically) convex if all
shortest paths between its members do not leave the subset. The geodetic
convex hull of a node set W is the smallest convex superset of W. The
hull number of a graph is then defined as the size of the smallest node
subset S (called hull set) whose convex hull contains all graph nodes. In
contrast to the geometric setting, where the point subset on the boundary
of the convex hull can be computed in polynomial time, it is NP-hard
to decide whether a graph has a hull set of size at most $s \in \mathbb{N}$. We
establish novel theoretical bounds for graph parameters related to convex
graph structures, and also design practical algorithms for upper and
lower bounding the hull number. We evaluate the quality of our bounds
as well as the performance of the proposed algorithms on road networks
and wireless sensor networks of varying size.

Keywords: Hull number · Graph contour · Geodetic iteration number

1 Introduction

There exist different notions of a hull of a graph. For embedded graphs, the
convex hull of the nodes [12] or the polygonal hull of the edges [14] are utilized
for applications such as clustering, shape recognition, or area-of-interest visualization. In this paper, we focus on geodetic convex hulls, though, which do not
depend on any embedding but are based solely on the toplogy of the graph [13].
Here, given a connected, undirected and unweighted graph $G(V, E)$, we denote
with $I(a, b)$ the set of nodes on shortest paths between $a \in V$ and $b \in V$. This
set is also called the interval of a and b. The interval of a node set $S \subseteq V$ is the
union of the intervals of all pairs of nodes $a, b \in S$, that is, $I[S] := \bigcup_{a,b \in S} I(a, b)$.
A set S is called convex if $I[S] = S$. This implies that all shortest paths between
nodes in S are fully contained in S. The (geodetic) convex hull $h(W)$ of a node
set $W \subseteq V$ is then defined as the smallest convex superset S of W. This notion
is inspired by the geometric concept of convexity, where a point set is convex if
and only if it contains all straight line segments between its members.

N. Balachandran and R. Inkulu (Eds.): CALDAM 2022, LNCS 13179, pp. 181–194, 2022.
https://doi.org/10.1007/978-3-030-95018-7_15

Similar to the geometric setting, a geodetic convex hull may be represented by its extreme points. For a convex point set in the plane, an extreme point is a point that does not lie on any open line segment between two other points in the set. For a convex node set, a node is extreme if it is not contained as an inner node in a shortest path between other extreme nodes. More formally, for a convex node set S in a graph, we call a node set W a hull set of S if $I^k[W] = S$ for some $k \in \mathbb{N}$, where $I^k[W] = I[W]$ for $k = 1$ and $I[I^{k-1}[W]]$ for $k > 1$. The other way around, for any set W, its convex hull $h(W)$ can be computed by applying this iterated interval operation until $I^{k+1}[W] = I^k[W]$ and hence $I^k[W] = h(W)$ holds. The respective value of k is called the geodetic iteration number of W, denoted by $gin(W)$. Figure 1 illustrates the concept of iterated interval computation.

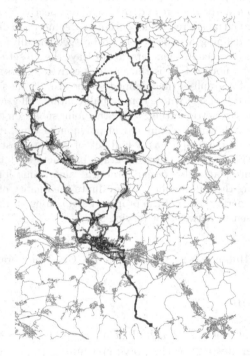

Fig. 1. Cutout of a road network with the geodetic convex hull of a set W of three nodes (red). Each color encodes an iteration step in which nodes on shortest paths between nodes in the current set W are added to W. The final set (obtained after $gin(W) = 7$ iterations) contains 7886 nodes. (Color figure online)

The size of a smallest set S such that $I^k[S] = V$ for some $k \in \mathbb{N}$ is called the hull number of the graph, abbreviated as $hn(G)$. For geometric convex hulls as well as polygonal hulls, the set of extreme points or nodes on the boundary can be computed in polynomial time.

Unfortunately, deciding whether $hn(G) \leq s$ for $s \in \mathbb{N}$ poses an NP-hard decision problem, even in several restricted graph classes as e.g. bipartite graphs [1]. Nevertheless, geodetic convex hulls and convex node sets have important applications in data structure design [11], connectivity analysis [15], route planning

[20], and graph similarity assessment [6]. Moreover, the hull number and related parameters are relevant for theoretical investigations as e.g. the characterization of classes of chordal graphs [8] or parameterizability of certain optimization problems [16].

The goal of this paper is to establish new (theoretical) bounds for the geodetic iteration number as well as the hull number, and to enable the efficient computation of concise hull sets in practice.

1.1 Related Work

The computation of boundary nodes of a graph is relevant for a wide range of applications. One particular well-studied application field is the analysis of wireless sensor networks. Here nodes represent sensors which are able to detect signals and communicate with other sensors in their proximity. These connections are then modelled as undirected egdes. The goal is typically to compute the boundary of the network which inlcudes the detection of so called holes (areas that are not well covered by sensors). A recent survey [5] discusses a rich set of approaches for boundary detection. Most of these papers are focused on algorithms that work well in practice, though, and rarely provide clear definitions of boundary nodes or holes (or only ones that depend on selected thresholds). While there are some approaches which come with quality guarantees [10,19], those usually strongly depend on certain model assumptions.

Convexity structures and hull sets are well-defined in any graph and are fundamental concepts in graph theory [7]. The hull number (the size of the smallest hull set of a graph) turned out to be NP-hard to compute even in bipartite graphs [1] and chordal graphs [2], though.

The boundary of the graph $\delta(G)$ and the contour of the graph $Ct(G)$ (see Sect. 2 for definitions) were both shown to constitute a (not necessarily minium sized) hull set [4,17]. But they come with different geodetic iteration numbers. While $gin(\delta(G)) = 1$ in all graphs, $gin(Ct(G)) = 1$ only holds in certain graph classes (as e.g. chordal graphs) but examples with up to $gin(Ct(G)) = 3$ are known in general graphs [17]. It is an open question whether graphs with $gin(Ct(G)) > 3$ exist. Further theoretical results and related notions of convexity are discussed in a survey by Brevsat et al. [3] and a book by Pelayo [18].

1.2 Contribution

The following theoretical and practical results for hull computation in graphs are presented in this paper:

- We first study the geodetic iteration number of a special node set, called the graph contour $Ct(G)$. It is known that the contour is always a hull set of the graph, but it is also relevant to identify the value of k such that $I^k[Ct(G)] = V$, that is, to compute $gin(Ct(G))$. We prove several new upper bounds for k that relate the geodetic iteration number to the graph diameter as well as the contour size.

- For the hull number $hn(G)$, we provide novel lower and upper bounding techniques based on structural insights. The upper bound is based on a heuristic that computes a feasible hull set (with bounded gin) in linear time.
- In the experimental evaluation, we demonstrate the quality of our bounds as well as the applicability of our novel algorithms on road networks and wireless sensor networks. As one important result, we show that our heuristic for computing a hull set is fast in practice and produces solutions that are significantly smaller than the graph contour.

2   Preliminaries

Throughout the paper, we assume to be given a connected, undirected and unweighted graph $G(V, E)$ with $|V| = n$ nodes and $|E| = m$ edges. With $d(v, w)$ we denote the minimum hop distance between nodes $v \in V$ and $w \in V$. For each node $v \in V$, the eccentricity $ecc(v)$ is defined as the length of the longest shortest path emerging from v. More formally, $ecc : V \to \mathbb{N}$ with $ecc(v) = \max_{w \in V} d(v, w)$. Based on this notion, we define the following graph parameters and special node sets (see Fig. 2 for illustrations):

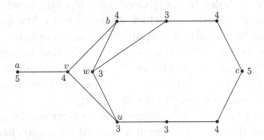

Fig. 2. Small graph with eccentricity values depicted in purple. The diameter is 5 and the radius is 3. The contour is formed by the node set $\{a, b, c\}$, the boundary by all nodes except of v. Note that this is an example of $gin(Ct(G)) = 2$ as $w \notin I[Ct(G)]$. The path u, v, a is a trail in G. (Color figure online)

- The *diameter* $diam(G)$ of a graph is the largest shortest path distance in G, that is, $diam(G) := \max_{v \in V} ecc(v)$.
- The *radius* $rad(G)$ of a graph is the smallest distance value such that some node v can reach all other nodes in G with a shortest path of at most that length, hence $rad(G) := \min_{v \in V} ecc(v)$.
- The *contour* $Ct(G)$ is the set of nodes with maximum eccentricity among their neighbors: $Ct(G) := \{v \in V | \forall w \in N(v) : ecc(v) \geq ecc(w)\}$ where $N(v) := \{w \in V | \{v, w\} \in E\}$.
- A *trail* is a path $p = v_1, v_2, \ldots, v_l$ in G with $ecc(v_{i+1}) = ecc(v_i) + 1$ for $i = 1, \ldots, l - 1$. The nodes $v_2, \ldots, v_l$ are then said to be on a trail from v_1 and belong to the *trail set* $tr(v_1)$.

For disambiguation, we further include the definition of the *boundary* $\delta(G)$ of a graph: $\delta(G) := \{v \in V | \exists u \in V : \forall w \in N(v) : d(u,w) \leq d(u,v)\}$. The boundary is always a hull set of G with $gin(\delta(G)) = 1$ and a superset of the contour [4].

3 Bounds for the Gin of the Graph Contour

The contour $Ct(G)$ of a graph G always yields a hull set. The geodetic iteration number of the contour $gin(Ct(G))$ denotes the smallest value k such that $I^k[Ct(G)] = V$. It was conjectured for over a decade that $gin(Ct(G)) \leq 2$ holds in all graphs. But a counter-example with $gin(Ct(G)) = 3$ was provided in [17]. It still is an open question whether $gin(Ct(G))$ is a always a small constant. A trivial upper bound is $gin(Ct(G)) \leq n$ as in each iteration at least one node has to be added to the set. In the following, we prove several non-trivial upper bounds for $gin(Ct(G))$ by establishing connections to the graph diameter and the contour size.

Theorem 1. *For any graph G, the geodetic iteration number of the contour $Ct(G)$ is upper bounded by $diam(G) - rad(G)$.*

Proof. We prove the following statement by induction over k:

$$I^k[Ct(G)] \supseteq \{v \in V | ecc(v) \geq diam(G) - k\}$$

Hence after at most $k = diam(G) - rad(G)$ iterations, all nodes $v \in V$ are contained in $I^k[Ct(G)]$. For $k = 0$, it is required that all nodes with an eccentricity equal to the diameter of the graph are contained in the contour. This is true by definition of the contour (as $ecc(v) \leq diam(G)$ for all $v \in V$). Next, we consider $I^{k+1}[Ct(G)]$ and a node $v \notin I^k[Ct(G)]$ with $ecc(v) = diam(G) - (k+1)$. As from $v \notin I^k[Ct(G)]$ it follows $v \notin Ct(G)$, there has to exist a neighboring node $w \in N(v)$ with $ecc(w) = ecc(v) + 1$. Let z be a node with $d(w,z) = ecc(w)$. Then both, w and z, have a higher eccentricity than v and are hence contained in $I^k[Ct(G)]$ according to the induction hypothesis. As $ecc(v) = ecc(w) - 1$, it follows that $d(v,z) \leq ecc(v) = ecc(w) - 1$ and therefore $d(w,v) + d(v,z) \leq d(w,z)$. Obviously, the inequality has to be tight. It follows that v is on a shortest path from w to z; and with $w, z \in I^k[Ct(G)]$, we hence conclude that $v \in I^{k+1}[Ct(G)]$. □

Note that this bound is tight for the example graph in Fig. 2, as there we have $diam(G) - rad(G) = 2 = gin(Ct(G))$.

Corollary 1. *Theorem 1 in combination with the simple fact that $rad(v) \geq diam(G)/2$ yields $gin(Ct(G)) \leq diam(G)/2$.*

We next want to prove connections between $gin(Ct(G))$ and the eccentricity of the contour nodes as well as the size of the contour. Our proofs are based on a lemma from Mezzini [17] which is rephrased below.

Lemma 1. *If $I^2[Ct(G)] \neq V$, then there exist nodes $a_1, \ldots, a_6$ such that*

- $a_1 \in V \setminus I^2[Ct(G)]$
- $a_2 \in tr(a_1) \cap Ct(G)$
- $a_3 \notin Ct(G),\ d(a_2, a_3) = ecc(a_2)$
- $a_4 \in tr(a_3) \cap Ct(G)$

- $a_5 \notin Ct(G),\ d(a_4, a_5) = ecc(a_4)$
- $a_6 \in tr(a_5) \cap Ct(G)$
- $ecc(a_6) \geq ecc(a_1) + 3$

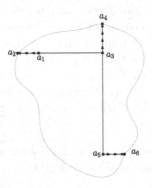

Fig. 3. Schematic depiction of the relationship between the nodes described in Lemma 1. The turquoise cycle illustrates the graph contour with nodes a_2, a_4 and a_6 being part of the contour. Straight lines indicate shortest paths between nodes, and path sections with arrows correspond to trails on which the eccentricity of the nodes increases by one along each directed edge.

An illustration of the configuration described in Lemma 1 is provided in Fig. 3. We will next prove a generalization of this lemma.

Theorem 2. *For any node $a_1 \notin I^k[Ct(G)]$ for $k \geq 2$, nodes $a_2, \ldots, a_6$ with properties as described in Lemma 1 exist.*

Proof. Let $a_1 \notin I^k[Ct(G)]$ for a $k \geq 2$. Accordingly, $a_1 \notin Ct(G)$ and hence a_1 has a neighboring node with eccentricity $ecc(a_1) + 1$. As this applies to all nodes that are not part of the contour, there exists a path from any node $v \in V \setminus Ct(G)$ to a contour node such that the eccentricity of the nodes along the path increases by one in each step. Therefore, $a_2 \in tr(a_1) \cap Ct(G)$ has to exist. Of course, there needs to be a node a_3 such that the eccentricity of a_2 is realized, that is, $d(a_2, a_3) = ecc(a_2)$. Now, we further conclude that there exists a shortest path from a_2 to a_3 that traverses a_1, as we know that $d(a_1, a_3) \leq ecc(a_1)$ and $ecc(a_1) = ecc(a_2) - d(a_2, a_1)$ based on $a_2 \in tr(a_1)$ and therefore $d(a_2, a_1) + d(a_1, a_3) \leq ecc(a_2)$. It follows that a_3 cannot be part of the contour, as otherwise a_1 would be included in $I[Ct(G)]$. The node a_4 exists for the same reasons as a_2, now based on $a_3 \notin Ct(G)$. For node a_5, we repeat the argument used to show the existence of a_3 but now $a_5 \in Ct(G)$ would lead to $a_1 \in I^2[Ct(G)]$ which still is a contradiction to the choice of a_1. The node a_6 then exists for the same reasons as a_2 and a_4 based on the observation that $a_5 \notin Ct(G)$. It remains to show that the nodes are all distinct and that $ecc(a_6) \geq ecc(a_1) + 3$. As $a_1 \notin Ct(G)$ and $a_2 \in tr(a_1) \cap Ct(G)$, we clearly have $ecc(a_2) > ecc(a_1) > 0$ and with that also $ecc(a_2) \geq 2$. The latter excludes $a_3 = a_2$ and $a_3 = a_1$. The node a_4 then need to have a higher eccentricity than all nodes with smaller index and is hence distinct from all of them. Analogue arguments apply to a_5 and a_6 and we conclude $ecc(a_6) > ecc(a_5) \geq ecc(a_4) > ecc(a_3) \geq ecc(a_2) > ecc(a_1) > 0$. These inequalities imply that $ecc(a_6) \geq ecc(a_1) + 3$.

Observation 1. *It follows directly from the proof of Theorem 2 that $a_5 \in I^k[Ct(G)]$ implies $a_1 \in I^{k+2}[Ct(G)]$, because $a_2, a_4 \in Ct(G)$, a_3 is on a shortest path between a_4 and a_5, and a_1 is on a shortest path between a_2 and a_3.*

Let now $\xi(G) := \min_{v \in Ct(G)} ecc(v)$ be the minimum eccentricity of a contour node in G. Then the following relationship holds.

Theorem 3. *For any graph G, the geodetic iteration number of the contour $Ct(G)$ is upper bounded by $diam(G) - \xi(G) + 1$.*

Proof. In the proof of Theorem 2, we observed that $ecc(a_5) > ecc(a_2)$. With $a_2 \in Ct(G)$, we have $ecc(a_2) \geq \xi(G)$. Based on Theorem 1, we can now conclude that $a_5 \in I^k[Ct(G)]$ with $diam(G) - k = ecc(a_2) + 1$. Combined with Observation 1, we have $a_1 \in I^{k+2}[Ct(G)]$ with $k + 2 = diam(G) - ecc(a_2) + 1 \leq diam(G) - \xi(G) + 1$.

Again, the graph in Fig. 2 is a tight example, as there we have $diam(G) - \xi(G) + 1 = 5 - 4 + 1 = 2 = gin(Ct(G))$. But Theorem 3 provides a stronger upper bound than Theorem 1 whenever the minimum eccentricity of the contour nodes is larger than the minimum eccentricity of all nodes plus one.

Theorem 4. *For any graph G, the geodetic iteration number of the contour $Ct(G)$ is upper bounded by $|Ct(G)|$.*

Proof. If there are only two contour nodes, the lemma is trivially true. Now let $c_1, \ldots, c_s$ with $s = |Ct(G)| \geq 3$ be the contour nodes, sorted increasingly by eccentricity. Hence we have $ecc(c_{s-1}) = ecc(c_s) = diam(G)$. We define the node sets A_i for $i = 1, \ldots, s - 1$ as $A_i := \{v \in V | ecc(c_i) \leq ecc(v) < ecc(c_{i+1})\}$ as well as $A_0 := \{v \in V | ecc(v) < ecc(c_1)\}$ and $A_s := \{v \in V | ecc(v) \geq ecc(c_s)\}$. We observe that all nodes in A_i for $i \geq s - 3$ are contained in $I^2[Ct(G)]$ as for $a_1 \in A_i$ there do not exist three contour nodes with higher and pairwise different eccentricity that can take the roles of a_2, a_4 and a_6 described in Lemma 1. Now consider $a_1 \in A_i$ for any $i < s - 3$. Then the smallest possible indices of the contour nodes a_2, a_4, a_6 in the sorted order are $i + 1, i + 2, i + 3$. Based on $ecc(a_5) \geq ecc(a_4)$, it follows that a_5 has to be contained in set A_j for some $j \geq i + 2$. Together with Observation 1, that tells us that a_1 is added to the iterated contour set at most two rounds after a_5, we conclude that the nodes in any set $A_{i \geq 0}$ are added after at most s iterations and hence $gin(Ct(G)) \leq |Ct(G)|$.

4 Bounding the Hull Number

In this section, we investigate upper and lower bounds for the hull number of a graph with a focus on bounds that can be efficiently computed in practice.

Upper Bounds. Simple theoretical upper bounds for the hull number as $hn(G) \leq n - diam(G) + 1$ were discussed in [7]. The size of the graph contour also constitutes a valid upper bound (with unclear a priori $gin(Gt(G))$ value), and so does the size of the boundary (with $gin(\delta(G)) = 1$). But the computation of the contour as well as the boundary requires the knowledge of all pairwise distances between graph nodes and therefore takes time $\Theta(n^2 + nm)$ which is not practical for large input networks. However, we can also compute an upper bound with as single BFS run in $\mathcal{O}(n + m)$ as detailed out in the following observation.

Observation 2. *For a given graph $G(V, E)$ and any node $v \in V$ the set $S = v \cup L(v)$, where $L(v)$ denotes the set of leaves in a BFS-tree rooted at v, is a hull set with $gin(S) = 1$.*

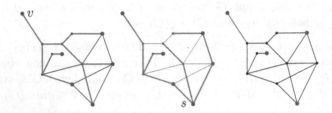

Fig. 4. Example of BFS based hull set computation. Left image: The leaf nodes in BFS tree from v (thick green edges) form together with v a hull set S (red nodes) with $gin(S) = 1$. Middle image: In the BFS tree from s there are two paths which contain three current hull nodes. Hence the middle nodes (purple) can be pruned. Right image: Valid reduced hull set (red nodes) S with $gin(S) \leq 2$. (Color figure online)

This simple observation also reveals a connection between $hn(G)$ and the maximum leaf number $ml(G)$, the largest number of leaves in any spanning tree of G, which is used in FPT algorithm design of e.g. coloring problems [9].

Corollary 2. $hn(G) \leq ml(G) + 1$

The bound is tight for complete graphs K_n with $n \geq 2$ where $hn(G) = n$ and $ml(G) = n - 1$.

The BFS bound from Observation 2 may be further improved in practice by iterating the following process: Select any node $s \in S$ and compute the respective BFS tree. If in an interval $I(s, s')$ with $s' \in S$ there are other nodes from S, those can be pruned from S. Figure 4 shows a successful example of this pruning strategy (with paths instead of intervals for clarity). We note, that with every iteration of BFS based pruning the geodetic iteration number might increase by at most 1. Hence, using B BFS runs in total, we get a running time of $\mathcal{O}(B(n + m))$ and end up with a valid hull set S with $gin(S) \leq B$.

Lower Bounds. Clearly, for any graph with more than one node, $hn(G) \geq 2$ has to hold, as there need to be at least two nodes in a hull set S to ensure $|I[S]| > 1$. In [1], it was proven that the hull number of a graph can be computed based on considering its two-connected components splitted at their cut nodes. To improve on that, we will now consider general node subsets $W \subset V$ together with their set $c(W)$ of cut nodes (the subset of W with a neighbor outside of W) and show that under certain conditions, W has to contain a node from the hull set.

Lemma 2. *If* $W \setminus h(c(W)) \neq \emptyset$ *for a node subset* W *of graph* G, *then every hull set of* G *needs to contain at least one node from* W.

Proof. Assume for contradiction that S is a hull set, $S \cap W = \emptyset$ and $W \setminus h(c(W)) \neq \emptyset$. Based on S being a hull set, we know that $h(S) \supset W$. Obviously, shortest paths between nodes $a, b \in V \setminus W$ that intersect W have to enter and exit W via nodes in $c(W)$, and their intersections with W have to be shortest paths as well. As $S \cap W = \emptyset$, it follows that $h(S) \cap W \subseteq h(c(W))$. But as we know that $W \setminus h(c(W)) \neq \emptyset$ this poses a contradiction to $h(S) \supset W$. Accordingly, there has to exist a hull set node in $S \cap W$.

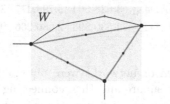

Fig. 5. Node subset W (indicated by the turquoise box) and its cut nodes $c(W)$ (large black dots at the border of the box). The shortest paths between the cut nodes all have length 2 and are drawn in green. The nodes on these paths belong to $h(c(W))$. However, the red nodes are not on any shortest path between nodes in $h(c(W))$. Hence, a valid hull set has to contain at least one of them. (Color figure online)

Figure 5 provides a small example instance for which the condition specified in Lemma 2 is met. Based on this lemma, a lower bound for $hn(G)$ can be obtained by first selecting a set of pairwise intersection free induced subgraphs $W_1, \ldots, W_l$ and then counting the W_i for which the lemma applies.

5   Experimental Results

Algorithms were implemented in C++. Experiments were conducted on a single core of an Intel(R) i5-8250U CPU clocked at @1.60GHz with 32 GB of RAM. We use two different graph types in the experiments: real-world road networks and simulated sensor networks.

Road Networks. The road networks are connected subgraphs of the OSM Germany graph[1], which we consider as undirected and unweighted. We note, that these graphs contain many nodes of degree-1 (dead-ends). This is very beneficial for our hull set computation algorithms, as all of them need to be contained

---

[1] https://i11www.iti.kit.edu/resources/roadgraphs.php

in any feasible hull set. Therefore, we also consider the corresponding network instances in which we recursively delete all nodes of degree-1 until the minimum degree in the graph is 2. Those should pose more difficult instances. Table 1 provides an overview of the characteristics of the used graphs.

Table 1. Road network benchmark data. The last column ($\delta 1$) denotes the percentage of degree-1 nodes.

Name	n	m	$\delta 1$
ROAD1$_{\geq 1}$	99,127	105,517	7%
ROAD1$_{\geq 2}$	73,185	79,575	0%
ROAD2$_{\geq 1}$	290,659	300,967	8%
ROAD2$_{\geq 2}$	180,875	191,183	0%
ROAD3$_{\geq 1}$	990,732	1,057,821	6%
ROAD3$_{\geq 2}$	783,079	850,168	0%
ROAD4$_{\geq 1}$	3,638,604	3,794,477	8%
ROAD4$_{\geq 2}$	2,530,393	2,686,266	0%

Sensor Networks. The sensor networks were obtained by choosing n random node positions in the unit square and then connecting node pairs with an edge if their Euclidean distance is at most $r = c/\sqrt{n}$ for some constant c. We used $c = 2$ and $c = 3$ to end up with sensor networks similar to those used in other simulations. Figure 6 shows two examples of the resulting networks. We use the following nomenclature for the generated graphs: SN[n]c[c]. Hence the two graphs in Fig. 6 would be referred to as SN500c2 and SN500c3, respectively. Presented results for sensor networks are always averaged over 10 random networks of the specified size with the same c value.

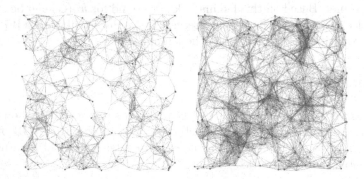

Fig. 6. Sensor networks with 500 nodes together with heuristic hull sets (red). In the sparser graph on the left ($c = 2$, average degree 6), a hull of size 50 is depicted. In the denser graph on the right ($c = 3$, average degree 13), a hull set of size 37 is shown. (Color figure online)

5.1 Graph Hull Sets and Gin

We discussed two methods to compute a valid hull set for a given graph G: computing the graph contour $Ct(G)$ and a BFS-based heuristic.

Contour Computation. We first evaluate the size of the contour and other relevant parameters on our benchmark instances. As their computation times are quadratic in n, we only consider the two smaller road network instances here, as well as sensor networks with up to 16,000 nodes (which already is a larger number of nodes than considered in most sensor network simulations). The results are summarized in Table 2. Interestingly, for all our tested instances the value $gin(Ct(G))$ turned out to be equal to 2. There are significant differences in the quality of our upper bounds, though. For road networks, the diameter is rather large and also significantly larger than the radius or $\xi(G)$. Hence our bounds turn out to be loose on such networks. But $\xi(G)$ is also larger than the radius by more than 1, showing that Theorem 3 indeed might provide stronger bounds than Theorem 1. For the sensor networks, our bounds are better due to the significantly smaller diameters. But we also see that $\xi(G)$ is always very close to the radius, hence Theorems 3 and 1 produce similar bounds.

Table 2. Experimental results for parameter and contour computation for selected road and sensor networks.

Name	$diam(G)$	$rad(G)$	$\xi(G)$	$|Ct(G)|$	Time
ROAD1$_{\geq 1}$	1406	727	735	12614	30 min
ROAD1$_{\geq 2}$	1322	664	679	7567	16 min
ROAD2$_{\geq 1}$	2956	1487	1508	33991	5 h
ROAD2$_{\geq 2}$	2898	1449	1486	11682	2 h
SN1000c2	28	15	17	183	0.1 s
SN1000c3	17	9	10	133	0.2 s
SN4000c2	55	28	29	630	2.9 s
SN4000c3	34	18	19	369	4.3 s
SN16000c2	112	56	57	2294	99.9 s
SN16000c3	67	34	35	1382	136.1 s

BFS-Based Heuristic. Next, we turn to our BFS-based heuristic, where the running time is linear as long as the number of iterations B is kept constant. We use $B = 10$ in the evaluation. The respective results for road networks are summarized in Table 3. For sensor networks of varying size, the lower bound and hull sizes are depicted in Fig. 7, once for $c = 2$ and once for $c = 3$.

Table 3. LResults for heuristic hull set computation on the road network instances with $B = 10$. Timings are in seconds.

Name	LB	HS	apx	Time
ROAD1$_{\geq 1}$	7103	10381	1.5	0.72
ROAD1$_{\geq 2}$	998	4239	4.2	0.40
ROAD2$_{\geq 1}$	23218	30363	1.3	4.98
ROAD2$_{\geq 2}$	1831	8306	4.5	1.86
ROAD3$_{\geq 1}$	59514	98114	1.6	13.69
ROAD3$_{\geq 2}$	8986	45910	4.9	8.54
ROAD4$_{\geq 1}$	288324	409867	1.4	44.64
ROAD4$_{\geq 2}$	23477	121278	5.2	18.72

For the road network instances, we observe that (as expected) the lower and upper bound are much closer for the instances which include dead-ends. But in these graphs also the hull set sizes are significantly larger. In the pruned graphs, the quality guarantee is about a factor of 5. The computation time for the lower bound was always comparable to that of the heuristic, and on average over 60% of the investigated subgraphs (chosen as Voronoi cells induced by the heuristic solution) that contributed to the lower bound value had a cut size of 3 or more. That means that without Lemma 2, the lower bounds would have been significantly weaker. For sensor networks, we see that the hull set sizes for $c = 2$ are larger than for $c = 3$ but at the same time the lower bounds are much better. This makes sense as with $c = 3$ induced subgraphs typically have many

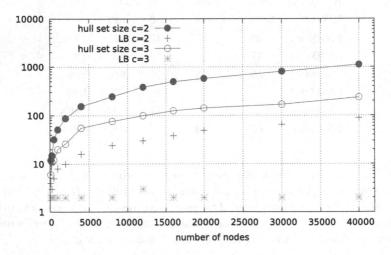

Fig. 7. Results for heuristic hull set computation for sensor networks with $B = 10$ in dependency of the number of nodes. Note the logscale on the y-axis.

cut nodes, and given the high ambiguity of shortest paths in these networks, the certificate from Lemma 2 is rarely issued. For $c = 2$, the quality guarantee ranges from a factor of 3 to 12, showing that our computed hull sets are sensible. Note that here 100% of the induced subgraphs that contributed to the lower bound had a cut size of 3 or more. Hence without Lemma 2, we could not have gotten a value larger than the trivial lower bound of 2.

Remarkably, across all instances - road networks and sensor networks alike - the size of the BFS-based hull set is always significantly smaller than the contour size. This is true even when we use $B = 2$ and thus have matching *gin* values. There, we get reductions around 10%–50%. Accordingly, our heuristic is not only significantly faster than the contour computation, but also yields better results and further allows us to trade running time (and gin value) for solution size.

6   Conclusions and Future Work

We demonstrated in this paper that hull sets of good quality (with bounded gin) can be computed efficiently even in large networks. It still would be interesting to design or rule out approximation algorithms for the hull number. Furthermore, our new upper bounds for the gin of the graph contour might help to search for instances with $gin(Ct(G)) > 3$, which are currently unknown (to exist). The established bounds imply, for example, that a graph with $gin(Ct(G)) = 4$ has to have a contour of size at least 4 and a diameter of at least 8.

Acknowledgement. Funded by the Deutsche Forschungsgemeinschaft (DFG, German Research Foundation) – Project-ID 50974019 – TRR 161.

References

1. Araujo, J., Campos, V., Giroire, F., Nisse, N., Sampaio, L., Soares, R.: On the hull number of some graph classes. Theoret. Comput. Sci. **475**, 1–12 (2013)
2. Bessy, S., Dourado, M.C., Penso, L.D., Rautenbach, D.: The geodetic hull number is hard for chordal graphs. SIAM J. Discret. Math. **32**(1), 543–547 (2018)
3. Brešar, B., Kovše, M., Tepeh, A.: Geodetic sets in graphs. In: Dehmer, M. (ed.) Structural Analysis of Complex Networks, pp. 197–218. Springer, Boston (2011). https://doi.org/10.1007/978-0-8176-4789-6_8
4. Cáceres, J., Puertas, M.L., Hernando, C., Mora, M., Pelayo, I.M., Seara, C.: Searching for geodetic boundary vertex sets. Electron. Notes Discret. Math. **19**, 25–31 (2005)
5. Das, S., DebBarma, M.K.: A review on coverage-hole boundary detection algorithms in wireless sensor networks. Comput. Sist. **24**(1), 121–140 (2020)
6. De Araújo, P.H.M., Campêlo, M., Corrêa, R.C., Labbé, M.: The geodesic classification problem on graphs. Electron. Notes Theor. Comput. Sci. **346**, 65–76 (2019)
7. Everett, M.G., Seidman, S.B.: The hull number of a graph. Discret. Math. **57**(3), 217–223 (1985)
8. Farber, M., Jamison, R.E.: Convexity in graphs and hypergraphs. SIAM J. Algebraic Discrete Methods **7**(3), 433–444 (1986)

9. Fellows, M., Lokshtanov, D., Misra, N., Mnich, M., Rosamond, F., Saurabh, S.: The complexity ecology of parameters: an illustration using bounded max leaf number. Theory Comput. Syst. **45**(4), 822–848 (2009)
10. Funke, S., Klein, C.: Hole detection or: "how much geometry hides in connectivity?". In: Proceedings of the Twenty-Second Annual Symposium on Computational Geometry, pp. 377–385 (2006)
11. Gallo, G.: An o (n log n) algorithm for the convex bipartite matching problem. Oper. Res. Lett. **3**(1), 31–34 (1984)
12. Green, P., Silverman, B.W.: Constructing the convex hull of a set of points in the plane. Comput. J. **22**(3), 262–266 (1979)
13. Harary, F., Nieminen, J.: Convexity in graphs. J. Differ. Geom. **16**(2), 185–190 (1981)
14. Lalem, F., et al.: LPCN: least polar-angle connected node algorithm to find a polygon hull in a connected euclidean graph. J. Netw. Comput. Appl. **93**, 38–50 (2017)
15. Linial, N., Lovasz, L., Wigderson, A.: Rubber bands, convex embeddings and graph connectivity. Combinatorica **8**(1), 91–102 (1988)
16. Marcilon, T., Sampaio, R.: The P3 infection time is W[1]-hard parameterized by the treewidth. Inf. Process. Lett. **132**, 55–61 (2018)
17. Mezzini, M.: On the geodetic iteration number of the contour of a graph. Discret. Appl. Math. **206**, 211–214 (2016)
18. Pelayo, I.M.: Geodesic Convexity in Graphs. Springer, New York (2013). https://doi.org/10.1007/978-1-4614-8699-2
19. Wang, Y., Gao, J., Mitchell, J.S.: Boundary recognition in sensor networks by topological methods. In: Proceedings of the 12th Annual International Conference on Mobile Computing and Networking, pp. 122–133 (2006)
20. Yan, D., Cheng, J., Ng, W., Liu, S.: Finding distance-preserving subgraphs in large road networks. In: 2013 IEEE 29th International Conference on Data Engineering (ICDE), pp. 625–636. IEEE (2013)

Voronoi Games Using Geodesics

Arun Kumar Das[1]([✉]), Sandip Das[1], Anil Maheshwari[2], and Sarvottamananda[3]

[1] Indian Statistical Institute, Kolkata, India
arund426@gmail.com, sandipdas@isical.ac.in
[2] Carleton University, Ottawa, Canada
anil@scs.carleton.ca
[3] Ramakrishna Mission Vivekananda Educational and Research Institute,
Howrah, India
sarvottamananda@rkmvu.ac.in

Abstract. In this paper, we study the single-round geodesic Voronoi games on various classes of polygons and polyhedra for two players. We prove some tight bounds on the payoffs, that is, the number of clients served by both the first player, Alice, and the second player, Bob, for orthogonal convex polygons and polyhedra for the $\mathbb{L}_1$ metric.

Keywords: Facility location · Computational geometry · Combinatorial optimization · Voronoi game · Convex polygon · Orthogonal polygon · Rectilinear polygon · Orthogonal convex polyhedron

1 Introduction

The *competitive facility location problem* is a fundamental geometric optimization problem. The *Voronoi games*, introduced by Ahn et al. [1], are a subclass of competitive facility location problems that deal with two or more players taking turns in placing their facilities while optimizing their payoffs. A *Voronoi game* $\mathcal{G}$ consists of a playing field $\mathcal{M}$ in which there is a client area $\mathcal{C} \subseteq \mathcal{M}$ that demands a service provided by two competitive players (named as Alice and Bob). In each round of the game, Alice places a set of facilities in $\mathcal{M}$ at a time, followed by Bob, to serve the clients. A client avails service from its nearest facility according to the distance metric considered in the model.

In this paper, we consider a *geodesic Voronoi game in the $\mathbb{L}_1$-space* for a polygon $\mathcal{P}$ and a fixed set of points $\mathcal{C}$ as clients in the plane. The polygonal region of $\mathcal{P}$ is supposedly owned by Alice. The clients in $\mathcal{C}$ are represented by points in the plane. Bob can only serve the clients in the exterior of the polygon $\mathcal{P}$ using exterior geodesic paths. Since Bob can only serve the clients exterior to the restricted region $\mathcal{P}$, and hence compete only for these exterior clients, we only consider the clients that are in the exterior of the polygon $\mathcal{P}$. Therefore, the metric for Bob is the external geodesic $\mathbb{L}_1$ distance. Here the interior and exterior of the polygons and polytopes are closed unless specifically mentioned.

The focus of this paper is on the *single-round* Voronoi game, where both the players place only one facility each. Alice places her facility $\mathcal{A}$ first in the

© Springer Nature Switzerland AG 2022
N. Balachandran and R. Inkulu (Eds.): CALDAM 2022, LNCS 13179, pp. 195–207, 2022.
https://doi.org/10.1007/978-3-030-95018-7_16

interior of an *orthogonal convex polygon* $\mathcal{P}$. A set $S \subseteq \mathbb{R}^d$ is *orthogonal convex* if its intersection with any orthogonal line, i.e., axis-parallel line, is convex [Unger [11]]. Then Bob places his facility in the exterior of $\mathcal{P}$. Alice and Bob serve the clients using unrestricted paths and external geodesic paths, respectively, that are shortest in the $\mathbb{L}_1$ metric. The *payoffs of Alice and Bob*, denoted by $\mathcal{S}_A$ and $\mathcal{S}_B$, respectively, also called their *scores*, are the total number of clients in their respective Voronoi regions. In case of a tie between the two nearest facilities, the client is equally served by both the players and counted as half in both the payoffs. The two optimization problems in the Voronoi game are to maximize the payoffs of Alice and Bob.

We use several interesting combinatorial and geometric techniques to prove the upper and lower bounds for the payoffs of Alice and Bob when the region possessed by Alice are *orthogonal convex polygons* (see Fig. 1) and *orthogonal convex polyhedra* in the $\mathbb{L}_1$-space.

1.1   Previous Results

Ahn et al. [1] described and solved the *Voronoi game in line segments and circles* in which Alice and Bob place an equal number of facilities, in one or more rounds, on line segments or circles. The payoff is the total length of their respective Voronoi cells. Cheong et al. [8] solved the *single-round Voronoi game in a square* in which Alice and Bob place multiple facilities at once, Alice before Bob, and the payoff is the total area of their respective Voronoi cells. Later, Fekete and Meijer [9] improved on the finer details of their solution and showed the intractability of the *Voronoi games in simple polygons with holes* and the payoff is the total polygonal area of their respective Voronoi cells.

Banik et al. [3] studied the *discrete version* of the Voronoi game on line segments with a finite number of point clients, and the payoff is the total number of clients in Alice and Bob's individual Voronoi cells. They proved bounds on the payoffs and designed optimal strategies for Alice and Bob. Banik et al. [2], and later, de Berg et al. [7], proposed algorithms for the discrete version of the *single-round Voronoi game in* $\mathbb{R}^1$ in which Alice and Bob place multiple facilities in a single round and the payoff is the number of clients. Banik et al. [4] studied the discrete version of the *single-round Voronoi game in* $\mathbb{R}^2$ in which Alice and Bob place new facilities among their previously owned facilities, and the payoff is the total number of clients in their respective Voronoi cells. They present polynomial-time algorithms for an optimal placement for Alice and Bob for each of the $\mathbb{L}_1$, $\mathbb{L}_2$ and $\mathbb{L}_\infty$ metrics. Later, Banik et al. [6] studied the discrete version of the *single-round Voronoi game in a simple polygon* in which the game arena is the internal geodesic space of a simple polygon, and the payoff is the total number of clients in their individual Voronoi cells. They devised polynomial-time algorithms for an optimal placement of both Alice and Bob for this game.

Recently, Banik et al. [5], introduced the *Voronoi game on a polygon* in which Alice and Bob play the Voronoi game on the boundary of simple or convex polygons and polyhedra, the metrics are internal and external geodesic $\mathbb{L}_2$ metric, and the payoff is the number of clients in their respective Voronoi regions. They proved tight upper and lower bounds for the payoffs of both for the single-round or k-round, simple or convex, polygons or polyhedra. They also devised algorithms for the optimal placements for Alice and Bob for the single round Voronoi game on a convex polygon.

1.2   New Results

In this paper, we study the same Voronoi game problems as in [5], termed the geodesic Voronoi games, but for the $\mathbb{L}_1$ metric in place of the Euclidean metric $\mathbb{L}_2$. The convexity is similarly replaced by orthogonal convexity. We take up questions regarding the Voronoi game, such as if Alice and Bob are guaranteed some payoff, and if not, whether there exist some preconditions that may guarantee a payoff, what are these preconditions, how much is the guaranteed payoff, etc. We assume both the players play with their optimal strategies in this model. We show that if the clients are only on the boundary of the given convex orthogonal polygon, then Alice is always guaranteed to get at least half of the clients in every case, whereas Bob can only ensure one-sixth of the total number of clients in the worst case (Sect. 3). We also prove that this bound for Alice is $\lceil \frac{n}{4} \rceil$ and for Bob is $\lceil \frac{n}{24} \rceil$, when we extend our model in 3-dimensional space (Sect. 4). We also show examples to ensure the tightness of the bounds. In addition, we outline several interesting properties of the Voronoi regions of Alice and Bob in this metric space (Sect. 2).

2   Preliminaries

Bob's maximization of his payoff in any single-round Voronoi game depends on where Alice places her facility first. Hence Bob's strategy is easier to understand. Bob's naive strategy will be to search among all possible feasible locations, say $\mathcal{F}$, and choose that location that maximizes his payoff. Alice's maximization of her payoff is comparatively more complex. Since Alice cannot predetermine Bob's choice afterwards, she has to preempt Bob's choice and needs to choose a location preparing for the worst; thus, she puts her facility among all possible locations in $\mathcal{F}$ where her minimum payoff is maximized. It is easy to see that the game, as described above is a constant-sum game, and hence the sum of the payoffs of Alice and Bob is the total number of clients.

Observation 1. Let $\mathcal{S}_{\mathcal{A}}^*$ and $\mathcal{S}_{\mathcal{B}}^*$ denote the final scores of Alice and Bob respectively in a single-round geodesic Voronoi game when they play optimally. Then, $\mathcal{S}_{\mathcal{A}}^* + \mathcal{S}_{\mathcal{B}}^* = n$.

2.1   Orthogonal Convex Polygons for the $\mathbb{L}_1$ Metric

Let $\mathcal{A}$ and $\mathcal{B}$ denote the optimal placement of the facilities by Alice and Bob, respectively, in the single-round geodesic Voronoi game $\mathcal{G}$. Let $\mathcal{P}$ denote the orthogonal convex polygon in the game. Let $d_{\mathcal{A}}$ and $d_{\mathcal{B}}$ be the distance metrics for Alice and Bob defined as before. First, we mention a property of geodesic paths on orthogonal polygons with the $\mathbb{L}_1$ metric, see Fig. 1.

Observation 2. The closed boundary of any orthogonal convex polygon can be non-uniquely partitioned into at least two and at most four monotonic $\mathbb{L}_1$-paths (staircases). The anti-clockwise boundary from the maximum x-coordinate to the maximum y-coordinate, on the top right of $\mathcal{P}$, is called the (++)-*quadrant* $\partial \mathcal{P}$ boundary. Likewise, we define the (-+)-*quadrant* $\partial \mathcal{P}$, (+-)-*quadrant* $\partial \mathcal{P}$ and (--)-*quadrant* $\partial \mathcal{P}$ boundaries. This partitioning can be proved by induction.

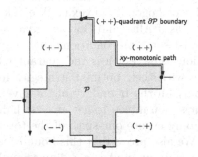

Fig. 1. An orthogonal convex polygon $\mathcal{P}$ with the boundary divided into four xy-monotonic parts.

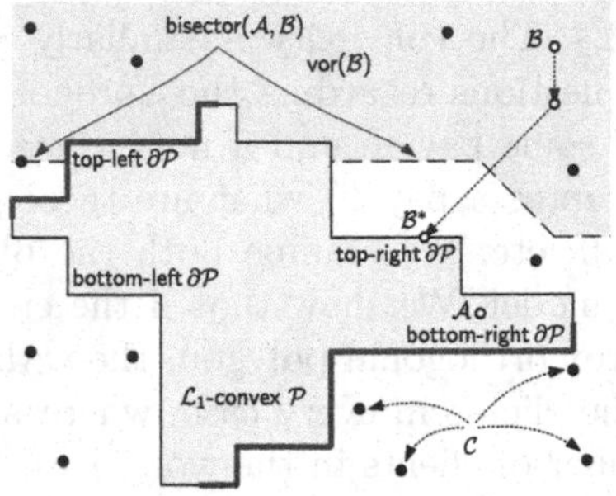

Fig. 2. Bob can move along the path BB^* to the boundary $\partial \mathcal{P}$ to increase his payoff.

Next, for geodesic Voronoi games, we make an important observation about boundary and non-boundary geodesic paths similar to [10].

Lemma 1. *Let s and t be any pair of points on the boundary $\partial \mathcal{P}$ of any orthogonal convex polygon $\mathcal{P}$. There exists an external geodesic $\mathbb{L}_1$-path between s and t that is a part of the boundary $\partial \mathcal{P}$.*

Proof. The external geodesic path can be deformed smoothly to a boundary path of same length due to the orthogonal convexity of $\mathcal{P}$ and Observation 2.

Corollary 1. *If an external geodesic path in the $\mathbb{L}_1$ metric, for an orthogonal convex polygon, does not touch the boundary $\partial \mathcal{P}$ of the polygon then the path length is the $\mathbb{L}_1$ distance between the end-points.*

Thus there exists external geodesic paths that are either completely free from the boundary $\partial \mathcal{P}$ of the orthogonal convex polygon $\mathcal{P}$ or contains only one connected part of the boundary $\partial \mathcal{P}$. Next, we characterize the optimal locations of Alice and Bob that maximize their payoffs. We note that whenever Alice places her facility in the region she owns, Bob has to place the facility as near to the region as possible. By moving nearer, Bob can gain more clients.

Lemma 2. *If Alice's facility $\mathcal{A}$ is in the (closed) interior of the orthogonal convex polygon $\mathcal{P}$ in the game $\mathcal{G}$, then there exists a point B^* on the boundary $\partial \mathcal{P}$ of $\mathcal{P}$ that maximizes Bob's payoff.*

Proof. Let $\mathcal{B}$ be any point in the exterior of the polygon $\mathcal{P}$. Without loss of generality, we assume that $\mathcal{B}$ is on the top-right of $\mathcal{P}$. We consider the geodesic bisector of $\mathcal{A}$ and $\mathcal{B}$ using their respective distance metrics. Because of the staircase nature of the first quadrant boundary, we can show that $\mathcal{B}$ can be moved successively vertically below or horizontally left, and then bottom-left direction at an angle of $\pi/4$ towards $\mathcal{P}$ such that the Voronoi region of $\mathcal{B}$ enlarges or remains the same. For example, in Fig. 2 Bob first moves vertically down and then obliquely towards the bottom-left. □

There are certain conditions when Alice has to place her facility in the orthogonal convex polygon $\mathcal{P}$. This happens when the clients are on the boundary of $\mathcal{P}$. The *orthogonal convex hull* of a set of points S is the smallest orthogonal convex polygon that contains S. We state a more general condition below.

Lemma 3. *There exists a point $\mathcal{A}^*$, that maximizes Alice's payoff, in the (closed) interior of the orthogonal convex hull of C in the game $\mathcal{G}$.*

Proof. For any point $\mathcal{A}$ in the exterior of an orthogonal convex hull there exists a point $\mathcal{A}'$ on the hull such that the distance from any point in the hull to $\mathcal{A}$ is not less than to $\mathcal{A}'$ ($\mathcal{A}'$ is the point nearest to $\mathcal{A}$ in one of the eight cardinal directions). Thus $\mathcal{A}'$ increases the payoff of Alice compared to $\mathcal{A}$. An interior $\mathcal{A}^*$ improves on $\mathcal{A}'$. □

Since the orthogonal convex hull of a set of points on the boundary of the orthogonal convex polygon $\mathcal{P}$ is contained in $\mathcal{P}$, we have the following lemma.

Lemma 4. *If all the clients C are on the boundary of the orthogonal convex polygon $\mathcal{P}$ in the game $\mathcal{G}$, then there exists an optimal facility location of Alice, that maximizes her payoff, in the (closed) interior of $\mathcal{P}$.*

The playing field $\mathcal{M}$ is partitioned by Alice and Bob by the clients that they serve, respectively grouped in their Voronoi regions. We state some conditions when these regions are connected or disconnected in the following lemmas. We denote the Voronoi regions of Alice's facility $\mathcal{A}$ and Bob's facility $\mathcal{B}$ as $\text{vor}(\mathcal{A})$ and $\text{vor}(\mathcal{B})$, respectively. We call them *Alice's and Bob's Voronoi regions*, respectively. See Figs. 3 and 4.

Lemma 5. *Let $d_{\mathcal{A}}$ and $d_{\mathcal{B}}$ be any two metrics over $\mathcal{M}$. Let Alice and Bob follow $d_{\mathcal{A}}$ and $d_{\mathcal{B}}$ metrics respectively for their payoffs. If $d_{\mathcal{A}}(x,y) \leq d_{\mathcal{B}}(x,y)$, for every $x, y \in \mathcal{M}$, then $\text{vor}(\mathcal{B})$ is connected in the Voronoi diagram of the set $\{\mathcal{A}, \mathcal{B}\}$.*

Proof. Assume $\text{vor}(\mathcal{B})$ is not connected. $\mathcal{B}$ is in $\text{vor}(\mathcal{B})$ as $d_{\mathcal{B}}(\mathcal{B}, \mathcal{B}) = 0$ and $d_{\mathcal{A}}(\mathcal{A}, \mathcal{B}) \geq 0$. We consider a point $p \in \text{vor}(\mathcal{B})$ not in the connected region to which $\mathcal{B}$ belongs. Since $p \in \text{vor}(\mathcal{B})$, $d_{\mathcal{B}}(\mathcal{B}, p) \leq d_{\mathcal{A}}(\mathcal{A}, p)$. Next, we consider a geodesic path π from $\mathcal{B}$ to p which connects the two disconnected regions of $\text{vor}(\mathcal{B})$ (and which may not be unique). We can select a point q in the path π not in $\text{vor}(\mathcal{B})$ due to the disconnectedness of $\text{vor}(\mathcal{B})$. Then, $d_{\mathcal{A}}(\mathcal{A}, q) < d_{\mathcal{B}}(\mathcal{B}, q)$. From the premise, $d_{\mathcal{A}}(p,q) \leq d_{\mathcal{B}}(p,q)$. Using the geodesic path π, these inequalities and the triangle inequality, we get $d_{\mathcal{B}}(\mathcal{B}, p) = d_{\mathcal{B}}(\mathcal{B}, q) + d_{\mathcal{B}}(p, q) > d_{\mathcal{A}}(\mathcal{A}, q) + d_{\mathcal{A}}(p, q) \geq d_{\mathcal{A}}(\mathcal{A}, p)$, which leads to a contradiction. □

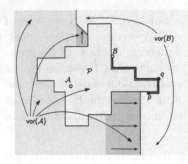

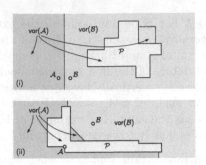

Fig. 3. Bob's Voronoi region is connected in geodesic Voronoi game $\mathcal{G}$ in $\mathbb{L}_1$. Bob's Voronoi region on the boundary $\partial \mathcal{P}$ is also connected.

Fig. 4. Alice Voronoi region is disconnected in (i) and Bob's Voronoi region on boundary of $\mathcal{P}$ is disconnected in (ii) in $\mathcal{G}$.

The premise of Lemma 5 holds true for any geodesic Voronoi game, therefore Bob's Voronoi region is connected in $\mathcal{G}$. Alice's Voronoi region is also connected conditionally. Bob's Voronoi region on the boundary $\partial \mathcal{P}$ is also connected when Bob places his facility on the boundary. We state these facts below.

Lemma 6. *Bob's Voronoi region is connected in the geodesic Voronoi game $\mathcal{G}$.*

Proof. $d_{\mathcal{A}}(x, y) \leq d_{\mathcal{B}}(x, y)$, $x, y \in \mathbb{R}^2$. □

Lemma 7. *If Alice's facility $\mathcal{A}$ is in the interior of $\mathcal{P}$ in the geodesic Voronoi game $\mathcal{G}$ then Alice's Voronoi region is connected and may be disconnected otherwise.*

Proof. The proof is similar to the proof of Lemma 5 above if we exchange $\mathcal{A}$ and $\mathcal{B}$. However, the substituted premise $d_{\mathcal{B}}(x, y) \leq d_{\mathcal{A}}(x, y)$ does not hold for every $x, y \in \mathcal{M}$. We give the proof sketch below. We assume that $\mathcal{A}$ is in the interior of $\mathcal{P}$ and vor($\mathcal{A}$) is disconnected. Let $p \in$ vor($\mathcal{A}$) be disconnected from $\mathcal{A}$. We consider the geodesic path from $\mathcal{A}$ to p and the point q to be the last exit point of that path from vor($\mathcal{B}$). $d_{\mathcal{A}}(p, q) = d_{\mathcal{B}}(p, q)$ follows from Corollary 1 as both are $\mathbb{L}_1$ distances. $d_{\mathcal{A}}(\mathcal{A}, p) = d_{\mathcal{A}}(\mathcal{A}, q) + d_{\mathcal{A}}(p, q) = d_{\mathcal{B}}(\mathcal{B}, q) + d_{\mathcal{B}}(p, q) \geq d_{\mathcal{B}}(\mathcal{B}, p)$, which contradicts $p \in$ vor($\mathcal{A}$). See Fig. 4 for an idea on how vor($\mathcal{A}$) might be disconnected when Alice's facility $\mathcal{A}$ isn't in the interior of $\mathcal{P}$. □

Lemma 8. *Let $\mathcal{P}$ be the orthogonal polygon in a geodesic Voronoi game $\mathcal{G}$ for the $\mathbb{L}_1$ metric. If Bob's facility $\mathcal{B}$ is on the boundary of $\mathcal{P}$, then the Voronoi region of Bob on the boundary, $\mathcal{P} \cap$ vor($\mathcal{B}$), is connected and maybe disconnected otherwise.*

Proof. If $\mathcal{B}$ is on the boundary, then for any point $p \in \mathcal{P} \cap$ vor($\mathcal{B}$), the geodesic path from $\mathcal{B}$ to p will be on the boundary and in $\mathcal{P} \cap$ vor($\mathcal{B}$). Therefore $\mathcal{P} \cap$ vor($\mathcal{B}$) will be connected. See Fig. 4 for an idea on how vor($\mathcal{B}$) on $\partial \mathcal{P}$ might be disconnected when Bob's facility $\mathcal{B}$ isn't on the boundary. □

We note that if Bob's Voronoi region on the boundary of the $\mathbb{L}_1$-convex polygon $\mathcal{P}$ is connected, Alice's Voronoi region is also connected.

2.2 Orthogonal Convex Polyhedra for the $\mathbb{L}_1$ Metric

First, we make some observations on the structure of the axis-parallel orthogonal convex polyhedra.

Observation 3. Similar to the case of plane, the surface $\partial \mathcal{P}$ of the axis parallel orthogonal convex polyhedra can be divided into eight parts (staircase like) in $\mathbb{R}^3$. We call them $(+++)$-*octant* $\partial \mathcal{P}$, $(-++)$-*octant* $\partial \mathcal{P}$, ..., $(---)$-*octant* $\partial \mathcal{P}$ surfaces.

We consider a geodesic Voronoi game $\mathcal{G}$ for an axis parallel orthogonal convex polyhedra $\mathcal{P}$ and clients $\mathcal{C}$. In $\mathbb{R}^{d>2}$, the condition that Alice's optimal facility location has to be inside the orthogonal convex hull of $\mathcal{C}$ does not hold. We generalize Lemma 3 as following for $\mathbb{R}^d$.

Observation 4. Any point $\mathcal{A}^*$ that maximizes Alice's payoff in a geodesic Voronoi game $\mathcal{G}$ for an orthogonal convex polytope in $\mathbb{R}^{d \geq 2}$ for $\mathbb{L}_1$, is in the (closed) interior of the smallest hypercube that contains $\mathcal{C}$.

We use the properties of the $\mathbb{L}_1$ metric to prove Observation 4. Contrary to our expectations, even if Alice is in the interior of $\mathcal{P}$, neither the optimal placement of Bob that maximizes his payoff might be on $\partial \mathcal{P}$ nor the Voronoi region of Alice on $\partial \mathcal{P}$ might be connected. See Fig. 5 where both Alice's as well as Bob's optimal facility location is in the exterior.

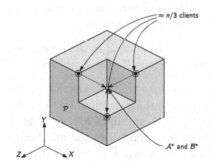

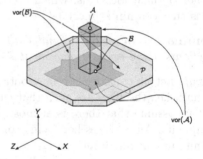

Fig. 5. Both Bob's and Alice's optimal facility location is at the corner extended on the exterior in the geodesic Voronoi game for $\mathbb{L}_1$ in $\mathbb{R}^3$.

Fig. 6. Alice's Voronoi region on the surface $\partial \mathcal{P}$ is disconnected in the geodesic Voronoi game for $\mathbb{L}_1$. Alice's facility $\mathcal{A}$ may be slightly below or above the surface to be in the strict interior or strict exterior, respectively.

Lemma 9. *There exists a geodesic Voronoi game in $\mathbb{R}^{d \geq 3}$ for an orthogonal convex polytope $\mathcal{P}$, such that there exists no optimal placement for Alice or Bob on the surface $\partial \mathcal{P}$, even if the clients $\mathcal{C}$ are on the surface $\partial \mathcal{P}$.*

Next, we consider the connectivity of the Voronoi regions of Alice and Bob. We have already shown that Bob's Voronoi region, if Bob is on the surface ∂P, is connected on the surface. However, since Bob's optimal placement may be on the exterior of the polytope, his Voronoi region on the surface might be disconnected. Moreover, even if Bob is on the surface and Alice anywhere, Alice's Voronoi region on the surface may be disconnected (see Fig. 6).

Lemma 10. *Alice's Voronoi region may be disconnected on the surface ∂P of the orthogonal convex polyhedron P in a geodesic Voronoi game in $\mathbb{R}^{d \geq 3}$, whether A is in the interior, exterior or on the boundary of P. This also holds for the Alice's optimal facility location A^*.*

3 Bounds for Orthogonal Convex Polygons

Let Alice and Bob play a single-round Voronoi game in the $\mathbb{L}_1$ metric with a set of n clients in the plane. First, we prove some bounds for the unrestricted case in which Alice, like Bob, does not own any region. Subsequently, we prove the bounds for the geodesic Voronoi games for orthogonal convex polygons with the clients on the boundary. We note that the unrestricted case is sometimes referred as the *discrete Voronoi game* in the literature.

Observe that in the unrestricted single-round Voronoi game in the $\mathbb{L}_1$ metric, Bob is guaranteed to serve at least $\frac{n}{2}$ clients by placing his facility precisely at the exact location as that of Alice's. On the other hand, as shown in the lemma below, Alice is also guaranteed to serve $\frac{n}{2}$ clients, if she places her facility at one of her optimal locations, which are the intersections of vertical and horizontal lines that contain at most $\lfloor \frac{n}{2} \rfloor$ clients in each of their open halves.

Lemma 11. *In a single round Voronoi game in the $\mathbb{L}_1$ metric in the plane with n clients, there exists a placement A^* that ensures a payoff of $\frac{n}{2}$ for Alice.*

Proof. Let us consider the horizontal and vertical lines that divide the clients in half, i.e., the lines are such that each open half contains at most $\lfloor \frac{n}{2} \rfloor$ clients. Let us assume that there is at most one point on both the lines for the sake of simplicity. Alice places her facility on the intersection point. Let the number of clients in four quadrants be x_{++}, x_{+-}, x_{-+} and x_{--}. Then, $x_{++} + x_{+-} = x_{+-} + x_{--} = x_{--} + x_{-+} = x_{-+} + x_{++} = \lfloor \frac{n}{2} \rfloor$. The Voronoi region of Alice, wherever Bob places his facility, will either contain two neighboring quadrants, contain one quadrant and share its two neighboring quadrants or share all four quadrants. In either case, it will ensure a payoff of $\frac{n}{2}$, including the share of the client at the facility location itself, if any. This follows from the equalities above. The case when there are clients on the lines can be analyzed similarly.   □

Thus, in every Voronoi game in the plane for the $\mathbb{L}_1$ metric without restrictions, the optimal payoff of Alice and Bob is always equal.

Theorem 1. *The optimal payoffs of Alice and Bob in any single-round Voronoi game in the plane for the $\mathbb{L}_1$ metric without restrictions for a set of n clients is $\frac{n}{2}$.*

3.1   Orthogonal Convex Polygon with Clients on Boundary

Let $\mathcal{G}$ be a geodesic Voronoi game such that the clients in $\mathcal{C}$ are located on the boundary $\partial \mathcal{P}$ of the orthogonal convex polygon $\mathcal{P}$ that Alice owns. The metric is $\mathbb{L}_1$ as before. Alice and Bob can place their facilities anywhere in plane. However, we have already shown in Lemma 4 that there always exists an optimal placement of Alice inside $\mathcal{P}$, so we only consider Alice's placements inside $\mathcal{P}$. On the other hand, Bob can get any client only if he places his facility outside $\mathcal{P}$, so we only consider Bob's placements outside $\mathcal{P}$. We note that, since Bob's distances to clients are not less than the distances in the equivalent unrestricted game without $\mathcal{P}$. So, following Lemma 11, Alice naturally has a guaranteed payoff of $\frac{n}{2}$. Surprisingly, however, we can show that there exist games such that Alice gets no more than $\frac{n}{2}$ clients. We prove this tight bound below.

Lemma 12. *Alice's optimal payoff is at least $\frac{n}{2}$ in any single-round geodesic Voronoi game for orthogonal convex polygon $\mathcal{P}$, clients $\mathcal{C}$ on $\partial \mathcal{P}$ and the $\mathbb{L}_1$ metric.*

Proof. The proof follows from Lemma 11 and the fact that d_A remains same whereas d_B increases in the argument there. $\square$

We next show that this bound is tight.

Lemma 13. *There exists a single-round geodesic Voronoi game for a orthogonal convex polygon $\mathcal{P}$ and n clients on $\partial \mathcal{P}$ in the $\mathbb{L}_1$ metric such that Alice's optimal payoff is $\frac{n}{2}$.*

Proof. We construct a game where Alice owns a square region. We place $\lfloor \frac{n}{2} \rfloor$ clients on one vertex, the other $\lfloor \frac{n}{2} \rfloor$ clients on the diagonally opposite vertex and a last remaining one client, if it exists, on one of the other two vertices. Wherever Alice places her facility, Bob can get at least $\frac{n}{2}$ clients by placing a facility on this last vertex (even if n is even and there is no remaining client). $\square$

Next, we show a non-trivial bound for Bob. See Figs. 7 and 8.

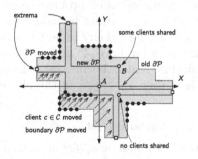

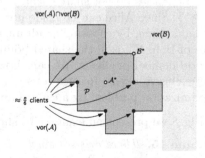

Fig. 7. Why Bob's payoff is at least $\frac{n}{6}$?   **Fig. 8.** Bob's payoff is at most $\lceil \frac{n}{6} \rceil$.

Lemma 14. *Bob's optimal payoff is at least $\lceil \frac{n}{6} \rceil$ in any single-round geodesic Voronoi game for any orthogonal convex polygon $\mathcal{P}$ and n clients on $\partial \mathcal{P}$ in the $\mathbb{L}_1$ metric.*

Proof. As a result of Lemma 4 we can assume that Alice's location $\mathcal{A}^*$ is in the interior of $\mathcal{P}$. We note that we can move the boundary of $\mathcal{P}$ together with the clients and $\mathcal{B}^*$ in each of the four quadrant boundaries mentioned previously towards $\mathcal{A}^*$ at an angle $\pi/4$ with axes, such that the distances to Bob from every client remain same, but to Alice, it might decrease. We do not move the four x and y extremes. We can show this using the properties of the $\mathbb{L}_1$ metric and the geodesics. We move till the polygon remains simple, i.e., the boundaries do not cross the other boundaries. If we show that Bob gets $\lceil \frac{n}{6} \rceil$ clients for this new polygon then it gets $\lceil \frac{n}{6} \rceil$ clients for the polygon $\mathcal{P}$. Let us rename the new polygon as $\mathcal{P}$ for this proof. See Fig. 7 for an illustration.

We observe that out of the four quadrant boundaries of $\mathcal{P}$, only two, and only the opposite ones, can intersect the horizontal and vertical lines passing through $\mathcal{A}^*$. This is because we have two axes and four intersections around $\mathcal{A}^*$ by the orthogonal convex polygon boundary (we argue using a combined reasoning of pigeon hole principle and partition function $p(4) = 5$). We consider Bob's placement in each quadrant boundary. If a quadrant boundary, ∂Q, intersects both the axes, we can show that Bob's best possible location for ∂Q gets at least half of the clients in ∂Q. Otherwise, if a quadrant boundary, ∂Q, does not intersect both the axes, Bob's best possible location for ∂Q can get all the clients in ∂Q, since ∂Q will be fully inside the Bob's Voronoi region, i.e., $\partial Q \in \text{vor}(\mathcal{B})$. Thus, Bob has four candidate locations to consider optimality, at most two candidates getting at least half, and at least two candidates getting all of their respective quadrant boundary clients. This is because of the pigeonhole principle, as each axis can intersect only two quadrant boundaries due to orthogonality.

The problem of maximizing Bob's payoff can be written as a min-max optimization problem: say, without loss of generality and in the worst case, $\mathcal{S}_{\mathcal{B}}^* \geq \min_{n_{++}, n_{+-}, n_{--}, n_{-+}} \max\{n_{++}, n_{+-}/2, n_{--}, n_{-+}/2\}$, where $n_{++} + n_{+-} + n_{--} + n_{-+} = n$ and each $n_{++}, n_{+-}, n_{--}, n_{-+} \geq 0$. The numbers $n_{++}, n_{+-}, n_{--}, n_{-+}$ are the number of clients in the respective quadrant boundaries (we assume for simplicity that the boundaries do not share clients). We solve this optimization problem to get $\mathcal{S}_{\mathcal{B}}^* \geq \frac{n}{6}$. Since Alice can be made not to touch any two quadrant boundaries at the same time, by choosing quadrant boundaries carefully, then the best candidate of Bob, $\mathcal{B}^*$, is in the shared boundary that does not have $\mathcal{A}^*$ and contains an odd number of clients. We can show, by counting, that Bob will have at least one exclusive client. In such a case we can show that $\mathcal{S}_{\mathcal{B}}^* \geq \lceil \frac{n}{6} \rceil$ in the above maximization problem by a careful analysis. $\square$

In Fig. 8 we present an example to show that this bound is tight.

Lemma 15. *There exists a single round geodesic Voronoi game for an orthogonal convex polygon $\mathcal{P}$ and n clients on $\partial \mathcal{P}$ in the $\mathbb{L}_1$ metric such that Bob's optimal payoff is $\lceil \frac{n}{6} \rceil$.*

We summarize the bounds in the theorem below.

Theorem 2. *Alice's optimal payoff is at least $\frac{n}{2}$ and Bob's optimal payoff is at least $\lceil \frac{n}{6} \rceil$ in any single-round geodesic Voronoi game $\mathcal{G}$ in the $\mathbb{L}_1$ metric for an orthogonal convex polygon $\mathcal{P}$ and n clients on the boundary $\partial \mathcal{P}$ of the polygon. Both the bounds are tight.*

As the matters stand, we can also show that similar tight bounds hold if the clients are unrestricted in the plane.

4   Bounds for Orthogonal Convex Polyhedra

This section extends the results on the bounds on orthogonal polygons to 3-space. We observe that the geodesic Voronoi games for orthogonal convex polyhedra differ significantly from the games for orthogonal polygons in the $\mathbb{L}_1$ metric. Let $\mathcal{G}$ be a geodesic Voronoi game and let $\mathcal{P}$ and $\mathcal{C}$ be the orthogonal convex polyhedra and the set of n clients, respectively.

In the unrestricted case, similar to the case of the plane, Alice is guaranteed a fraction of clients, and dissimilar to the case of the plane, Alice wins only a small fraction of clients. We compute the three orthogonal planes parallel to xy-plane, yz-plane and xz-plane, respectively, that contain at most $\lfloor \frac{n}{2} \rfloor$ clients in their open halves. Alice is guaranteed a payoff of $\lceil \frac{n}{4} \rceil$ if she puts her facility at the intersection point of these three planes.

Theorem 3. *Alice's optimal payoff is at least $\lceil \frac{n}{4} \rceil$ and Bob's optimal payoff is at least $\frac{n}{2}$ in any single-round Voronoi game $\mathcal{G}$ in the $\mathbb{L}_1$ metric for n clients. Both the bounds are tight.*

Proof. We assume, for the sake of simplicity that the orthogonal planes contain at the most one point. We can show that the eight quadrants have the number of clients satisfying the following relations for some m: $n_{+++} = n_{---} + m$, $n_{+--} = n_{-++} + m$, $n_{-+-} = n_{+-+} + m$, and $n_{--+} = n_{++-} + m$. The best placement for Bob will be infinitesimal near Alice's placement. Then $vor(\mathcal{A})$ either shares whole space, shares four octants and contains two, contains four octants on one side of one of the three orthogonal planes mentioned in the discussion, or contains four octants that are three neighbors of the remaining one. In either case, we can prove that Alice wins at least $\lceil \frac{n}{4} \rceil$ clients.

We can construct an example where the bounds are tight such that $m = \lfloor \frac{n}{4} \rfloor$. We distribute equal number of clients at uniform intervals in four rays in four quadrants $x = y = z > 0$, $x = -y = -z > 0$, $-x = y = -z > 0$ and $-x = -y = z > 0$, and put at least one client at the origin, if $n \bmod 4 \neq 0$.   □

4.1   Orthogonal Convex Polyhedra with Boundary Clients

Consider the case of the geodesic Voronoi game when the clients are located on the boundary of convex polyhedron $\mathcal{P}$. We again analyze the intersection point of orthogonal planes mentioned previously for Alice's optimal location. From those arguments, we have that Alice's payoff is at least $\frac{n}{4}$.

The lower bound is not tight. We construct a geodesic Voronoi game in which Alice's payoff is not more than $\lceil \frac{11n}{26} \rceil$, implying that the lower bound is not more than $\lceil \frac{11n}{26} \rceil$. We state this in the following lemma.

Lemma 16. *There exists a single round geodesic Voronoi game in the $\mathbb{L}_1$ metric for a convex polyhedron $\mathcal{P}$ with n clients on the boundary $\partial \mathcal{P}$ such that Alice's optimal payoff is $\lceil \frac{11n}{26} \rceil$.*

Proof. We prove this by distributing points (i.e., clients) on the center of faces of a cube and its corners, forming a tetrahedral. We place $5n/26$ clients on each of the four tetrahedral corners and $n/26$ clients on each of the six centers of faces. See, Fig. 9 for an illustration. □

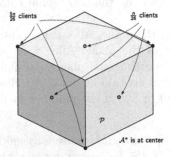

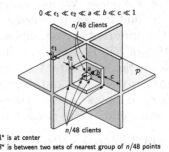

Fig. 9. Alice's payoff is at most $\lceil \frac{11n}{26} \rceil$ in the geodesic Voronoi game for $\mathbb{L}_1$.

Fig. 10. Bob's payoff is at most $\lceil \frac{n}{24} \rceil$ in the geodesic Voronoi game for $\mathbb{L}_1$.

Next we show that Bob's payoff is at least $\lceil \frac{n}{24} \rceil$ in the lemma below.

Lemma 17. *Bob's optimal payoff is at least $\lceil \frac{n}{24} \rceil$ in any single-round geodesic Voronoi game for an orthogonal convex polyhedron $\mathcal{P}$ and n clients on the boundary $\partial \mathcal{P}$ in the $\mathbb{L}_1$ metric.*

Proof. Let Alice place her facility at $\mathcal{A}$. We consider the eight octant surfaces $\partial \mathcal{P}$ of the orthogonal convex polyhedron $\mathcal{P}$ as mentioned in the Observation 3. We can prove that three points on each octant surface $\partial \mathcal{P}$ are sufficient so that $\partial \mathcal{P}$ is included in vor$(\mathcal{B})'s$ for any $\mathcal{A}$. Thus a total of 24 $\mathcal{B}$'s are sufficient to cover the whole of $\partial \mathcal{P}$ in vor$(\mathcal{B})$'s. Bob places his facility in that location for which vor$(\mathcal{B})$ contains the maximum number of clients. □

We also show that this lower bound of $\lceil \frac{n}{24} \rceil$ is tight by constructing a geodesic Voronoi game where Bob does not win more than $\lceil \frac{n}{24} \rceil$ clients.

Lemma 18. *There exists a single round geodesic Voronoi game in the $\mathbb{L}_1$ metric for a convex polyhedron $\mathcal{P}$ with clients on the boundary $\partial \mathcal{P}$ such that Bob's optimal payoff is $\lceil \frac{n}{24} \rceil$.*

Proof. We construct a cube with double hollowed corners along with a smaller cube in the center and distribute $n/48$ clients each at the six corners nearer to the center, as in Fig. 10. $\square$

We summarize our results on Voronoi games on orthogonal convex polyhedra with boundary clients in the theorem below.

Theorem 4. *Alice's optimal payoff is at least $\lceil \frac{n}{4} \rceil$ and Bob's optimal payoff is at least $\lceil \frac{n}{24} \rceil$ in any single round Voronoi game $\mathcal{G}$ in the $\mathbb{L}_1$ metric for an orthogonal convex polyhedron $\mathcal{P}$ and $n \geq 2$ clients on the boundary $\partial \mathcal{P}$ of $\mathcal{P}$. The lower bound of Bob's optimal payoff is tight. Moreover, the tight lower bound of Alice's optimal payoff is at most $\lceil \frac{11n}{26} \rceil$.*

References

1. Ahn, H.K., Cheng, S.W., Cheong, O., Golin, M., van Oostrum, R.: Competitive facility location: the Voronoi game. Theoret. Comput. Sci. **310**(1), 457–467 (2004)
2. Banik, A., Bhattacharya, B.B., Das, S.: Optimal strategies for the one-round discrete Voronoi game on a line. J. Comb. Optim. **26**(4), 655–669 (2012). https://doi.org/10.1007/s10878-011-9447-6
3. Banik, A., Bhattacharya, B.B., Das, S., Das, S.: The 1-dimensional discrete Voronoi game. Oper. Res. Lett. **47**(2), 115–121 (2019)
4. Banik, A., Bhattacharya, B.B., Das, S., Mukherjee, S.: The discrete Voronoi game in $\mathcal{R}^2$. Comput. Geom. **63**, 53–62 (2017)
5. Banik, A., Das, A.K., Das, S., Maheshwari, A., Sarvottamananda, S.: Optimal strategies in single round Voronoi game on convex polygons with constraints. Theoretical Computer Science (to appear in the special issue of COCOA 2020) (TBA)
6. Banik, A., Das, S., Maheshwari, A., Smid, M.: The discrete Voronoi game in a simple polygon. Theoret. Comput. Sci. **793**, 28–35 (2019)
7. de Berg, M., Kisfaludi-Bak, S., Mehr, M.: On one-round discrete Voronoi games. In: Lu, P., Zhang, G. (eds.) 30th International Symposium on Algorithms and Computation (ISAAC 2019). Leibniz International Proceedings in Informatics (LIPIcs), vol. 149, pp. 37:1–37:17. Schloss Dagstuhl-Leibniz-Zentrum fuer Informatik, Dagstuhl (2019)
8. Cheong, O., Har-Peled, S., Linial, N., Matousek, J.: The one-round Voronoi game. Discrete Comput. Geom. **31**(1), 125–138 (2004)
9. Fekete, S.P., Meijer, H.: The one-round Voronoi game replayed. Comput. Geom. **30**(2), 81–94 (2005)
10. Lee, D.T., Preparata, F.P.: Euclidean shortest paths in the presence of rectilinear barriers. Networks **14**(3), 393–410 (1984)
11. Unger, S.H.: Pattern detection and recognition. Proc. IRE **47**(10), 1737–1752 (1959)

Algorithms and Optimization

Approximation and Parameterized Algorithms for Balanced Connected Partition Problems

Phablo F. S. Moura[1] , Matheus Jun Ota[2](✉) , and Yoshiko Wakabayashi[3]

[1] Departamento de Ciência da Computação, Universidade Federal de Minas Gerais, Belo Horizonte, Brazil
phablo@dcc.ufmg.br
[2] Department of Combinatorics and Optimization, University of Waterloo, Waterloo, Canada
mjota@uwaterloo.ca
[3] Instituto de Matemática e Estatística, Universidade de São Paulo, São Paulo, Brazil
yw@ime.usp.br

Abstract. For a given integer $k \geq 2$, partitioning a connected graph into k vertex-disjoint connected subgraphs of similar (or fixed) orders is a classical problem that has been intensively investigated since late seventies. A connected k-partition of a graph is a partition of its vertex set into classes such that each one induces a connected subgraph. Given a connected graph $G = (V, E)$ and a weight function $w : V \to \mathbb{Q}_{\geq}$, the balanced connected k-partition problem looks for a connected k-partition of G into classes of roughly the same weight. To model this concept of balance, we seek connected k-partitions that either maximize the weight of a lightest class (MAX-MIN BCP_k) or minimize the weight of a heaviest class (MIN-MAX BCP_k). These problems, known to be NP-hard, are equivalent only when $k = 2$. We present a simple pseudo-polynomial $\frac{k}{2}$-approximation algorithm for MIN-MAX BCP_k that runs in time $\mathcal{O}(W|V||E|)$, where $W = \sum_{v \in V} w(v)$; then, using a scaling technique, we obtain a (polynomial) $(\frac{k}{2} + \varepsilon)$-approximation with running-time $\mathcal{O}(|V|^3|E|/\varepsilon)$, for any fixed $\varepsilon > 0$. Additionally, we propose a fixed-parameter tractable algorithm for the unweighted MAX-MIN BCP (where k is part of the input) parameterized by the size of a vertex cover.

Keywords: Balanced connected partition · Approximation algorithm · Parameterized algorithm

1 Introduction

The problem of partitioning a connected graph into a given number $k \geq 2$ of connected subgraphs with prescribed orders was first studied by Lovász [16] and

Research partially supported by grant #2015/11937-9, São Paulo Research Foundation (FAPESP). Moura is supported by FAPEMIG (Proc. APQ-01040-21) and Pró-Reitoria de Pesquisa da Universidade Federal de Minas Gerais. Wakabayashi is supported by CNPq (Proc. 306464/2016-0 and 423833/2018-9).

© Springer Nature Switzerland AG 2022
N. Balachandran and R. Inkulu (Eds.): CALDAM 2022, LNCS 13179, pp. 211–223, 2022.
https://doi.org/10.1007/978-3-030-95018-7_17

Győri [15] in the late seventies. Let $[k]$ denote the set $\{1, 2, \ldots, k\}$, for every integer $k \geq 1$. A *connected k-partition* of a connected graph $G = (V, E)$ is a partition of V into nonempty classes $\{V_i\}_{i=1}^k$ such that, for each $i \in [k]$, the subgraph $G[V_i]$ is connected, where $G[V_i]$ denotes the subgraph of G induced by the set of vertices V_i.

We denote by (G, w) a pair consisting of a connected graph $G = (V, E)$ and a function $w \colon V \to \mathbb{Q}_\geq$ that assigns non-negative weights to the vertices of G. For each $V' \subseteq V$, we define $w(V') = \sum_{v \in V'} w(v)$. Furthermore, if $G' = (V', E')$ is a subgraph of G, we write $w(G')$ instead of $w(V')$. If $\mathcal{P} = \{V_i\}_{i \in [k]}$ is a connected k-partition of G, then $w^+(\mathcal{P})$ stands for $\max_{i \in [k]} \{w(V_i)\}$, and $w^-(\mathcal{P})$ stands for $\min_{i \in [k]} \{w(V_i)\}$.

The concept of balance of the classes of a connected partition can be expressed in different ways. In this work, we consider two related variants whose objective functions express this concept.

Problem. Min-Max Balanced Connected k-Partition (MIN-MAX BCP$_k$)
INSTANCE: a connected graph $G = (V, E)$, and a weight function $w \colon V \to \mathbb{Q}_\geq$.
FIND: a connected k-partition $\mathcal{P}$ of G.
GOAL: minimize $w^+(\mathcal{P})$.

Analogously, MAX-MIN BCP$_k$ has the same set of instances as MIN-MAX BCP$_k$, but it seeks a connected k-partition $\mathcal{P}$ that maximizes $w^-(\mathcal{P})$.

The problems MIN-MAX BCP$_2$ and MAX-MIN BCP$_2$ are equivalent, that is, an optimal solution for one of the versions is also an optimal solution for the other version (but they may have different optimal values). However, for $k > 2$, equivalence do not hold for MIN-MAX BCP$_k$ and MAX-MIN BCP$_k$.

For all problems mentioned here the input graph G is always simple and connected (and possibly with further properties). We also use the convention that n (resp. m) is the number of vertices (resp. edges) of the graph under consideration.

Throughout this paper we assume that $k \geq 2$. When k is in the name of the problem, we are considering that k is fixed. The problems in which k is part of the instance are denoted similarly but without specifying k in the name (e.g. MAX-MIN BCP). The unweighted (or cardinality) versions of the problems refer to the case in which all vertices have equal weight, which may be assumed to be 1. We denote the corresponding problems as 1-MIN-MAX BCP$_k$, 1-MAX-MIN BCP$_k$, 1-MIN-MAX BCP and 1-MAX-MIN BCP.

In this paper, we show approximation algorithms for MIN-MAX BCP$_k$, but mention approximation results for both MIN-MAX BCP$_k$ and MAX-MIN BCP$_k$. We observe that whenever we refer to an approximation algorithm, we mean that it runs in polynomial time on the size of the instance. If an approximation ratio α can be guaranteed for an algorithm, but it may run in pseudo-polynomial time, we refer to it as a *pseudo-polynomial α-approximation*. This is not a usual terminology, but it will be appropriate for our purposes.

Problems of finding balanced connected partitions can model a rich collection of applications in logistics, image processing, data base, operating systems, cluster analysis and robotics [4,17,18,22].

Dyer and Frieze [12] proved that 1-MAX-MIN BCP_k is NP-hard on bipartite graphs. Furthermore, 1-MAX-MIN BCP_k has been shown by Chlebíková [10] to be NP-hard to approximate within an absolute error guarantee of $n^{1-\varepsilon}$, for all $\varepsilon > 0$. For the weighted versions, Becker et al. [3] proved that MAX-MIN BCP_2 is NP-hard on grid graphs. Wu [21] showed that MAX-MIN BCP_k is NP-hard on interval graphs for every k. Chataigner et al. [7] proved that MAX-MIN BCP_k is strongly NP-hard, even on k-connected graphs. Hence, unless P = NP, the problem MAX-MIN BCP_k does not admit a fully polynomial-time approximation scheme (FPTAS). They also showed that, when k is part of the instance, the problem MIN-MAX BCP cannot be approximated within a ratio better than 6/5.

For MAX-MIN BCP_k (resp. MIN-MAX BCP_k), Perl and Schach [20] (resp. Becker, Schach, and Perl [5]) designed polynomial-time algorithm when the input graph is a tree. Also for trees, Frederickson [14] proposed linear-time algorithms for both MAX-MIN BCP_k and MIN-MAX BCP_k. Polynomial-time algorithms were also derived for MAX-MIN BCP_2 on graphs with at most two cut-vertices [1,10]. For MAX-MIN BCP_k on ladders, a polynomial-time algorithm was obtained by Becker et al. [2].

Chlebíková [10] designed a (4/3)-approximation algorithm for MAX-MIN BCP_2. In 2020, Chen et al. [8] observed that the algorithm obtained by Chlebíková has approximation ratio 5/4 for MIN-MAX BCP_2 (but requires another analysis). These authors also obtained approximation algorithms with ratio 3/2 and 5/3 for MIN-MAX BCP_3 and MAX-MIN BCP_3, respectively. In 2012, Wu [21] designed a FPTAS for MAX-MIN BCP_2 restricted to interval graphs. When k is part of the input, very recently Casel et al. [6] derived a very involved 3-approximation algorithm for both MAX-MIN BCP and MIN-MAX BCP, based on the crown decomposition of the graph. For recent exact algorithms based on mixed integer linear programs for these problems we refer the reader to Miyazawa et al. [19].

1.1   Our Contribution

We show an approximation algorithm for MIN-MAX BCP_k, $k \geq 3$, that was inspired by the $k/2$-approximation algorithm, designed by Chen et al. [9] for (the unweighted version) 1-MIN-MAX BCP_k. The algorithm we present here has basically the same approximation ratio: namely, $k/2 + \varepsilon$, for any arbitrarily small $\varepsilon > 0$. When the weights assigned to the vertices of the input graph are bounded by a polynomial on the order of the graph, it achieves the ratio $k/2$. The additional constant ε in the ratio $k/2$ comes from a scaling technique used to deal with weights that might be very large. These results are presented in Sect. 2. We note that a 3/2-approximation algorithm for MIN-MAX BCP_3 was obtained by Chen et al. [8], but its analysis and implementation are slightly more complicated than the algorithm we show here.

In Sect. 3, we prove that 1-MAX-MIN BCP is fixed-parameter tractable when the parameter is the size of a vertex cover of the input graph. The proposed algorithm is based on an integer linear program that has a doubly exponential dependency on the size of a vertex cover. To the best of our knowledge, no

FPT algorithm for balanced connected partition problems is described in the literature. We believe that the strategy used to model connected partitions may be useful to show that other problems involving connectivity constraints are fixed-parameter tractable when the parameter is the size of a vertex cover.

2   Approximation Algorithm for Min-Max BCP_k

Chen et al. [9] devised an algorithm for 1-MIN-MAX BCP_k with approximation ratio $k/2$. This algorithm iteratively applies two simple operations, namely PULL and MERGE, to reduce the size of the largest class. In what follows, we show how to generalize such operations for the *weighted* case to design a $(\frac{k}{2} + \varepsilon)$-approximation for MIN-MAX BCP_k, for any $\varepsilon > 0$. First we discuss the algorithm for the case $k = 3$, and then we show how to use the connected 3-partition produced by this algorithm to obtain a connected k-partition for any $k \geq 4$.

Throughout this section, (G, w) denotes an instance of MIN-MAX BCP_k, as defined previously. Moreover, we assume without loss of generality that w is an integer-valued function (otherwise, we may simply multiply all weights by the least common multiple of the denominators).

The following trivial fact is used to show the approximation ratio of the algorithms for MIN-MAX BCP_k proposed here. When convenient, we denote by $\mathrm{OPT}_k(I)$ the value of an optimal solution for an instance I of MIN-MAX BCP_k.

Fact 1. *Any optimal solution for an instance $I = (G, w)$ of* MIN-MAX BCP_k *has value at least $w(G)/k$, that is, $\mathrm{OPT}_k(I) \geq w(G)/k$.*

For $k \geq 3$, let $\mathcal{G}_k$ be the class of connected graphs G containing a cut-vertex v such that $G - v$ has at least $k - 1$ components. We denote by $c(H)$ the number of components of a graph H. The next lemma provides a lower bound for the value of an optimal solution of MIN-MAX BCP_k on instances (G, w) with $G \in \mathcal{G}_k$.

Lemma 1. *Let $I = (G, w)$ be an instance of* MIN-MAX BCP_k *in which $G \in \mathcal{G}_k$, and v is a cut-vertex of G such that $c(G - v) = \ell \geq k - 1$. Let $\mathcal{C} = \{C_i\}_{i \in [\ell]}$ be the set of the components of $G - v$. Suppose further that $w(C_i) \leq w(C_{i+1})$ for every $i \in [\ell - 1]$. Then every connected k-partition $\mathcal{P}$ of G satisfies $w^+(\mathcal{P}) \geq w(v) + \sum_{i \in [\ell - k + 1]} w(C_i)$. In particular, $\mathrm{OPT}_k(I) \geq w(v) + \sum_{i \in [\ell - k + 1]} w(C_i)$.*

Proof. Consider a connected k-partition $\mathcal{P}$ of G, and let V^* be the class in $\mathcal{P}$ that contains v. Let $q^* := |\{C \in \mathcal{C} : V(C) \subseteq V^*\}|$ and $q := |\{C \in \mathcal{C} : V(C) \not\subseteq V^*\}|$.

Hence, $q^* + q = \ell$ and $q \leq k - 1$. Therefore, $q^* = \ell - q \geq \ell - k + 1$. Since $w(C_1) \leq w(C_2) \leq \ldots \leq w(C_\ell)$, we conclude that $w(V^*) \geq w(v) + \sum_{i \in [\ell - k + 1]} w(C_i)$, and thus, $w^+(\mathcal{P}) \geq w(V^*)$. Clearly, it holds that $\mathrm{OPT}_k(I) \geq w(v) + \sum_{i \in [\ell - k + 1]} w(C_i)$.

$\square$

We now present an algorithm for MIN-MAX BCP_3 that generalizes the algorithm proposed by Chen et al. [9] for the unweighted version of this problem.

We adopt the same notation used by these authors to refer to the core operations of the algorithm. The strategy used in the algorithm is to start with an arbitrary connected 3-partition and improve it by applying successively (while it is possible) the operations MERGE and PULL, defined in what follows.

We say that a connected 3-partition $\{V_1, V_2, V_3\}$ of G is *ordered* if $w(V_1) \leq w(V_2) \leq w(V_3)$. The input for PULL and MERGE is an ordered connected 3-partition $\{V_1, V_2, V_3\}$. As these operations may be applied several times, a reordering of the classes is performed at the end, if necessary. In this context, we say that an ordered 3-partition $\mathcal{P} = \{V_1, V_2, V_3\}$ is *better* than an ordered 3-partition $\mathcal{Q} = \{X_1, X_2, X_3\}$ if $w(V_3) < w(X_3)$. Two classes V_i and V_j are adjacent if there is an edge in G joining these classes. For $X \subset V$, we denote by $N(X)$ the set of vertices in $G - X$ that are adjacent to a vertex of X.

- MERGE($\mathcal{P}$)
  - *Input*: an ordered connected 3-partition $\mathcal{P} = \{V_1, V_2, V_3\}$ of G.
  - *Preconditions*: (a) $w(V_3) > w(G)/2$; (b) $|V_3| \geq 2$; (c) V_1 and V_2 are adjacent.
  - *Output*: a connected 3-partition $\{V_1 \cup V_2, V_3', V_3''\}$, where $\{V_3', V_3''\}$ is an arbitrary connected 2-partition of $G[V_3]$. Reorder the classes if necessary, and return an ordered partition.

The 3-partition returned by MERGE is better than the input partition since $w(V_3') < w(V_3)$, $w(V_3'') < w(V_3)$ and $w(V_1) + w(V_2) < w(G)/2 < w(V_3)$.

Note that a depth-first search suffices to check the preconditions. Moreover, a connected 2-partition of $G[V_3]$ can be easily obtained from any spanning tree of this graph. Hence, MERGE can be executed in $\mathcal{O}(|V| + |E|)$.

- PULL($\mathcal{P}, U, i$)
  - *Input*: an ordered connected 3-partition $\mathcal{P} = \{V_1, V_2, V_3\}$ of G, a nonempty subset U of vertices, and $i \in \{1, 2\}$.
  - *Preconditions*: (a) $w(V_3) > w(G)/2$; (b) $U \subsetneq V_3$, $G[V_i \cup U]$ and $G[V_3 \setminus U]$ are connected; (c) $w(V_i \cup U) < w(V_3)$.
  - *Output*: a connected 3-partition $\{V_j, V_i \cup U, V_3 \setminus U\}$ where $j \in \{1, 2\} \setminus \{i\}$. Reorder the classes if necessary, and return an ordered partition.

Note that PULL($\mathcal{P}, U, i$) outputs a partition that is better than $\mathcal{P} = \{V_1, V_2, V_3\}$, since $w(V_3 \setminus U) < w(V_3)$, $w(V_j) < w(V_3)$ and $w(V_i \cup U) < w(V_3)$. Moreover, it is only executed when a set U satisfying the preconditions is given. Thus, this operation can be executed in $\mathcal{O}(|V|)$ time. One may show that the time complexity to find such a set $U \subsetneq V_3$ (if it exists) is $\mathcal{O}(|V||E|)$. Let us denote by PULLCHECK the algorithm that receives as input an ordered connected 3-partition $\mathcal{P} = \{V_1, V_2, V_3\}$ of (G, w), and $i \in \{1, 2\}$, then outputs either a set $U \subset V_3$ that satisfies the preconditions of PULL w.r.t. i, or the empty set $\emptyset$ (if no such U exists).

Algorithm 1. MIN-MAX-BCP3

    Input: An instance (G, w) of MIN-MAX BCP$_3$
    Output: A connected 3-partition of G
    Routines: MERGE, PULL and PULLCHECK.
1: **procedure** MIN-MAX-BCP3(G, w)
2:     Let $\mathcal{P} = \{V_1, V_2, V_3\}$ be an ordered connected 3-partition of G; $W = w(G)$
3:     **while** $w(V_3) > W/2$ **do**
4:         **if** V_1 and V_2 are adjacent **and** $|V_3| \geq 2$ **then**
5:             $\mathcal{P} \leftarrow$ MERGE$(\mathcal{P})$ # $\mathcal{P} = \{V_1, V_2, V_3\}$
6:         **else if** PULLCHECK$(\mathcal{P}, i)$ returns a nonempty set U for $i \in [2]$ **then**
7:             $\mathcal{P} \leftarrow$ PULL$(\mathcal{P}, U, i)$ # $\mathcal{P} = \{V_1, V_2, V_3\}$
8:         **else**
9:             **break**
10:     **return** $\mathcal{P}$

Lemma 2. *Algorithm 1 on input (G, w), where $G = (V, E)$ and w is an integer-valued function, finds a connected 3-partition of G in $\mathcal{O}(w(G)|V||E|)$ time.*

Proof (sketch). Each time a MERGE or a PULL operation is executed, the weight of the heaviest class decreases. Thus, at most $w(G)$ calls of such operations are performed by the algorithm. Note that both MERGE and PULL operations take $\mathcal{O}(|V| + |E|)$ time. The routine PULLCHECK has time complexity $\mathcal{O}(|V||E|)$. It follows that Algorithm 1 has time complexity $\mathcal{O}(w(G)|V||E|)$.   □

It is clear that when Algorithm 1 halts and returns a partition $\mathcal{P}$, one of the two cases occurs: (a) either the loop condition in line 3 failed, and in this case, $\mathcal{P}$ has value $w^+(\mathcal{P}) \leq w(G)/2$, or (b) neither MERGE nor PULL operations could be performed (and $w^+(\mathcal{P}) > w(G)/2$). In what follows, we prove that in case (b) the input graph has a particular "star-like" structure which allows us to conclude that the solution produced by the algorithm is optimal.

Lemma 3. *Let $\mathcal{P} = \{V_1, V_2, V_3\}$ be an ordered connected 3-partition produced by Algorithm 1, and let $G_i = G[V_i]$, for $i = 1, 2, 3$. If $|V_3| \geq 2$ and $w(V_3) > w(G)/2$, the following hold:*

(i) *$w(V_1) < w(G)/4$, and V_1 and V_2 are not adjacent; and*
(ii) *there exists $u \in V_3$ such that u is a cut-vertex of G, $\{G_1, G_2\} \subseteq \mathcal{C}$, $w(C) \leq w(V_1) \leq w(V_2)$ for each $C \in \mathcal{C} \setminus \{G_1, G_2\}$, where $\mathcal{C}$ is the set of components of $G - u$. Moreover, if $|\mathcal{C}| = 3$ then $w(u) > w(G)/4$.*

Theorem 1. *Algorithm 1 is a pseudo-polynomial approximation with ratio $\frac{3}{2}$ for MIN-MAX BCP$_3$ which runs in $\mathcal{O}(w(G)|V||E|)$ time on an instance (G, w), where $G = (V, E)$.*

Proof. Let $\mathcal{P} = \{V_1, V_2, V_3\}$ be an ordered 3-partition of G, returned by the algorithm; and let $G_i = G[V_i]$, for $i = 1, 2, 3$. By Lemma 2, $\mathcal{P}$ is indeed a connected 3-partition of G and it can be computed in time $\mathcal{O}(w(G)|V||E|)$.

If $w(V_3) \leq w(G)/2$, then it follows directly from Fact 1 that $w^+(\mathcal{P}) = w(V_3) \leq \frac{3}{2}\text{OPT}_3(G, w)$.

Suppose now that $w(V_3) > w(G)/2$. If V_3 is a singleton $\{u\}$, then $w(u) \leq \text{OPT}_3(G, w)$ and $\mathcal{P}$ is optimal. Otherwise, the algorithm terminated because neither MERGE nor PULL operation can be performed on $\mathcal{P}$. By Lemma 3(ii), there exists $u \in V_3$ such that u is a cut-vertex of G, $\{G_1, G_2\} \subseteq \mathcal{C}$, and $w(C) \leq w(V_1) \leq w(V_2)$ for each $C \in \mathcal{C} \setminus \{G_1, G_2\}$, where $\mathcal{C}$ is the set of components of $G - u$. By Lemma 1, we have $w^+(\mathcal{P}) = w(V_3) = w(u) + \sum_{C \in \mathcal{C} \setminus \{G_1, G_2\}} w(C) \leq \text{OPT}_3(G, w)$. Therefore, in this case the partition $\mathcal{P}$ produced by the algorithm is an optimal solution for the instance (G, w) of MIN-MAX BCP$_3$. $\square$

In what follows, we show how to extend the result obtained for MIN-MAX BCP$_3$ to obtain results for MIN-MAX BCP$_k$, for all $k \geq 4$. For simplicity, we say that a vertex u satisfying condition (ii) of Lemma 3 is a *star-center*. Moreover, when u is a star-center, we label the ℓ components of $G - u$ as $\mathcal{C} = \{C_1, C_2, \ldots, C_\ell\}$, where $C_\ell = G[V_2]$, $C_{\ell-1} = G[V_1]$ and $w(C_i) \leq w(C_{i+1})$ for all $i \in [\ell-1]$. The next algorithm uses a routine called GETSINGLETONS which receives as input a connected graph $G = (V, E)$, a connected k'-partition $\mathcal{P}$ of G, and an integer $q \geq 0$ such that $k' + q \leq |V|$, then it produces a connected $(k' + q)$-partition of G in time $\mathcal{O}(|V||E|)$ (where q of the classes in the partition are singletons).

Algorithm 2. MIN-MAX-BCPk ($k \geq 3$)

    Input: An instance $(G = (V, E), w)$ of MIN-MAX BCP$_k$, $3 \leq k \leq |V|$
    Output: A connected k-partition of G
    Routines: MIN-MAX-BCP3, GETSINGLETONS

1: **procedure** MIN-MAX-BCP$k(G, w)$
2:     $\mathcal{P} \leftarrow$ MIN-MAX-BCP3(G, w) # $\mathcal{P} = \{V_1, V_2, V_3\}$
3:     **if** $w^+(\mathcal{P}) \leq w(G)/2$ **or** $|V_3| = 1$ **then**
4:         $\mathcal{P}' \leftarrow$ GETSINGLETONS$(G, w, k - 3, \mathcal{P})$
5:     **else**
6:         Let u be the star-center and let $\mathcal{C} = \{C_i\}_{i \in [\ell]}$ be the components of $G - u$.
7:         **if** $\ell > k - 1$ **then**
8:             Let $t = \ell - k + 1$ and $V' = (\bigcup_{i \in [t]} V(C_i)) \cup \{u\}$.
9:             $\mathcal{P}' \leftarrow \{V', V(C_{t+1}), \ldots, V(C_{\ell-1}), V(C_\ell)\}$
10:        **else**
11:           $\mathcal{P} \leftarrow \{\{u\}\} \cup \{C_i\}_{i \in [\ell]}$
12:           $\mathcal{P}' \leftarrow$ GETSINGLETONS$(G, w, k - 1 - \ell, \mathcal{P})$
13:     **return** $\mathcal{P}'$

Theorem 2. *For each integer $k \geq 3$, Algorithm 2 is a pseudo-polynomial $\frac{k}{2}$-approximation for the problem MIN-MAX BCP$_k$ that runs in $\mathcal{O}(w(G)|V||E|)$ time on an instance (G, w), where $G = (V, E)$.*

Algorithm 2 is a (polynomial) $\frac{k}{2}$-approximation if the weights assigned to the vertices are bounded by a polynomial on the order of the graph. When the weights are arbitrary, we apply a scaling technique and use the previous algorithm as a subroutine to obtain a polynomial algorithm for MIN-MAX BCP_k with approximation ratio $(\frac{k}{2} + \varepsilon)$, for any fixed $\varepsilon > 0$.

Algorithm 3. ε-MIN-MAX-BCPk $(k \geq 3)$

   Input: An instance $(G = (V, E), w)$ of MIN-MAX BCP_k, $3 \leq k \leq |V|$
   Output: A connected k-partition of G
   Routine: A pseudo-polynomial α-approximation algorithm $\mathcal{A}$ for MIN-MAX BCP_k

1: **procedure** ε-MIN-MAX-BCP$k(G, w)$
2:     $\theta \leftarrow \max_{v \in V} w(v)$
3:     $\lambda \leftarrow \frac{\varepsilon\theta}{|V|}$
4:     **for** $v \in V$ **do**
5:         $\widehat{w}(v) \leftarrow \left\lceil \frac{w(v)}{\lambda} \right\rceil$

6:     $\mathcal{P} \leftarrow \mathcal{A}(G, \widehat{w})$
7:     **return** $\mathcal{P}$

Theorem 3. *Let $k \geq 3$ be an integer, and let $I = (G = (V, E), w)$ be an instance of MIN-MAX BCP_k. If there is a pseudo-polynomial α-approximation algorithm $\mathcal{A}$ for MIN-MAX BCP_k that runs in $\mathcal{O}(w(G)^c |V||E|)$ time for some constant c, then Algorithm 3 is an $\alpha(1 + \varepsilon)$-approximation for MIN-MAX BCP_k that runs in $\mathcal{O}(|V|^{2c+1}|E|/\varepsilon^c)$ time.*

Corollary 1. *For each integer $k \geq 3$ and $\varepsilon' > 0$, there is a $(\frac{k}{2} + \varepsilon')$-approximation for MIN-MAX BCP_k that runs in $\mathcal{O}(|V|^3|E|/\varepsilon')$ time on a input $(G = (V, E), w)$.*

Proof. The result follows from Theorem 3, by taking Algorithm 3 with $\varepsilon = \varepsilon'/(k/2)$ and Algorithm 2 as the routine $\mathcal{A}$ it requires. The approximation ratio $k/2$ of Algorithm 2 is guaranteed by Theorem 2.   □

An algorithm analogous to Algorithm 3 can be designed for MAX-MIN BCP_k. In this case, change line 2 to $\theta \leftarrow \min_{v \in V} w(v)$, change line 5 to $\widehat{w}(v) \leftarrow \left\lfloor \frac{w(v)}{\lambda} \right\rfloor$, and consider a routine that is a pseudo-polynomial α-approximation for MAX-MIN BCP_k. Then, a theorem similar to Theorem 3 can be obtained for MAX-MIN BCP_k.

3   Parameterized Algorithm for 1-MAX-MIN BCP

This section is devoted to the design of a fixed-parameter tractable (FPT) algorithm for 1-MAX-MIN BCP when parameterized by the vertex cover. In this problem, we are given an unweighted graph G, a positive integer k, and a vertex cover X of G. The objective is to find a connected k-partition of G that maximizes the size of the smallest class. Let us consider a fixed instance (G, k) of 1-MAX-MIN BCP and a vertex cover X of G.

Let us denote by I the stable set $V(G) \setminus X$. Recall that we assume $k \leq |V(G)| = |X| + |I|$. If $k > |X|$, then there are at least $k - |X|$ classes of size exactly 1 contained in I, and so an optimal solution (which has value equal to 1) can be easily computed. If $|X| = 1$, then G is a star, and so it is trivial to compute an optimal solution. From now on, we assume that $k \leq |X|$ and $|X| \geq 2$.

Before presenting the details of the proposed algorithm, we show a lemma that guarantees the existence of an optimal solution in which each class intersects the given vertex cover X.

Lemma 4. *Let (G, k) be an instance of 1-MAX-MIN-BCP and let X be a vertex cover of G. Then, there exists an optimal connected k-partition $\{V_i\}_{i \in [k]}$ of G such that $V_i \cap X \neq \emptyset$ for all $i \in [k]$.*

We remark that the proof of the above lemma follows from the fact that, if there exists an optimal solution for 1-MAX-MIN-BCP where a class is contained in $V(G) \setminus X$, then the cost of such a solution is 1. Therefore, any connected k-partition of G is an optimal solution. However, the same observation is not valid for 1-MIN-MAX-BCP as one may easily construct a (bipartite) graph G and a vertex cover X of G such that every optimal 3-connected partition of G has a class which does not intersect X.

We next use hypergraphs to model the constraints of our ILP formulation for 1-MAX-MIN BCP. A hyperpath of length m between two vertices u and v in a hypergraph H is a set of hyperedges $\{e_1, \ldots, e_m\} \subseteq E(H)$ such that $u \in e_1$, $v \in e_m$, and $e_i \cap e_{i+1} \neq \emptyset$ for each $i \in \{1, \ldots, m - 1\}$. A set of hyperedges $F \subseteq E(H)$ is a (u, v)-cut if there is no hyperpath between u and v in $H - F$.

For each $S \subseteq X$, we define $I(S) = \{v \in I : N(v) = S\}$. Let $u, v \in X$ be a pair of non-adjacent vertices in G, and let $\Gamma_X(u, v)$ be the set of all separators of u and v in $G[X]$. Consider a separator $Z \in \Gamma_X(u, v)$, and denote by $\mathcal{C}(Z)$ the set of components of $G[X \setminus Z]$. Let H_Z denote the hypergraph with vertices $\mathcal{C}(Z)$ such that, for each $S \subseteq X$ with $I(S) \neq \emptyset$, there is a hyperedge $\{C \in \mathcal{C}(Z) : S \cap V(C) \neq \emptyset\}$ in H_Z. We denote by $\Lambda_Z(u, v)$ the set of all (C_u, C_v)-cuts in H_Z, where C_u and C_v are the components of $G[X \setminus Z]$ containing u and v, respectively.

Suppose that u and v belong to a same class of a connected partition of G. Hence, there exists a path P linking u and v in G. Note that either P intersects Z (i.e. $V(P) \cap Z \neq \emptyset$), or P contains a vertex in the stable set I (i.e. $V(P) \cap I \neq \emptyset$). In the latter case, one may easily see that P guarantees the existence of a hyperpath Q which connects the vertices C_u and C_v in the hypergraph H_Z. Therefore, the hyperedges of Q must cross every (C_u, C_v)-cut in H_Z. The hypergraph H_Z is illustrated in Fig. 1.

For each $v \in X$ and $i \in [k]$, there is a binary variable $x_{v,i}$ that equals 1 if and only if v belongs to the i-th class of the partition. Moreover, for every $S \subseteq X$ and $i \in [k]$, there is an integer variable $y_{S,i}$ that equals the amount of vertices in $I(S)$ that are assigned to the i-th class. The intuition behind the y-variables is that all vertices in $I(S)$, for a fixed $S \subseteq X$, play essentially the same role in a connected partition. The idea of using integer variables to count indistinguishable vertices in a stable set appeared before in Fellows et al. [13].

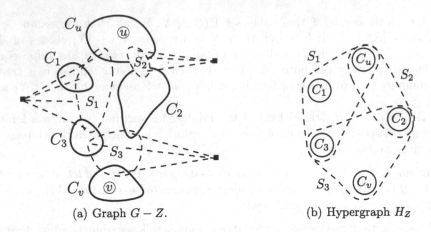

(a) Graph $G - Z$.   (b) Hypergraph H_Z

Fig. 1. Illustration of the hypergraph construction. Continuous lines indicate the components in $\mathcal{C}(Z)$. Subsets S of X are represented with dashed lines and their corresponding vertices $I(S)$ are depicted with squares. In this example, $\{S_1, S_2\}$ and $\{S_3\}$ are (C_u, C_v)-cuts in H_Z.

Let $\eta = 2^{|X|}$ (number of subsets of X), and let $\mathcal{B}(G, X, k)$ be the set of vectors in $\mathbb{R}^{(|X|+\eta)k}$ that satisfy the following inequalities (1)–(7). To shorten the description of inequalities (3) in the next ILP, we denote by Ψ the set

$$\{(uv, Z, F): \{u, v\} \subseteq X, uv \notin E, Z \in \Gamma_X(u, v), \text{ and } F \in \Lambda_Z(u, v)\}.$$

$$\sum_{v \in X} x_{v,i} + \sum_{S \subseteq X} y_{S,i} \leq \sum_{v \in X} x_{v,i+1} + \sum_{S \subseteq X} y_{S,i+1} \qquad \forall i \in [k-1], \quad (1)$$

$$\sum_{i \in [k]} x_{v,i} = 1 \qquad \forall v \in X, \quad (2)$$

$$x_{u,i} + x_{v,i} - \sum_{z \in Z} x_{z,i} - \sum_{S \in F} y_{S,i} \leq 1 \qquad \forall (uv, Z, F) \in \Psi, i \in [k], \quad (3)$$

$$y_{S,i} \leq |I(S)| \left(\sum_{v \in S} x_{v,i} \right) \qquad \forall S \subseteq X, i \in [k], \quad (4)$$

$$\sum_{i \in [k]} y_{S,i} = |I(S)| \qquad \forall S \subseteq X, \quad (5)$$

$$x_{v,i} \in \{0, 1\} \qquad \forall v \in X \text{ and } i \in [k], \quad (6)$$

$$y_{S,i} \in \mathbb{Z}_{\geq} \qquad \forall S \subseteq X \text{ and } i \in [k]. \quad (7)$$

Inequalities (1) establish a non-decreasing ordering of the classes according to their sizes. Inequalities (2) and (5) guarantee that every vertex of the graph belongs to exactly one class (i.e. the classes define a partition). Due do Lemma 4, we may consider only partitions such that each of its classes intersects X. Thus, whenever a vertex in the stable set I is chosen to belong to some class, at least one of its neighbors in X has to be in the same class. This explains the meaning of inequalities (4). Inequalities (3) guarantee that each class of the partition induces a connected subgraph. The following lemma shows that the formulation correctly models the problem.

Lemma 5. *Let G be a connected graph, let $k \geq 2$ be an integer, and let X be a vertex cover of G. The problem* 1-MAX-MIN BCP *on instance (G, k) is equivalent to*

$$\max \left\{ \sum_{v \in X} x_{v,1} + \sum_{S \subseteq X} y_{S,1} \colon (x, y) \in \mathcal{B}(G, X, k) \right\}.$$

An instance of an INTEGER LINEAR PROGRAMMING problem consists of a matrix $A \in \mathbb{Z}^{p \times q}$, a vector $b \in \mathbb{Z}^p$ and a vector $c \in \mathbb{Z}^q$. The objective is to find a vector $x \in \mathbb{Z}^q$ that satisfies $Ax \leq b$, and maximizes $c^T x$. Let us denote by L the size of the binary representation of an instance (A, b, c) of the problem. We next present the maximization version of the theorem showed by Cygan et al. [11] on the existence of an FPT algorithm for an INTEGER LINEAR PROGRAMMING problem parameterized by the number of variables.

Theorem 4 (Cygan et al. [11]). *Let $I = (A, b, c)$ be an instance of an INTEGER LINEAR PROGRAMMING problem with size L and q variables. Then I can be solved using $\mathcal{O}(q^{2.5q+o(q)} \cdot (L + \log M_x) \log(M_x M_c))$ arithmetic operations and space polynomial in $L + \log M_x$, where M_x is an upper bound on the absolute value a variable can take in a solution, and M_c is the largest absolute value of a coefficient in the vector c.*

Theorem 5. *The problem* 1-MAX-MIN BCP, *parameterized by the size of a vertex cover of the input graph, is fixed-parameter tractable.*

Proof. Let (G, k) be an instance of 1-MAX-MIN BCP, and X a vertex cover of G. From Lemma 5, we have that $\max\{\sum_{v \in X} x_{v,1} + \sum_{S \subseteq X} y_{S,1} \colon (x, y) \in \mathcal{B}(G, X, k)\}$ is equivalent to solving instance (G, k). Note that the size of the corresponding ILP is $2^{2^{\mathcal{O}(|X|)}} \log |V(G)|$. By Theorem 4, this ILP can be solved in time $2^{2^{\mathcal{O}(|X|)}} |V(G)|^{\mathcal{O}(1)}$. Therefore, 1-MAX-MIN BCP is fixed parameter-tractable when parameterized by the size of a vertex cover of the input graph. $\square$

4   Concluding Remarks

Problems on balanced connected partitions of graphs have been largely investigated since late seventies. Many variants of these problems, either of existential or optimization nature (with different objective functions), have been considered, most of them known to be computationally hard.

One of these intriguing existential problems, in which the input graph is k-connected and one is interested in finding a connected k-partition into classes of prescribed sizes, was solved by Lovász [16] and Györi [15]. It has been shown that such a partition does exist, but it has not been settled whether it can be found in polynomial time.

The variants we have considered here (MIN-MAX BCP_k and MAX-MIN BCP_k), and the corresponding versions in which the number of classes k is part of the input, have gained much attention more recently in terms of approximation algorithms. We note that, although for the latter variant an inapproximability threshold (of $6/5$) has been proved in 2007, for the variants in which k is fixed such thresholds are not known. The parameterized algorithm shown here seems to be the first of this nature for this class of problems.

References

1. Alimonti, P., Calamoneri, T.: On the complexity of the max balance problem. In: Argentinian Workshop on Theoretical Computer Science (WAIT 1999), pp. 133–138 (1999)
2. Becker, R., Lari, I., Lucertini, M., Simeone, B.: A polynomial-time algorithm for max-min partitioning of ladders. Theory Comput. Syst. **34**(4), 353–374 (2001). https://doi.org/10.1007/s00224-001-0008-8
3. Becker, R.I., Lari, I., Lucertini, M., Simeone, B.: Max-min partitioning of grid graphs into connected components. Networks **32**(2), 115–125 (1998)
4. Becker, R.I., Perl, Y.: Shifting algorithms for tree partitioning with general weighting functions. J. Algorithms **4**(2), 101–120 (1983)
5. Becker, R.I., Schach, S.R., Perl, Y.: A shifting algorithm for min-max tree partitioning. J. ACM **29**(1), 58–67 (1982)
6. Casel, K., Friedrich, T., Issac, D., Niklanovits, A., Zeif, Z.: Balanced crown decomposition for connectivity constraints. In: Mutzel, P., Pagh, R., Herman, G. (eds.) 29th Annual European Symposium on Algorithms (ESA 2021). Leibniz International Proceedings in Informatics (LIPIcs), vol. 204, pp. 26:1–26:15. Schloss Dagstuhl - Leibniz-Zentrum für Informatik, Dagstuhl (2021)
7. Chataigner, F., Salgado, L.R.B., Wakabayashi, Y.: Approximation and inapproximability results on balanced connected partitions of graphs. Discret. Math. Theor. Comput. Sci. **9**(1), 177–192 (2007)
8. Chen, G., Chen, Y., Chen, Z.-Z., Lin, G., Liu, T., Zhang, A.: Approximation algorithms for the maximally balanced connected graph tripartition problem. J. Comb. Optim. 1–21 (2020). https://doi.org/10.1007/s10878-020-00544-w
9. Chen, Y., Chen, Z.-Z., Lin, G., Xu, Y., Zhang, A.: Approximation algorithms for maximally balanced connected graph partition. Algorithmica **83**(12), 3715–3740 (2021). https://doi.org/10.1007/s00453-021-00870-3
10. Chlebíková, J.: Approximating the maximally balanced connected partition problem in graphs. Inf. Process. Lett. **60**(5), 225–230 (1996)
11. Cygan, M., et al.: Miscellaneous. In: Parameterized Algorithms, pp. 129–150. Springer, Cham (2015). https://doi.org/10.1007/978-3-319-21275-3_6
12. Dyer, M., Frieze, A.: On the complexity of partitioning graphs into connected subgraphs. Discret. Appl. Math. **10**(2), 139–153 (1985)
13. Fellows, M.R., Lokshtanov, D., Misra, N., Rosamond, F.A., Saurabh, S.: Graph layout problems parameterized by vertex cover. In: Hong, S.-H., Nagamochi, H., Fukunaga, T. (eds.) ISAAC 2008. LNCS, vol. 5369, pp. 294–305. Springer, Heidelberg (2008). https://doi.org/10.1007/978-3-540-92182-0_28

14. Frederickson, G.N.: Optimal algorithms for tree partitioning. In: Proceedings of the Second Annual ACM-SIAM Symposium on Discrete Algorithms, SODA 1991, pp. 168–177. Society for Industrial and Applied Mathematics, USA (1991)
15. Györi, E.: On division of graph to connected subgraphs. In: Combinatorics (Proc. Fifth Hungarian Colloq., Koszthely, 1976). Colloq. Math. Soc. János Bolyai, vol. 18, pp. 485–494 (1978)
16. Lovász, L.: A homology theory for spanning tress of a graph. Acta Mathematica Academiae Scientiarum Hungarica 30, 241–251 (1977)
17. Lucertini, M., Perl, Y., Simeone, B.: Most uniform path partitioning and its use in image processing. Discret. Appl. Math. 42(2), 227–256 (1993)
18. Maravalle, M., Simeone, B., Naldini, R.: Clustering on trees. Comput. Stat. Data Anal. 24(2), 217–234 (1997)
19. Miyazawa, F.K., Moura, P.F., Ota, M.J., Wakabayashi, Y.: Partitioning a graph into balanced connected classes: formulations, separation and experiments. Eur. J. Oper. Res. 293(3), 826–836 (2021)
20. Perl, Y., Schach, S.R.: Max-min tree partitioning. J. ACM 28(1), 5–15 (1981)
21. Wu, B.Y.: Fully polynomial-time approximation schemes for the max-min connected partition problem on interval graphs. Discrete Math. Algorithms Appl. 04(01), 1250005 (2012)
22. Zhou, X., Wang, H., Ding, B., Hu, T., Shang, S.: Balanced connected task allocations for multi-robot systems: an exact flow-based integer program and an approximate tree-based genetic algorithm. Expert Syst. Appl. 116, 10–20 (2019)

Algorithms for Online Car-Sharing Problem

Xiangyu Guo[1] and Kelin Luo[2(✉)]

[1] Department of Computer Science and Engineering, University at Buffalo,
Buffalo, NY, USA
xiangyug@buffalo.edu
[2] Department of Mathematics and Computer Science,
Eindhoven University of Technology, Eindhoven, The Netherlands
k.luo@tue.nl

Abstract. In the online car-sharing (a.k.a. ride-sharing) problem, we are given a set of m available car, and n requests arrive sequentially in T periods, in which each request consists of a pick-up location and a drop-off location. In each period, we must *immediately* and *irrevocably* assign free cars to serve arrived requests, such that two requests share one car. The goal is to find an online algorithm to process all requests while minimizing the total travel distance of cars.

We give the first algorithm for this problem under the adversarial model and the random arrival model. For the adversarial model, we give a $2T + 1/2$-competitive algorithm, then we show this can be further improved to $2T$-competitive by a carefully designed edge cost function. This almost matches the known $2T - 1$ lower bound in this model. For the random arrival model, our algorithm is $3H_T - 1/2 + o(1)$-competitive, where H_T is the T-th harmonic number. All the above three results are based on one single algorithm that runs in $O(n^3)$ time.

Keywords: Car-sharing · Online matching · Competitive analysis

1 Introduction

In a car-sharing system, a company offers cars to customers in which each car serves at most two requests (see [17]). A typical scenario is the following: There are a number of available cars with current location information, and requests with pick-up and drop-off locations arrive over time. The car-sharing company has to serve these requests with available cars. The company wishes to minimize the total driving distance because it reflects the costs (serving time or fuel consumption) of serving requests. From a societal point of view, this objective also helps to reduce emissions and protect the environment.

This project has received funding from the European Union's Horizon 2020 research and innovation programme under the Marie Skłodowska-Curie grant agreement number 754462.

© Springer Nature Switzerland AG 2022
N. Balachandran and R. Inkulu (Eds.): CALDAM 2022, LNCS 13179, pp. 224–236, 2022.
https://doi.org/10.1007/978-3-030-95018-7_18

Formally, the online car-sharing (OCS) problem can be described as follows: Given a set C of cars $\{c_k : k = 1, 2, ..., |C|\}$, with car k initially at location c_k; Each customer request i consists of a *pick-up* location s_i and a *drop-off* location t_i. The request set R is revealed online in T different time periods; During the d-th period, a subset of requests is revealed together as a *group*, and the algorithm must immediately and irrevocably pair them and assign each request pair $\{i, j\}$ to an unmatched car k. The travel distance of car k serving requests i and j, is the minimum distance of starting from location c_k to visit the locations s_i, s_j, t_i, and t_j, such that s_i is visited before t_i and s_j is visited before t_j. The goal of the online algorithm is to minimize the total travel distance of cars involved in the assignment.

Related Work. **The offline car-sharing problem.** if $T = 1$, the online car-sharing problem becomes the offline car-sharing problem. All requests in R are revealed at once. The objective is to assign all requests to the cars such that each car serves exactly two requests while minimizing total travel distance. Bei and Zhang [2] first studied this problem and proved that the offline car-sharing problem is NP-hard, and they also gave a 2.5-approximation algorithm. Recently, Luo and Spieksma [11] gave a new algorithm with an improved approximation ratio of 2; For the special case where the pick-up and drop-off location coincides, their algorithm achieves 7/5-approximation. Similar problems have also been studied in the data mining community under various contexts: one closest is the work by [8] where the problem is called"food delivery problem", with flow time objective.

The Online Minimum Metric Bipartite Matching Problem. Set $T = n/2$ in our problem, i.e. request pairs arrive one by one. If the pick-up and drop-off locations of any two requests in a pair are the same, then our problem is reduced to the *online minimum metric bipartite matching problem*. For the adversarial model, Kalyanasundaram and Pruhs [9] proved that no deterministic algorithm can achieve a competitive ratio better than $2T - 1$, which implies a same lower bound for our problem in the adversarial model. Furthermore, they gave an optimal algorithm, Permutation algorithm, matching this lower bound. In the restricted version when all locations are on a line, Koutsoupias and Nanavati [10] showed that the work function method has a competitive ratio between $O(\log T)$ and $O(T)$. Later, Nayyar and Raghvendra [12] gave an input-sensitive analysis of the algorithm in [14]. They showed that for any metric space $\mathbb{M}$ and car locations in S, the competitive ratio of the algorithm is $O(\mu_{\mathbb{M}}(S) \log^2 T)$ where $\mu_{\mathbb{M}}(S)$ is the maximum ratio of the traveling salesman tour and the diameter of a subset of cars among all subsets of S. In particular, if S is a set of points on a line, then the competitive ratio is $O(\log^2 T)$. For the random arrival model, Raghvendra [14] came up with a $2H_T - 1 + o(1)$-competitive algorithm, matching the lower bound. This lower bound also directly implies a $2T - 1$ lower bound for our problem in the random arrival model.

The Dial-a-Ride Problem (DaRP). Let $\lambda \in \mathbb{N}_{>0}$ be an input parameter, if each car can carry up to λ requests, and can be *reused* unlimited times, then the car-sharing problem becomes the classical dial-a-ride problem. The DaRP has been studied extensively in both the online and offline setting [1, 3–7,13]. Most research works focused on the single-vehicle case though: The best offline algorithm achieves approximation ratio $\min\{\tilde{O}(\sqrt{n}), \tilde{O}(\sqrt{\lambda})\}$ [4,7], while the best online algorithm has competitive ratio 4 (with exponential-time computation) [3]. It would be interesting to extend our algorithm to the case when each car can serve $\lambda > 2$ requests. On the application side, DaRP has been studied extensively by the operation research and data mining communities. Some recent results are [15,16,18], where they give algorithms solving large-scale DaRP with various constraints or objectives. Notably, the algorithm of [18] also has approximation guarantee matching the best theoretical results [4,7] mentioned before.

Our Results. We consider two models: the adversarial model and the random arrival model.

- **The (adaptive) adversarial model.** In this model, at each period d the adversary can choose any request subset from the remaining of R to release as the arriving group, based on all the decisions the algorithm has made up to that period. We use *competitive ratio* $\alpha \geq 1$ to measure the performance of our algorithm. Let $W(M)$ be our algorithm's assignment cost and let $W(M^*)$ be the minimum-possible cost (in other words, M^* is an optimal offline assignment). If $W(M) \leq \alpha W(M^*)$ for any C and R and the arrival order in R, then we say our algorithm is α-competitive in the adversarial model.
- **The random arrival model.** In this model, the adversary got to choose the requests for all groups *at the very beginning*, but the groups will arrive in an uniformly random order. If $\mathbb{E}[W(M)] \leq \alpha W(M^*)$ for any C and R (with expectation taken on the arrival order of R), then we say our algorithm is α-competitive in the random arrival model.

Motivated by the Permutation algorithm [9] and Match-and-Assign algorithm [2], we give an $O(n^3)$-time algorithm named Online-Match-and-Assign (OMA) that works in both the adversarial model and the random arrival model. The algorithm essentially computes two matchings: it first pairs arrived requests by computing a minimum-cost matching on them, then assigns these request pairs to cars by computing another minimum-cost matching. By carefully choosing the *cost function* used when computing the two matchings, we obtain the following results:

1. For the adversarial model, we first show that the OMA algorithm achieves $2T + 1/2$-competitiveness with a natural cost function defined by the metric. When $T = 1$, the result matches the approximation ratio 2.5 of the offline car-sharing problem [2]. We then tweak the cost function and show that OMA achieves an improved ratio of $2T$. Note that this also implies a 2-approximation algorithm for the offline car-sharing problem, matching the best approximation ratio so far [11].

2. For the random arrival model. We show the OMA algorithm is $3H_T - 1/2 + o(1)$-competitive (H_T is the Tth harmonic number), while the tightest known lower bound is $2H_T - 1 - o(1)$ [14].

Lastly, we point out that our algorithm can be easily adapted to allow *exclusive requests* (e.g. customers that don't want to share ride with others): we can create a virtual request co-locating with each exclusive request, and pair them together. Then the above competitive ratios still hold.

2 Preliminaries

2.1 Problem Setting and Notations

We now define the Car-Sharing Problem formally. We are given a metric space $(\mathcal{X}, \text{dist})$. There is a set C of m cars numbered from 1 to m, each of capacity 2, and the k-th car is identified by its initial location $c_k \in \mathcal{X}$. There is also a set R of n rider requests. The i-th request is denoted as a tuple $(s_i, t_i) \in \mathcal{X} \times \mathcal{X}$, where s_i is the pick-up location and t_i is the drop-off location of the rider. R is further partitioned into T *groups* $G_1, \ldots, G_T$, where each group $G_d(d \in [T])$ contains requests arrived in a same short time period and should be handled separately from those in other groups. We shall use $R_d = \bigcup_{i \in [d]} G_i$ to denote the set of all requests seen till period d.

Now we describe the desired output for the Car-Sharing Problem. Without loss of generality, we can assume in each group there are an even number of requests. We want to assign cars to requests in a way that minimizes the number of cars used as well as the total distance all cars travel. Specifically, a solution for the Car-Sharing problem can be represented as two matchings (MR, M): $\text{MR} = \biguplus_{d \in [T]} Q_d \subset R \times R$ is a perfect matching over all requests R, and it is also the disjoint union of $Q_d, d \in [T]$, where $Q_d \in G_d \times G_d$ is a perfect matching on G_d. MR represents a *pairing* of requests: every request pair $(i, j) \in \text{MR}$ will share a car. The second matching $M \in C \times \text{MR}$ represents an *assignment* from cars to paired requests. We can w.l.o.g. assume all pairs in MR are matched, i.e., every request pair has its exclusive car.

We can now define the cost of a solution. Let $(a_1, a_2, \ldots, a_K) \in \mathcal{X}^K$ be any sequence of locations in $\mathcal{X}$, we use $\text{dist}(a_1, a_2, \ldots, a_K) := \text{dist}(a_1, a_2) + \text{dist}(a_2, a_3) + \cdots + \text{dist}(a_{K-1}, a_K)$ to denote the total travel distance of moving a_1 to a_K and visit each a_i in order. Then suppose requests i, j are assigned to car k, the travel distance $w(k, \{i, j\})$ for car k to serve request pair $\{i, j\}$ is defined as

$$w(k, \{i, j\}) := \min\{\text{dist}(c_k, s_i, t_i, s_j, t_j), \text{dist}(c_k, s_i, s_j, t_i, t_j),$$
$$\text{dist}(c_k, s_i, s_j, t_j, t_i), \text{dist}(c_k, s_j, t_j, s_i, t_i),$$
$$\text{dist}(c_k, s_j, s_i, t_j, t_i), \text{dist}(c_k, s_j, s_i, t_i, t_j)\}. \qquad (1)$$

Now the cost of a solution (MR, M) can be denoted as

$$W(M) = \sum_{(k, \{i, j\}) \in M} w(k, \{i, j\}), \qquad (2)$$

and the goal for the Car-Sharing Problem is to find a solution with the minimum $W(M)$.

In the Online Car-Sharing (OCS) problem, the set C of cars is known a priori, while the request set R is revealed sequentially in T *periods*: in the d-th period the group G_d arrives, and we need to assign some cars to serve this group immediately and irrevocably. We will also use $\mathsf{OCS}_{s=t}$ to denote the special version when every request only has one location, i.e. $t_i = s_i$ for every request i.

Summary of notations. We use notation $v_{k,i} := \mathsf{dist}(c_k, s_i)$ to denote the distance between car k and request i. We define u_{ij} to be the shortest path length that starts at s_i and serves request pair $\{i, j\}$:

$$u_{ij} = \min\{\mathsf{dist}(s_i, t_i, s_j, t_j), \mathsf{dist}(s_i, s_j, t_i, t_j), \mathsf{dist}(s_i, s_j, t_j, t_i)\}. \tag{3}$$

Similarly, u_{ji} is the shortest length of such paths starting at s_j. Usually $u_{ij} \neq u_{ji}$, but in the special case $\mathsf{OCS}_{s=t}$ we always have $u_{ij} = u_{ji} = \mathsf{dist}(s_i, s_j)$.

Proposition 1. *For u, v, w defined as above, we have*

1a $w(k, \{i, j\}) = \min\{v_{k,i} + u_{ij}, v_{k,j} + u_{ji}\}$
1b For any two requests i and j, we have

$$\max\{u_{ij}, u_{ji}\} - \min\{u_{ij}, u_{ji}\} \leq \mathsf{dist}(s_i, s_j)$$

1c For any assignment $(k, \{i, j\})$, we have

$$\min\{v_{k,i}, v_{k,j}\} + \min\{u_{ij}, u_{ji}\} \leq w(k, \{i, j\}) \leq \max\{v_{k,i}, v_{k,j}\} + \min\{u_{ij}, u_{ji}\}$$

1d For any assignment $(k, \{i, j\})$, w.l.o.g, suppose $u_{ij} \geq u_{ji}$, we have

$$w(k, \{i, j\}) = \min\left\{v_{k,i} + \frac{u_{ij} - u_{ji}}{2}, v_{k,j} - \frac{u_{ij} - u_{ji}}{2}\right\} + \frac{u_{ij} + u_{ji}}{2}$$

Table 1 gives an overview of important notations used in this paper.

Table 1. Overview of important notations

Notation	Definition
$C, |C| = m$	Set of cars
c_k	Initial location of car $k \in C$
$R, |R| = n$	Set of requests
(s_i, t_i)	Pick-up location and drop-off location for the ith request
T	The total number of periods
R_d	The set of all requests arrived till period $d \in [T]$
MR	Pairing of requests
M	Assignment from C to MR
$w(k, \{i, j\})$	The min travel distance of serving request i, j using car k
$W(M)$	The sum of travel distances $W(M) = \sum_{(k, \{i, j\}) \in M} w(k, \{i, j\})$
$v_{k,i}$	$v_{k,i} := \mathsf{dist}(c_k, s_i)$ for $k \in C$ and $i \in R$
u_{ij} (resp. u_{ji})	Length of shortest path that serves request i, j, see (3)

2.2 τ-Net-Cost Augmenting Path

One building block for our algorithm is the τ-net-augmenting path proposed by Raghvendra [14]. Consider a complete bipartite graph G with non-negative edge costs w, and let F be a matching on this graph. An *alternating path* (or cycle) is a simple path (resp. cycle) with edges alternating between those in F and outside F. Suppose there are some vertices unmatched by F (which we call *free vertices*). An *augmenting path* P is an alternating path starting from one free vertex and ending at another free vertex. Given any such P, we can enlarge F by letting $F = F \oplus P$ (the symmetric difference of F and P). In the minimum cost matching problem, we want to match more vertices while having less edge cost. To characterize the cost incurred by augmenting F with P, we define the τ-net-cost as:

$$\Phi_\tau(F, P, w) = \tau \sum_{e \in P \setminus F} w(e) - \sum_{e \in P \cap F} w(e), \qquad (4)$$

where $\tau \geq 1$ is a parameter. When $\tau = 1$, this is just the net cost of augmenting F with P, and the well-known Hungarian algorithm augments F by implicitly computing the path with minimum 1-net-cost. For $\tau > 1$, Raghvendra [14] showed that one can still find the augmenting path with minimum τ-net-cost efficiently. We generalize [14]'s idea to finding a set of augmenting paths with small total τ-net-cost: let $\mathcal{P}$ be a set of augmenting paths and define the τ-net-cost of this set to be

$$\Phi_\tau(F, \mathcal{P}, w) = \sum_{P \in \mathcal{P}} \Phi_\tau(F, P, w). \qquad (5)$$

In Sect. 3 we give an algorithm to find an augmenting path set $\mathcal{P}$ with minimum Φ_τ, and use it to guide the search for a good assignment.

3   The Online-Match-and-Assign Algorithm

In this section, we describe our main algorithm for the online car-sharing problem: Online Match-and-Assign algorithm (OMA). OMA consists of two steps in each period: in the first step, the algorithm pairs the requests, and in the second step, run the min-tau-net-cost algorithm which assigns the request pairs to the cars. Both of the two steps are essentially computing minimum-cost matchings w.r.t. different *edge cost functions*. The algorithm is summarized in Algorithm 1. We give a brief explanation here.

The OMA algorithm maintains two matchings at every period d: an offline matching M'_d serving as an approximation for the optimal, and the actual online matching M_d. The algorithm updates these two matchings iteratively. Recall R_d is the set of all requests present till period d. In the first step, the algorithm computes a minimum-cost pairing Γ_d on newly arrived requests G_d w.r.t. the following edge cost $v_1 : G_d \times G_d \mapsto \mathbb{R}_{\geq 0}$:

$$v_1(\{i, j\}) := \min\{u_{ij}, u_{ji}\} \qquad (6)$$

where u_{ij} is defined in (3). The cost $v_1(\{i,j\})$ can be viewed as the shortest travel distance needed to serve request pair $\{i,j\}$. Let MR_d be the pairing on R_d after this step, we see that $\mathrm{MR}_d = \mathrm{MR}_{d-1} \cup \Gamma_d$. We use $v_1(\mathrm{MR}_d) := \sum_{\{i,j\} \in \mathrm{MR}_d} v_1(\{i,j\})$ to denote the total cost of edges in pairing MR_d. We also have the following simple proposition on the pairing cost:

Proposition 2. *For any feasible pairing* MR_d^* *until period* d,

$$v_1(\mathrm{MR}_d) \leq v_1(\mathrm{MR}_d^*).$$

After all newly arrived requests are paired up with each other, we run algorithm min-tau-net-cost$_{\tau,v_2}$ (see Algorithm 2) with parameter τ and edge cost function v_2 to assign cars to the new request pairs Γ_d. Here τ is the coefficient in τ-net-cost $\Phi_\tau(M, \mathcal{P}, v_2)$, and $v_2 \in \mathbb{R}_{\geq 0}^{C \times (R \times R)}$ is a cost measure for a car k to serve a request pair $\{i,j\}$. Specifically, for different arrival models we will choose different $v_2 \in \{\alpha, \beta, \gamma\}$ where:

$$\alpha(k, \{i,j\}) = \min\{v_{k,i}, v_{k,j}\} \tag{7}$$

$$\beta(k, \{i,j\}) = \min\left\{ v_{k,i} + \frac{u_{ij} - u_{ji}}{2}, v_{k,j} - \frac{u_{ij} - u_{ji}}{2} \right\} \quad \text{(assuming } u_{ij} > u_{ji}) \tag{8}$$

$$\gamma(k, \{i,j\}) = \max\{v_{k,i}, v_{k,j}\} \tag{9}$$

Among the three edge cost functions above, α can be understood as the distance of edge between a car k and a request pair $\{i,j\}$, while β can be thought as α compensated by the travel distance needed to serve i and j. We'll use $v_2 = \alpha$ or β for the adversarial arrival model, and $v_2 = \gamma$ for the random arrival model.

Algorithm 1. Online match and assign algorithm (OMA$_{\tau,v_2}$).
Parameters: $\tau \geq 1$, $v_2 \in \{\alpha, \beta, \gamma\}$

1: $\mathrm{MR}_0 \leftarrow \emptyset$ ▷ request pairing
2: $\mathrm{M}_0 \leftarrow \emptyset$ ▷ the online matching
3: $\mathrm{M}_0' \leftarrow \emptyset$ ▷ the offline matching
4: **for** requests that arrive in period d **do**
5:     $\Gamma_d \leftarrow$ minimum-cost matching on $R_d \setminus R_{d-1}$ w.r.t. cost v_1
6:     $\mathrm{MR}_d \leftarrow \mathrm{MR}_{d-1} \cup \Gamma_d$ ▷ current request pairing
7:     $(M_d', M_d) \leftarrow$ min-tau-net-cost$_{\tau,v_2}(C \cup \mathrm{MR}_d, M_{d-1}', M_{d-1})$ ▷ See Algorithm 2
8: **end for**
9: **return** $\mathrm{MR} = \mathrm{MR}_T$, $M' = M_T'$, $M = M_T$

The min-tau-net-cost algorithm is similar to the online minimum metric bipartite matching algorithm in [9,14]. The main difference here is that we need to compute multiple augmenting paths simultaneously while in [9,14] they only need one path at a time. In period d, we compute the minimum-τ-net-cost augmenting path set $\mathcal{P}$ w.r.t. the offline matching M_{d-1}', and update the offline

---

Algorithm 2. min-tau-net-cost$_{\tau,v_2}(C \cup \mathrm{MR}_d, M'_{d-1}, M_{d-1})$.
Parameters: $\tau \geq 1, v_2 \in \{\alpha, \beta, \gamma\}$

1: Compute the augmenting paths $\mathcal{P}$ with $\min \Phi_\tau(M'_{d-1}, \mathcal{P}, v_2)$
2: Update the offline matching with $\mathcal{P}$: $M'_d \leftarrow M'_{d-1} \oplus \mathcal{P}$.
3: Assign each request pair $\{i, j\}$ in $\mathrm{MR}_d \setminus \mathrm{MR}_{d-1}$ to its corresponding endpoint
   (*free car*) k along the augmenting path $P \in \mathcal{P}$ starting from $\{i, j\}$. Let Υ_d be the
   assignment of request pairs in $\mathrm{MR}_d \setminus \mathrm{MR}_{d-1}$.
4: Update the online matching: $M_d \leftarrow M_{d-1} \cup \Upsilon_d$.
5: **return** M'_d and M_d

---

matching $M'_d \leftarrow M'_{d-1} \oplus \mathcal{P}$; Then we assign the new request pairs using $\mathcal{P}$: each
$\{i, j\} \in \mathrm{MR}_d \setminus \mathrm{MR}_{d-1}$ is assigned to its corresponding endpoint (*free car*) k
along the augmenting path $P \in \mathcal{P}$ starting from $\{i, j\}$. Raghvendra [14] gave a
polynomial-time algorithm for the $|\mathcal{P}| = 1$ case, and we generalize it to get the
following theorem:

Theorem 1. *For every $d \in [T]$ and any $\tau \geq 1$, edge cost v_2, Step 1 of Algorithm 2 can be computed in $O(n^3)$ time to get a set $\mathcal{P}$ of augmenting paths such that*

1a $|\mathcal{P}| = |\mathrm{MR}_d \setminus \mathrm{MR}_{d-1}|$ and all paths in $\mathcal{P}$ are (node-)disjoint with each other.
1b Among all augmenting path sets that satisfy condition (1a), $\mathcal{P}$ is the one with minimum $\Phi_\tau(M, \mathcal{P}, v_2)$.

We note the main technical difficulty of the car-sharing problem: unlike in
the metric bipartite matching problem [9,14] where the cost of assigning a car
to a request is naturally defined by the distance between them, in our problem,
the "distance" from a car to a request pair is not well-defined; That's one reason
why our problem is inherently harder than matching: recall even the offline car-
sharing is already NP-hard. Although we have designed some edge cost functions
(e.g., (7) (8) (9)), they don't necessarily satisfy the triangle inequality, which
poses additional difficulty in the analysis.

In the next section, we analyze the algorithm performance with different τ
and v_2. For the adversarial arrivals model in Sect. 4.1, we use $\tau = 1, v_2 \in \{\alpha, \beta\}$
and the whole algorithm is referred to as $\mathsf{OMA}_{1,\alpha}$ or $\mathsf{OMA}_{1,\beta}$, respectively; while
for the random arrival model in Sect. 4.2, we use some $\tau > 1$ and $v_2 = \gamma$, with
the algorithm referred to as $\mathsf{OMA}_{\tau,\gamma}$.

4   Algorithm Analysis

In this section, we analyze the performance of the OMA algorithm in the adversarial model and the random arrival model.

4.1   Adversarial Order of Arrivals

In this section, we will present a detailed analysis of the $(2T + 1/2)$-competitive algorithm $\mathsf{OMA}_{1,\alpha}$. Using β in place of α we can slightly improve the competitive ratio to $2T$. The analysis of $\mathsf{OMA}_{1,\beta}$ is overall identical but technically more involved, so to keep a better flow of presentation we leave it to the supplementary.

In $\mathsf{OMA}_{1,\alpha}$, we run the OMA algorithm with $v_2 = \alpha$ (see Algorithm 1): in each period d, we compute the augmenting paths $\mathcal{P}$ with $\min \Phi_1(M'_{d-1}, \mathcal{P}, \alpha)$, then update the offline matching to $M'_d = \mathcal{P} \oplus M'_{d-1}$. First, we claim that M'_d is a minimum perfect matching in the weighted bipartite graph $G = (C \cup \mathrm{MR}_d, v_2(k, \{i, j\}))$: because an alternating path set $\mathcal{P}$ minimizing $\Phi_1(M'_{d-1}, \mathcal{P}, v_2)$ also minimizes $\Phi_1(M'_{d-1}, \mathcal{P}, v_2) + \sum_{(k,\{i,j\}) \in M'_{d-1}} v_2(k, \{i, j\})$, which is exactly $\sum_{(k,\{i,j\}) \in M'_d} v_2(k, \{i, j\})$. Therefore, we have the following proposition.

Proposition 3. *For any edge set M, let $v_2(M) := \sum_{(k,\{i,j\}) \in M} v_2(k, \{i, j\})$ denote the total v_2 for M. Let $\tilde{M}^*$ be an optimal assignment for request pairing MR with respect to cost function $v_2 \in \{\alpha, \beta\}$. Then for any period $d \geq 1$, we have*

$$v_2(M'_d) \leq v_2(M') = v_2(\tilde{M}^*).$$

Recall now the cost α for car k to serve request pair $\{i, j\}$ is defined to be $\alpha(k, \{i, j\}) = \min\{v_{k,i}, v_{k,j}\}$. Although α doesn't satisfy the triangle inequality, the following lemma shows that α can still be bounded by alternating path length *plus* the pairing cost.

Lemma 1. *Let $\alpha(M) := \sum_{(k,\{i,j\}) \in M} \alpha(k, \{i, j\})$ for any matching M. In period d, we have*

$$\alpha(M_d \setminus M_{d-1}) \leq \alpha(M'_{d-1}) + \alpha(M'_d) + \sum_{\{i,j\} \in \mathrm{MR}_{d-1}} \mathrm{dist}(s_i, s_j).$$

Let M^* be an optimal assignment and MR^* be the corresponding pairing. The following lemma shows that the pairing MR used by our algorithm actually induces a good matching when using α as the edge cost function.

Lemma 2. *(Lemma 4 in [2]) Let $\tilde{M}^*$ be the minimum perfect matching in the weighted bipartite graph $G = (C \cup \mathrm{MR}, \alpha(k, \{i, j\}))$ where $\alpha(k, \{i, j\}) = \min\{v_{k,i}, v_{k,j}\}$. We have:*

$$\alpha(\tilde{M}^*) \leq \sum_{(k,\{i,j\}) \in M^*} \frac{v_{k,i} + v_{k,j}}{2}.$$

Now we can prove the competitive-ratio for $\mathsf{OMA}_{1,\alpha}$.

Theorem 2. *For OCS in the adversarial model, $\mathsf{OMA}_{1,\alpha}$ algorithm is $(2T + 1/2)$-competitive.*

Proof. According to definition (2), we have:

$$W(M) = \sum_{(k,\{i,j\})\in M} w(k,\{i,j\}) \le \sum_{(k,\{i,j\})\in M} (\max\{v_{k,i}, v_{k,j}\} + \min\{u_{ij}, u_{ji}\})$$

$$\le \sum_{(k,\{i,j\})\in M} (\min\{v_{k,i}, v_{k,j}\} + \mathsf{dist}(s_i, s_j) + \min\{u_{ij}, u_{ji}\})$$

$$\le \sum_{d\in[T]} \sum_{(k,\{i,j\})\in M_d\backslash M_{d-1}} (\min\{v_{k,i}, v_{k,j}\} + \mathsf{dist}(s_i, s_j))$$

$$+ \sum_{\{i,j\}\in MR} \min\{u_{ij}, u_{ji}\}.$$

The first inequality is by Proposition 1c and the second one is by triangle inequality. Then, by Lemma 1 and the definition of α, we have,

$$\sum_{(k,\{i,j\})\in M_d\backslash M_{d-1}} \min\{v_{k,i}, v_{k,j}\} \le \alpha(M'_{d-1}) + \alpha(M'_d) + \sum_{\{i,j\}\in MR_{d-1}} \mathsf{dist}(s_i, s_j).$$

Recall that $v_1(\{i,j\}) := \min\{u_{ij}, u_{ji}\}$ and $v_1(MR) := \sum_{\{i,j\}\in MR} v_1(\{i,j\})$, thus

$$W(M) \le \sum_{d\in[T]} \left(\alpha(M'_{d-1}) + \alpha(M'_d) + \sum_{\{i,j\}\in MR_d} \mathsf{dist}(s_i, s_j) \right) + v_1(MR).$$

By Proposition 3 we have $\alpha(M'_d) \le \alpha(M') = \alpha(\tilde{M}^*)$; Furthermore, by definition for all $1 \le d \le T$ there is $\sum_{\{i,j\}\in MR} \mathsf{dist}(s_i, s_j) > \sum_{\{i,j\}\in MR_d} \mathsf{dist}(s_i, s_j)$. Combine the two we get

$$W(M) \le (2T-1)\alpha(\tilde{M}^*) + T \sum_{\{i,j\}\in MR} \mathsf{dist}(s_i, s_j) + v_1(MR)$$

$$\le (2T-1) \sum_{(k,\{i,j\})\in M^*} \frac{v_{k,i} + v_{k,j}}{2} + T \sum_{\{i,j\}\in MR} \mathsf{dist}(s_i, s_j) + v_1(MR)$$

$$\le (2T-1) \sum_{(k,\{i,j\})\in M^*} (\min\{v_{k,i}, v_{k,j}\} + \mathsf{dist}(s_i, s_j)/2)$$

$$+ T \sum_{\{i,j\}\in MR} \mathsf{dist}(s_i, s_j) + v_1(MR)$$

$$\le (2T-1) \sum_{(k,\{i,j\})\in M^*} \min\{v_{k,i}, v_{k,j}\} + ((2T-1)/2 + T + 1)v_1(MR^*)$$

$$\le (2T + 1/2)\, W(M^*).$$

The second inequality follows from Lemma 2. The third inequality follows from the triangle inequality. The fourth inequality holds because $\min\{u_{ij}, u_{ji}\} \ge \mathsf{dist}(s_i, s_j)$ and $v_1(MR^*) \ge v_1(MR) \ge \sum_{\{i,j\}\in MR} \mathsf{dist}(s_i, s_j)$ (by Proposition 2 and triangle inequality). The last inequality follows from Proposition 1c.   □

Replacing the edge cost function α with β (see (8)), we can remove the additive $1/2$ to get a $2T$-competitive algorithm, matching the best known offline ($T = 1$) approximation ratio. The analysis of $\mathsf{OMA}_{1,\beta}$ is quite similar to $\mathsf{OMA}_{1,\alpha}$.

Theorem 3. *For* OCS *in the adversarial model, the* $\mathsf{OMA}_{1,\beta}$ *algorithm is* $2T$-*competitive.*

For the special case where each request's pick-up location s_i coincides with its drop-off location t_i, we can show a slightly tighter ratio.

Theorem 4. *For* $\mathsf{OCS}_{s=t}$ *in the adversarial model,* $\mathsf{OMA}_{1,\alpha}$ *algorithm is* $(2T - 1/2)$-*competitive.*

Kalyanasundaram and Prushs [9] proved that, for the minimum metric bipartite matching problem no deterministic algorithm can achieve a competitive ratio smaller than $2T - 1$. This directly implies a $2T - 1$ lower bound for OCS and $\mathsf{OCS}_{s=t}$, because online minimum metric bipartite matching problem is a special case of $\mathsf{OCS}_{s=t}$. Also note that when $T = 1$, OCS becomes the offline car-sharing problem, and we get a competitive ratio of 2, which is also the best-known approximation ratio so far [11].

4.2   Random Order of Arrivals

In this section, we analyze the performance of $\mathsf{OMA}_{\tau,\gamma}$ for OCS in the random arrival model. As in $\mathsf{OMA}_{1,\alpha}$, we augment M'_{d-1} by finding a set of augmenting paths $\mathcal{P}$, and let $M'_d = \mathcal{P} \oplus M'_{d-1}$. The main differences from the adversarial model are: (1) we use a different edge cost function $v_2(k, \{i, j\}) = \gamma(k, \{i, j\}) = \max\{v_{k,i}, v_{k,j}\}$, and (2) M'_d ($d \geq 2$) is not necessarily the minimum weight perfect matching in the weighted graph $G = (C \cup \mathrm{MR}_d, \gamma(k, \{i, j\}))$. We first bound the increased cost by the augmenting path length.

Lemma 3. *Let* $\gamma(M) := \sum_{(k,\{i,j\}) \in M} \gamma(k, \{i, j\})$ *for any assignment M. In period d, let $\mathcal{P}$ be the alternating paths with respect to M'_{d-1}, we have*

$$\gamma(M_d \setminus M_{d-1}) \leq \gamma(\mathcal{P} \setminus M'_{d-1}) + \gamma(\mathcal{P} \cap M'_{d-1}).$$

We also have the following lemma from [14] that relates the matching cost with the total τ-net-cost of augmenting paths produced over time:

Lemma 4. (*Lemma 7. (ii) in* [14]) *Let $\tau > 1$. Let $\mathcal{P}_1, \mathcal{P}_2, ..., \mathcal{P}_T$ be the augmenting path sets computed by our algorithm in that order. Then, the τ-net-cost of these paths relates to the cost of the online matching as follows:*

$$\sum_{d=1}^{T} \Phi_\tau(\mathcal{P}_d) \geq \frac{\tau+1}{2}\gamma(M') + \frac{\tau-1}{2}\gamma(M).$$

Now we prove the competitive ratio for $\mathsf{OMA}_{\tau,\gamma}$.

Theorem 5. *In the random arrival model, the* OMA$_{\tau,\gamma}$ *algorithm is* $3H_T - 1/2 + o(1)$*-competitive.*

There also exists a $2H_n - 1 - o(1)$ lower bound for the minimum metric bipartite matching problem in the random arrival model [14]. Like in the case of adversarial model, this implies a same $2H_T - 1 - o(1)$ lower bound for OCS and OCS$_{s=t}$ in the random arrival model.

5 Conclusion

We gave the first algorithm for the online car-sharing problem in the adaptive adversarial model and the random arrival model. Our algorithm achieves near-optimal competitive ratio in both models. One immediate open problem is to allow each car to serve $\lambda > 2$ requests. It's natural to think along the same approach of this paper: i.e., first "cluster" the requests according to some criteria, then assign cars to the resulted clusters by solving certain min-cost matching. However, the competitive ratio will likely depend on λ: one feature that makes our problem hard is the rigid requirement that every car serves *exactly* λ requests, and this often implies solving hard problems like λ-dimensional matching. Another direction worth exploration is to consider different objectives, e.g., customer waiting time (a.k.a. flow time), which has apparent practical importance.

References

1. Ascheuer, N., Krumke, S.O., Rambau, J.: Online dial-a-ride problems: minimizing the completion time. In: Annual Symposium on Theoretical Aspects of Computer Science, pp. 639–650 (2000)
2. Bei, X., Zhang, S.: Algorithms for trip-vehicle assignment in ride-sharing. In: McIlraith, S.A., Weinberger, K.Q. (eds.) Proceedings of the Thirty-Second AAAI Conference on Artificial Intelligence, (AAAI-18), the 30th Innovative Applications of Artificial Intelligence (IAAI-18), and the 8th AAAI Symposium on Educational Advances in Artificial Intelligence (EAAI-18), New Orleans, Louisiana, USA, 2–7 February 2018, pp. 3–9. AAAI Press (2018)
3. Bienkowski, M., Kraska, A., Liu, H.H.: Traveling repairperson, unrelated machines, and other stories about average completion times. In: International Colloquium on Automata, Languages and Programming (ICALP) (2021)
4. Charikar, M., Raghavachari, B.: The finite capacity dial-a-ride problem. In: Proceedings 39th Annual Symposium on Foundations of Computer Science (Cat. No. 98CB36280), pp. 458–467 (1998)
5. Feuerstein, E., Stougie, L.: On-line single-server dial-a-ride problems. Theoret. Comput. Sci. **268**(1), 91–105 (2001)
6. Gørtz, I.L., Nagarajan, V., Ravi, R.: Minimum makespan multi-vehicle dial-a-ride. ACM Trans. Algorithms (TALG) **11**(3), 1–29 (2015)
7. Gupta, A., Hajiaghayi, M., Nagarajan, V., Ravi, R.: Dial a ride from k-forest. ACM Trans. Algorithms (TALG) **6**(2), 1–21 (2010)

8. Joshi, M., Singh, A., Ranu, S., Bagchi, A., Karia, P., Kala, P.: Batching and matching for food delivery in dynamic road networks. In: 2021 IEEE 37th International Conference on Data Engineering (ICDE), pp. 2099–2104. IEEE (2021)
9. Kalyanasundaram, B., Pruhs, K.: Online weighted matching. J. Algorithms **14**(3), 478–488 (1993)
10. Koutsoupias, E., Nanavati, A.: The online matching problem on a line. In: Solis-Oba, R., Jansen, K. (eds.) WAOA 2003. LNCS, vol. 2909, pp. 179–191. Springer, Heidelberg (2004). https://doi.org/10.1007/978-3-540-24592-6_14
11. Luo, K., Spieksma, F.C.R.: Approximation algorithms for car-sharing problems. In: Kim, D., Uma, R.N., Cai, Z., Lee, D.H. (eds.) COCOON 2020. LNCS, vol. 12273, pp. 262–273. Springer, Cham (2020). https://doi.org/10.1007/978-3-030-58150-3_21
12. Nayyar, K., Raghvendra, S.: An input sensitive online algorithm for the metric bipartite matching problem. In: Umans, C. (ed.) 58th IEEE Annual Symposium on Foundations of Computer Science, FOCS 2017, Berkeley, CA, USA, 15–17 October 2017, pp. 505–515. IEEE Computer Society (2017)
13. de Paepe, W., Lenstra, J.K., Sgall, J., Sitters, R.A., Stougie, L.: Computer-aided complexity classification of dial-a-ride problems. INFORMS J. Comput. **16**(2), 120–132 (2004)
14. Raghvendra, S.: A robust and optimal online algorithm for minimum metric bipartite matching. In: Jansen, K., Mathieu, C., Rolim, J.D.P., Umans, C. (eds.) Approximation, Randomization, and Combinatorial Optimization. Algorithms and Techniques, APPROX/RANDOM 2016, 7–9 September 2016, Paris, France. LIPIcs, vol. 60, pp. 18:1–18:16. Schloss Dagstuhl - Leibniz-Zentrum für Informatik (2016)
15. Ta, N., Li, G., Zhao, T., Feng, J., Ma, H., Gong, Z.: An efficient ride-sharing framework for maximizing shared route. IEEE Trans. Knowl. Data Eng. **30**(2), 219–233 (2017)
16. Tong, Y., Zeng, Y., Zhou, Z., Chen, L., Ye, J., Xu, K.: A unified approach to route planning for shared mobility. Proc. VLDB Endowment **11**(11), 1633 (2018)
17. Uber: How uberpool works (2021). https://www.uber.com/nl/en/ride/uberpool/
18. Zeng, Y., Tong, Y., Chen, L.: Last-mile delivery made practical: an efficient route planning framework with theoretical guarantees. Proc. VLDB Endowment **13**(3), 320–333 (2019)

Algebraic Algorithms for Variants of Subset Sum

Pranjal Dutta[1](✉) and Mahesh Sreekumar Rajasree[2]

[1] Chennai Mathematical Institute, Chennai, India
pranjal@cmi.ac.in
[2] Indian Institute of Technology, Kanpur, India
mahesr@cse.iitk.ac.in

Abstract. Given $(a_1, \ldots, a_n, t) \in \mathbb{Z}_{\geq 0}^{n+1}$, the Subset Sum problem (SSUM) is to decide whether there exists $S \subseteq [n]$ such that $\sum_{i \in S} a_i = t$. Bellman (1957) gave a pseudopolynomial time dynamic programming algorithm which solves the Subset Sum in $O(nt)$ time and $O(t)$ space.

In this work, we present *search* algorithms for variants of the Subset Sum problem. Our algorithms are parameterized by k, which is a given upper bound on the number of realisable sets (i.e. number of solutions, summing exactly t). We show that SSUM with a unique solution is already NP-hard, under randomized reduction. This makes the regime of parametrized algorithms, in terms of k, very interesting.

Subsequently, we present an $\tilde{O}(k \cdot (n+t))$ time deterministic algorithm, which finds the hamming weight of all the realisable sets for a subset sum instance. We also give a poly(knt)-time and $O(\log(knt))$-space deterministic algorithm that finds all the realisable sets for a subset sum instance. Our algorithms use analytic and number-theoretic techniques.

Keywords: Subset sum · Power series · Isolation lemma · Hamming weight · Interpolation · Logspace · Newton's identities

1 Introduction: Variants of Subset Sum

The Subset Sum problem (SSUM) is a well-known NP-complete problem [1, p. 226], where given $(a_1, \ldots, a_n, t) \in \mathbb{Z}_{\geq 0}^{n+1}$, the problem is to decide whether there exists $S \subseteq [n]$ such that $\sum_{i \in S} a_i = t$. In the recent years, provable-secure cryptosystems based on SSUM such as private-key encryption schemes [2], tag-based encryption schemes [3], etc. have been proposed. There are numerous improvements made in the algorithms that solve the SSUM problem in both the classical [4–8] and quantum world [9–11]. One of the first algorithms was due to Bellman [12] who gave a $O(nt)$ time (*pseudo-polynomial* time) algorithm which requires $\Omega(t)$ space. One can ask for a *search* version of this problem,

---

The full version is available at this link.

P. Dutta—Supported by Google PhD Fellowship.
M. S. Rajasree—Supported by Prime Minister's Research Fellowship.

© Springer Nature Switzerland AG 2022
N. Balachandran and R. Inkulu (Eds.): CALDAM 2022, LNCS 13179, pp. 237–251, 2022.
https://doi.org/10.1007/978-3-030-95018-7_19

i.e. to output all the solutions. Since there can be *exponentially* many solutions, it could take $\exp(n)$-time (and space), to output them. This motivates our first problem defined below.

Problem 1 (k−SSSUM). Given $(a_1, \ldots, a_n, t) \in \mathbb{Z}_{\geq 0}^{n+1}$, the k-solution SSUM($k-$SSSUM) problem asks to output all $S \subseteq [n]$ such that $\sum_{i \in S} a_i = t$ provided with the guarantee that the number of such subsets is at most k.

▶ Remark. We denote $1 -$ SSSUM as unique Subset Sum problem (uSSSUM). In stackexchange, a more restricted version was asked where it was assumed that $k = 1$, for *any* realizable t. Here we just want $k = 1$ for some fixed target value t and we do not assume anything for any other value t'.

Now, we consider a different restricted version of the $k -$ SSSUM, where we demand to output only the hamming weights of the k-solutions (we call it Hamming $- k -$ SSSUM, for definition see Problem 2). By hamming weight of a solution, we mean the number of a_i's in the solution set (which sums up to exactly t). In other words, if $\vec{a} \cdot \vec{v} = t$, where $\vec{a} = (a_1, \ldots, a_n)$ and $\vec{v} \in \{0,1\}^n$, we want $|v|_1$, the ℓ_1-norm of the solution vector.

Problem 2 (Hamming $- k -$ SSSUM). Given an instance of the $k -$ SSSUM, say $(a_1, \ldots, a_n, t) \in \mathbb{Z}_{\geq 0}^{n+1}$, with the promise that there are at most k-many $S \subseteq [n]$ such that $\sum_{i \in S} a_i = t$, Hamming $- k -$ SSSUM asks to output all the hamming weights (i.e., $|S|$) of the solutions.

It is obvious that solving $k-$SSSUM solves Problem 2. Importantly, the decision problem, namely the HWSSUM is already NP-hard. The HWSSUM problem is : given an instance $(a_1, \ldots, a_n, t, w) \in \mathbb{Z}_{\geq 0}^{n+2}$, decide whether there is a solution to the Subset Sum with hamming weight equal to w. Note that, there is a trivial Cook's reduction from the SSUM to the HWSSUM: SSUM decides 'yes' to the instance $(a_1, \ldots, a_n, t)$ iff at least one of the following HWSSUM instances $(a_1, \ldots, a_n, t, i)$, for $i \in [n]$ decides 'yes'. Therefore, the search-version of HWSSUM, the Hamming $- k -$ SSSUM problem, is already an interesting problem and worth investigating.

In this work, we give various deterministic algorithms for Problem 1-2. Our algorithms are algebraic and number theoretic in nature and mainly build upon the previous power series techniques, by Jin and Wu [6] and sparse interpolation [13].

1.1 Main Results

In this section, we briefly state our main results. The leitmotif of this paper is to give efficient algorithms for variants of SSUM, with a promise of a bounded number of solutions. Our first theorem gives an efficient pseudo-linear $\tilde{O}(n + t)$ time *deterministic* algorithm for Problem 2, for constant k.

Theorem 1 (Algorithm for hamming weight). *There is a $\tilde{O}(k(n+t))$-time deterministic algorithm for* Hamming $- k -$ SSSUM.

▶ **Remark (Optimality).** We emphasize the fact that Theorem 1 is likely to be *near*-optimal for bounded k, due to the following argument. An $O(t^{1-\epsilon})$ time algorithm for Hamming $-1-$ SSSUM can be directly used to solve $1-$ SSSUM, as discussed above. By using the *randomized* reduction (Theorem 3), this would give us a randomized $n^{O(1)}t^{1-\epsilon}$-time algorithm for SSUM. But, in [14] the authors showed that SSUM does not have $n^{O(1)}t^{1-\epsilon}$ time algorithm unless the Strong Exponential Time Hypothesis (SETH) is false.

Theorem 2 (Algorithms for finding solutions in low space). *There is a* poly(knt)*-time and* $O(\log(knt))$*-space deterministic algorithm which solves* $k-$ SSSUM.

▶ **Remark.** When considering low space algorithms outputting multiple values, the standard assumption is that the output is written onto a one-way tape which *does not* count into the space complexity; so an algorithm outputting $kn \log n$ bits (like in the above case) could use much less working memory than $kn \log n$; for a reference see McKay and Williams [15].

▶ **Comparison with the Trivial Algorithm.** Consider the usual search-to-decision reduction for subset sum: First try to include a_1 in the subset, and if it is feasible then we subtract t by a_1 and add a_1 into the solution, and then continue with a_2, and so on. This procedure finds a single solution, but if we implement it in a recursive way then it can find all the k solutions in $k \cdot n \cdot$ (time complexity for decision version) time; we can think about an n-level binary recursion tree where all the infeasible subtrees are pruned.

Theorem 1 Is Better than the Trivial. Since number of solutions is bounded by k, choosing a prime $p > n+t+k$ suffices in [6], to make the algorithm deterministic. Thus, the time complexity of the decision version is $\tilde{O}((n+t)\log k)$. Hence, from the above, the search complexity is $\tilde{O}(kn(n+t))$ which is *worse* than Theorem 1.

Theorem 2 Is Better than the Trivial. For solving the decision problem in low space, we simply use Kane's $O(\log(nt))$-space poly(nt)-time algorithm [16]. As explained (and improved) in [7], the time complexity is actually $O(n^3 t)$ and the extra space usage is $\tilde{O}(n)$ for remembering the recursion stack. Thus the total time complexity is $O(kn^4 t)$ and it takes $\tilde{O}(n) + O(\log t)$ space. While Theorem 2 takes $O(\log(knt))$ space and poly(knt) time. Although our time complexity is worse[1], when $k \leq 2^{O((n \log t)^{1-\epsilon})}$, for $\epsilon > 0$, our space complexity is *better*.

1.2   Technical Overview

All the algorithms presented in this paper consider that the number of solutions is *bounded* by a parameter k. This naturally raises the question whether the SSUM problem is hard, even when the number of solutions is bounded. We will show that this is true even for the case when $k = 1$, i.e., uSSSUM is NP-hard under *randomized* reduction.

---

[1] Thm. 2 is *not about* time complexity; as long as it is pseudopolynomial time it's ok.

Theorem 3 (Hardness of uSSSUM). *There exists a randomized reduction which takes a* SSUM *instance* $\mathcal{M} = (a_1, \ldots, a_n, t) \in \mathbb{Z}_{\geq 0}^{n+1}$, *as an input, and produces multiple* SSUM *instances* $SS_\ell = (b_1, \ldots, b_n, t^{(\ell)})$, *where* $\ell \in [2n^2]$, *such that if*

- $\mathcal{M}$ *is a YES instance of* SSUM $\implies \exists \ell$ *such that* SS_ℓ *is a YES instance of* uSSSUM;
- $\mathcal{M}$ *is a NO instance of* SSUM $\implies \forall \ell, SS_\ell$ *is a NO instance of* uSSSUM.

Proof. The core of the proof is based on the Lemma 1 (Isolation lemma). The reduction is as follows. Let $w_1, \ldots, w_n$ be chosen *uniformly at random* from $[2n]$. We define $b_i = 4n^2 a_i + w_i, \forall i \in [n]$ and the ℓth SSUM instance as $SS_\ell = (b_1, \ldots, b_n, t^{(\ell)} = 4n^2 t + \ell)$. Observe that all the new instances are different only in the target values $t^{(\ell)}$.

Suppose $\mathcal{M}$ is a YES instance, i.e., $\exists S \subseteq [n]$ such that $\sum_{i \in S} a_i = t$. Then, for $\ell = \sum_{i \in S} w_i$, the SS_ℓ is a YES instance, because

$$\sum_{i \in S} b_i - t^{(\ell)} = 4n^2 \left(\sum_{i \in S} a_i - t \right) - \left(\ell - \sum_{i \in S} w_i \right) = 0 \,.$$

If $\mathcal{M}$ is a NO instance, consider any ℓ and $S \subseteq [n]$. Since $\mathcal{M}$ is a NO instance, $4n^2 (\sum_{i \in S} a_i - t)$ is a non-zero multiple of $4n^2$, whereas $|\ell - \sum_{i \in S} w_i| < 4n^2$, which implies that

$$4n^2 (\sum_{i \in S} a_i - t) - (\ell - \sum_{i \in S} w_i) \neq 0 \implies \sum_{i \in S} b_i \neq t^{(\ell)} \,.$$

Hence, SS_ℓ is also a NO instance.

We now show that if $\mathcal{M}$ is a YES instance, then one of SS_ℓ is a uSSSUM. Let $\mathcal{F}$ contain all the solutions to the SSUM instance $\mathcal{M}$, i.e. $\mathcal{F} = \{S | S \subseteq [n], \sum_{i \in S} a_i = t\}$. Since w_i's are chosen uniformly at random, Lemma 1 says that there exists a *unique* $S \in \mathcal{F}$, such that $w(S) = \sum_{i \in S} w_i$, is *minimal* with probability at least $1/2$. Let us denote this minimal value $w(S)$ as ℓ^*. Then, SS_{ℓ^*} is uSSSUM because S is the only subset such that $\sum_{i \in S} w_i = \ell^*$. $\square$

Proof idea of Theorem 1. First we sketch the idea for $k = 1$. Suppose, we have a uSSSUM instance such that the hamming weight of the unique solution is w. Choose a prime $q = O(n + k + t)$ and a *primitive root* μ, i.e. $\text{ord}_q(\mu) = q - 1$ (for definition, see Definition 2). We can find them efficiently in $\tilde{O}(n + k + t)$ time.

Now, consider the following important polynomial $f(x) = \prod_{i=1}^{n} (1 + \mu \cdot x^{a_i})$. Observe that the coefficient of x^t in f is μ^w. Therefore, by using Lemma 6, we can find μ^w from $f(x)$ and extract w, since $\text{ord}_q(\mu) = q - 1 > n \geq w$. This solves Hamming $- 1 -$ SSSUM.

This idea can be extended to general k-SSSUM instance. Observe that, we cannot directly use the above trick, for a single polynomial $f(x)$, since, in this case, the coefficient of x^t is $\sum_{i \leq k} \lambda_i \cdot \mu^{w_i}$, where w_i are the hamming weights of the solution, which occur λ_i times. Eventually, we want to create a polynomial whose roots are of the form μ^{w_i}, so that we can first find the roots μ^{w_i} (over $\mathbb{F}_q$), and from them we can find w_i. To achieve that, we work with k-many polynomials $f_j := \prod_{i=1}^{n}(1 + \mu^j \cdot x^{a_i})$, for $j \in [k]$. Note that the coefficient of x^t in f_j is of the form $\sum_{i \leq k} \lambda_i \cdot \mu^{jw_i}$ (Claim 2). By Newton's Identities (Lemma 3) and Vieta's formulas (Lemma 4), we can now *efficiently* construct a polynomial whose roots are μ^{w_i}. For details, see Sect. 3.

Proof idea of Theorem 2. The above polynomial method *fails* to give a low space algorithm, since Lemma 6 requires $\Omega(t)$ space (eventually it needs to store all the coefficients mod x^{t+1}). Therefore, our proof idea of Theorem 2 is completely different from that of Theorem 1. Here, we work with a multivariate polynomial $f(x, y_1, \ldots, y_n) = \prod_{i=1}^{n}(1 + y_i x^{a_i})$ over $\mathbb{F}_q$, for a large prime $q = O(nt)$ and its multiple evaluations $f(\alpha, c_1, \ldots, c_n)$, where $(\alpha, c_1, \ldots, c_n) \in \mathbb{F}_q^{n+1}$.

Observe that, the coefficient of x^t in f is a multivariate polynomial $p_t(y_1, \ldots, y_n)$; each of its monomial carries the *necessary information* of a solution, for the instance $(a_1, \ldots, a_n, t)$. More precisely, S is a realisable set of $(a_1, \ldots, a_n, t) \iff \prod_{i \in S} y_i$ is a monomial in p_t. And, the sparsity (number of monomials) of p_t is at most k.

Therefore, it boils down to find the multivariate polynomial p_t. How easy it is to find p_t? Note that we cannot expect to find p_t, just by trivial multiplication as it would take $\tilde{O}(2^n t)$ time! Instead, our algorithm is a *reconstruction* algorithm, which *efficiently* reconstructs p_t, from multiple evaluations points $f(\alpha, c_1, \ldots, c_n)$, for $\alpha \in \mathbb{F}_q^*$. Eventually, we will use sparse interpolation [13] (see Theorem 6), which requires evaluations of the polynomial $p_t(y_1, \ldots, y_n)$ at multiple (polynomially many) points $(c_1, \ldots, c_n) \in \mathbb{F}_q^n$. To find $p_t(c_1, \ldots, c_n)$, we use Kane's identity (Lemma 2) which uses the evaluations $f(\alpha, c_1, \ldots, c_n)$, for $\alpha \in [1, q-1]$. Finding $p_t(c_1, \ldots, c_n)$ can be efficiently done in logspace. The rest (to reconstruct p_t) requires a brief space complexity analysis of [13]. For details, refer to Sect. 4.

1.3   Prior Works and Their Limitations

Before going into the details, we briefly review the state of the art of the problems (& its variants). After Bellman's $O(nt)$ dynamic solution [12], Pisinger [17] first improved it to $O(nt/\log t)$ on word-RAM models. Recently, Koiliaris and Xu gave a deterministic algorithm [18,19] in time $\tilde{O}(\sqrt{n}t)$, which is the best deterministic algorithm so far. Bringmann [5] & Jin and Wu [6] later improved the running time to randomized $\tilde{O}(n + t)$. All these algorithms require $\Omega(t)$ space. Moreover, most of the recent algorithms solve the decision versions. Here we remark that Abboud et al. [14] recently showed that SSUM has no $t^{1-\epsilon} n^{O(1)}$ time algorithm for any $\epsilon > 0$, unless the Strong Exponential Time Hypothesis (SETH) is false. Therefore, the $\tilde{O}(n + t)$ time bound is likely to be *near-optimal*.

In [18] (also see [19, Lemma 2]), the authors gave a deterministic $\tilde{O}(nt)$ algorithm that finds all the hamming weights for all realisable targets less than equal to t. Their algorithm *does not* depend on the number of solutions for a particular target. Compared to this, our Theorem 1 is *faster* when $k = o(n/(\log n)^c)$, for a large constant c. Similarly, with the 'extra' information of k, we give a *faster* deterministic algorithm (which even outputs all the hamming weights of the solutions) compared to $\tilde{O}(\sqrt{n}t)$ decision algorithm in [18,19] (which outputs all the realisable subset sums $\leq t$), when $k = o(\sqrt{n}/(\log n)^c)$, for a large constant c. Here we remark that the $O(nt)$-time dynamic programming algorithm [12] can be easily modified to find all the solutions, but this gives an $O(n(k + t))$-time (and space) algorithm solution.

On the other hand, there have been quite some work on solving SSUM in LOGSPACE. Lokshtanov and Nederlof [20], and Kane [16] (2010) gave $O(\log nt)$ space poly(nt)-time deterministic algorithm, which have been very recently improved to $\tilde{O}(n^2t)$-time and poly $\log(nt)$ space. On the other hand, Bringmann [5] gave a $nt^{1+\epsilon}$ time, $O(n \log t)$ space *randomized* algorithm, which have been improved to $O(\log n \log \log n + \log t)$ space in [7]. Again, most of the algorithms are decision algorithms and do not output the solution set. In contrast to this, our algorithm in Theorem 2 uses only $O(\log(knt))$ space and outputs all the solution sets, which is near-optimal.

Finally, we remark that in the proof of Theorem 1, we extend analytic tools from [6] to our advantage (see Lemma 6), yet our algorithm for Theorem 1 is *deterministic* (unlike in [6]).

2   Preliminaries and Notations

Notations. $\mathbb{Z}$ and $\mathbb{Q}$ denotes the set of all integers and rationals, respectively. For any integer $n > 0$, $[n]$ denotes the set $\{1, 2, \ldots, n\}$, while $2^{[n]}$ denotes the set of all subsets of $[n]$. log denotes $\log_2$. We also denote $\tilde{O}(g)$ to be $g \cdot \text{poly}(\log g)$.

Sparsity of a polynomial $f(x_1, \ldots, x_n) \in \mathbb{F}[x_1, \ldots, x_n]$ over a field $\mathbb{F}$, denotes the number of nonzero terms in f.

A weight function $w : [n] \longrightarrow [m]$, can be naturally extended to a set $S \in 2^{[n]}$, by defining $w(S) := \sum_{i \in S} w(i)$.

Definition 1 (Subset Sum problem (SSUM)). *Given* $(a_1, \ldots, a_n, t) \in \mathbb{Z}_{\geq 0}^{n+1}$, *the subset sum problem is to decide whether t is a realisable target with respect to* $(a_1, \ldots, a_n)$, *i.e., there exists $S \subseteq [n]$ such that $\sum_{i \in S} a_i = t$. Here, n is called the size, t is the target and any $S \subseteq [n]$ such that $\sum_{i \in S} a_i = t$ is a realisable set of the subset sum instance.*

Assumptions. Throughout the paper, we assume that $t \geq \max a_i$ for simplicity. Also, we work in the Turing model where basic operations like addition and multiplication over $\mathbb{F}_p$ are not unit-cost unlike Word Ram model considered in [6], for simplicity; in the word RAM model our results will give slightly better result shaving one $\log p$ factor.

Lemma 1 ([21, Isolation Lemma]). *Let n and N be positive integers, and let $\mathcal{F}$ be an arbitrary family of subsets of $[n]$. Suppose $w(x)$ is an integer weight* *given to each element $x \in [n]$ uniformly and independently at random from $[N]$. The weight of $S \in \mathcal{F}$ is defined as $w(S) = \sum_{x \in S} w(x)$. Then, with probability at least $1 - n/N$, there is a unique set $S' \in \mathcal{F}$ that has the minimum weight among all sets of $\mathcal{F}$.*

Lemma 2 (Kane's Identity [16]**).** *Let $f(x) = \sum_{i=0}^{d} c_i x^i$ be a polynomial of degree at most d with coefficients c_i being integers. Let $\mathbb{F}_q$ be the finite field of order $q = p^k > d + 2$. For $0 \leq t \leq d$, define*

$$r_t = \sum_{x \in \mathbb{F}_q^*} x^{q-1-t} f(x) = -c_t \in \mathbb{F}_q$$

Then, $r_t = 0 \iff c_t$ is divisible by p.

Lemma 3 (Newton's Identities). *Let $X_1, \ldots, X_n$ be $n \geq 1$ variables. Let $P_m(X_1, \ldots, X_n) = \sum_{i=1}^{n} X_i^m$, be the m-th power sum and $E_m(X_1, \ldots, X_n)$ be the m-th elementary symmetric polynomials i.e. $E_m(x_1, \ldots, x_n) = \sum_{1 \leq j_1 \leq \ldots \leq j_m \leq n} X_{j_1} \cdots X_{j_m}$, then*

$$m \cdot E_m(X_1, \ldots, X_n) = \sum_{i=1}^{m} (-1)^{i-1} E_{m-i}(X_1, \ldots, X_n) \cdot P_i(X_1, \ldots, X_n).$$

Lemma 4 (Vieta's formulas). *Let $f(x) = \prod_{i=1}^{n}(x - a_i)$ be a monic polynomial of degree n. Then, $f(x) = \sum_{i=0}^{n} c_i x^i$ where $c_{n-i} = (-1)^i E_i(a_1, \ldots, a_n), \forall 1 \leq i \leq n$ and $c_n = 1$.*

Lemma 5 (Polynomial division with remainder [22, **Theorem 9.6]).** *Given a d-degree polynomial f and a linear polynomial g over a finite field $\mathbb{F}_p$, there exists a deterministic algorithm that finds the quotient and remainder of f divided by g in $\tilde{O}(d \log p)$-time.*

Definition 2 (Order of a number mod p**).** *The order of a (mod p), denoted as $\text{ord}_p(a)$ is defined to be the smallest positive integer m such that $a^m \equiv 1$ mod p.*

Note that when p is prime, $\text{ord}_p(a)$ is clearly finite since $a^{p-1} \equiv 1$ mod p, from Fermat's Little Theorem. Emil Artin (1927, see [23]) conjectured that for any non-square $a \in \mathbb{Z} \setminus \{-1\}$, there exist infinitely many primes p such that a is a *primitive root* modulo p, i.e. $\text{ord}_p(a) = p - 1$. There has been impressive amount of work done to understand behaviour and distribution of $\text{ord}_p(a)$ [24–26]. In particular, we have the following.

Theorem 4 ([27]**).** *There exists a $\tilde{O}(p^{1/4+\epsilon})$ time algorithm to determinstically find a primitive root over $\mathbb{F}_p$.*

Theorem 5 ([28]**).** *For $n \geq 25$, there is a prime in the interval $[n, 6/5 \cdot n]$.*

Here is the most important lemma, which is an extension of [6, Lemma 4], where the authors considered the simplest form. In this paper, we need the extensions for the 'robust' usage of this lemma (in Sect. 3).

Lemma 6 (Coefficient Extraction Lemma). *Let $A(x) = \prod_{i \in [n]}(1 + W^b \cdot x^{a_i})$, for any non-negative integers a_i, b and $W \in \mathbb{Z}$. Then, for a prime $p > t$, one can compute $\mathrm{coef}_{x^r}(A(x)) \bmod p$ for all $0 \leq r \leq t$, in time $\tilde{O}((n + t\log(Wb))\log p)$.*

3   Proof of Theorem 1

We present an $\tilde{O}(k(n + t))$-time deterministic algorithm for outputting all the hamming weight of the solutions, given a Hamming $- k -$ SSSUM instance i.e. there are only at most k-many solutions to the SSUM instance $(a_1, \ldots, a_n, t) \in \mathbb{Z}_{\geq 0}^{n+1}$.

Proof of Theorem 1. We start with some notations that we will use throughout the proof.

▶ Basic notations. Assume that the SSUM instance $(a_1, \ldots, a_n, t) \in \mathbb{Z}_{\geq 0}^{n+1}$ has *exactly* m $(m \leq k)$ many solutions, and they have ℓ many *distinct* hamming weights $w_1, \ldots, w_\ell$; since two solutions can have same hamming weight, $\ell \leq m$. Moreover, assume that there are λ_i many solutions which appear with hamming weight w_i, for $i \in [\ell]$. Thus, $\sum_{i \in [\ell]} \lambda_i = m \leq k$.

▶ Choosing prime q and a primitive root μ. We will work with a fixed q in this proof, where $q > n + k + t := M$ (we will mention why such a requirement later). We can find a prime q in $\tilde{O}(n + k + t)$ time, since we can go over every element in the interval $[M, 6/5 \cdot M]$, in which we know a prime exists (Theorem 5) and primality testing is efficient [29]. Once we find q, we choose μ such that μ is a *primitive root* over $\mathbb{F}_q$, i.e. $\mathrm{ord}_q(\mu) = q - 1$. This μ can be found in $\tilde{O}((n + k + t)^{1/4+\epsilon})$ time using Theorem 4. Thus, the total time complexity of this step is $\tilde{O}(n + k + t)$.

▶ The polynomials. Define the k-many univariate polynomials as follows:

$$f_j(x) := \prod_{i \in [n]} \left(1 + \mu^j x^{a_i}\right), \, \forall j \in [k].$$

We remark that we do not know ℓ apriori, but we can find m efficiently.

Claim 1 (Finding the exact number of solutions). Given a Hamming$-k-$SSSUM instance, one can find the exact number of solutions, m, deterministically, in $\tilde{O}((n + t)\log(q))$ time.

Proof. Use [6] (see Lemma 6, for the general statement) which gives a deterministic algorithm to find the coefficient of x^t of $\prod_{i \in [n]} (1 + x^{a_i})$ over $\mathbb{F}_q$; this takes time $\tilde{O}((n + t)\log(q))$. □

Since we know the exact value of m, we will just work with f_j for $j \in [m]$, which suffices for our algorithmic purpose. Here is an important claim about coefficients of x^t in f_j's.

Claim 2. $C_j = \mathrm{coef}_{x^t}(f_j(x)) = \sum_{i \in [\ell]} \lambda_i \cdot \mu^{jw_i}$, for each $j \in [m]$.

Proof. If $S \subseteq [n]$ is a solution to the instance with hamming weight, say w, then this will contribute μ^{jw} to the coefficient of x^t of $f_j(x)$. Since, there are ℓ many weights $w_1, \ldots, w_\ell$ with multiplicity $\lambda_1, \ldots, \lambda_\ell$, the claim easily follows. $\square$

Using Lemma 6, we can find $C_j \mod q$ for each $j \in [m]$ in $\tilde{O}((n+t\log(\mu j))\log q)$ time, owing total $\tilde{O}(k(n+t))$, since $q = O(n+k+t)$, $\mu \le q-1$, and $\sum_{j \in [m]} \log j = \log(m!) \le \log(k!) = \tilde{O}(k)$.

Using the Newton's Identities (Lemma 3), we have the following relations, for $j \in [m]$:

$$E_j(\mu^{w_1}, \ldots, \mu^{w_\ell}) \equiv j^{-1} \cdot \left(\sum_{i=1}^{j} (-1)^{i-1} E_{j-i}(\mu^{w_1}, \ldots, \mu^{w_k}) \cdot P_i(\mu^{w_1}, \ldots, \mu^{w_\ell}) \right) \mod q.$$

$$(1)$$

In the above, by $E_j(\mu^{w_1}, \ldots, \mu^{w_\ell})$, we mean $E_j(\underbrace{\mu^{w_1}, \ldots, \mu^{w_1}}_{\lambda_1 \text{ times}}, \underbrace{\mu^{w_2}, \ldots, \mu^{w_2}}_{\lambda_2 \text{ times}},$
$\ldots, \underbrace{\mu^{w_\ell}, \ldots, \mu^{w_\ell}}_{\lambda_\ell \text{ times}})$, and similar for P_j. Since $q > k$, $j^{-1} \mod q$ exists, and thus
the above relations are valid. Here is another important and obvious observation, just from the definition of P_j's:

Observation 1. *For* $j \in [k]$, $C_j \equiv P_j(\mu^{w_1}, \ldots, \mu^{w_\ell}) \mod q$.

Note that we know $E_0 = 1$ and P_j's (and $j^{-1} \mod q$) are already computed. To compute E_j, we need to know $E_1, \ldots, E_{j-1}$ and additionally we need $O(j)$ many additions and multiplications. Suppose, $T(j)$ is the time to compute $E_1, \ldots, E_j$. Then, the trivial complexity is $T(m) \le \tilde{O}(k^2 \log q) + \tilde{O}(k(n+t))$. But one can do better than $\tilde{O}(k^2 \log q)$ and make it $\tilde{O}(k \log q)$ (i.e. solve the recurrence, using FFT), owing the total complexity to $T(m) \le \tilde{O}(k(n+t))$ (since $q = O(n+k+t)$).

Once, we have computed E_j, for $j \in [m]$, define a new polynomial

$$g(x) := \sum_{j=0}^{m} (-1)^j \cdot E_j(\mu^{w_1}, \ldots, \mu^{w_\ell}) \cdot x^j.$$

Using Lemma 4, it is immediate that $g(x) = \prod_{i=1}^{\ell} (x - \mu^{w_i})^{\lambda_i}$. Further, by definition, $\deg(g) = m$. From g, now we want to extract the roots, namely $\mu^{w_1}, \ldots, \mu^{w_\ell}$ over $\mathbb{F}_q$. We do this, by checking whether $(x - \mu^i)$ divides g, for $i \in [n]$ (since $w_i \le n$). Using Lemma 5, a single division with remainder takes $\tilde{O}(k \log q)$, therefore, the total time to find all the w_i is $\tilde{O}(nk \log q) = \tilde{O}(nk)$.

Here, we *remark* that we do not use the deterministic root finding or factoring algorithms (for e.g. [30, 31]), since it takes $\tilde{O}(mq^{1/2}) = \tilde{O}(k \cdot (k+t)^{1/2})$ time, which could be larger than $\tilde{O}(k(n+t))$.

▶ **Reason for choosing q and μ.** In the hindsight, there are three important properties of the prime q that will suffice to successfully output the w_i's using the above described steps:

1. Since, Lemma 6 *requires* to compute the inverses of numbers upto t, hence, we would want $q > t$.
2. While computing $E_j(\mu^{w_1}, \ldots, \mu^{w_k})$ using Lemma 3 in the above, one should be able to compute the inverse of all j's less than equal to m. So, we want $q > m,$.
3. To obtain w_i from $\mu^{w_i} \mod q$, we want $\text{ord}_q(\mu) > n$ (for definition see Definition 2). Since, $w_i \leq n$, this would ensure that we have found the correct w_i.

Here, we remark that we do not need to concern ourselves about the 'largeness' of the coefficients of C_j and make it nonzero $\mod q$, as required in [6]. For the first two points, it suffices to choose $q > k + t$. Since μ is a primitive root over $\mathbb{F}_q$, this guarantees that $\text{ord}_q(\mu) = q - 1 > n$ and thus we will find w_i from μ^{w_i} correctly.

▶ **Total time complexity.** The time complexity to find the correct m, q and μ is $\tilde{O}(n + k + t)$. Finding the coefficients of g takes $\tilde{O}(k(n + t))$ time and then finding w_i from g takes $\tilde{O}(nk \log q)$ time. Thus, the total time complexity remains $\tilde{O}(k(n + t))$. □

Remark 1. The above algorithm can be extended to find the multiplicities λ_i's in $\tilde{O}(k(n + t) + k^{3/2})$ time by finding the largest λ_i, by binary search, such that $(x - \mu^{w_i})^{\lambda_i}$ divides $g(x)$. Finding each λ_i takes $\tilde{O}(m \log q \log(\lambda_i))$ time over $\mathbb{F}_q$, for the same q as above, since the polynomial division takes $\tilde{O}(m \log q)$ time and binary search introduces a multiplicative $O(\log(\lambda_i))$ term. Since, $\sum_{i \in [\ell]} \log(\lambda_i) = \log\left(\prod_{i \in [\ell]} \lambda_i\right)$, using AM-GM, $\prod_{i \in [\ell]} \lambda_i \leq (m/\ell)^\ell$, which is maximized at $\ell = \sqrt{m} \leq \sqrt{k}$, implying $\sum_{i \in [\ell]} \log(\lambda_i) \leq O(\sqrt{k} \log k)$. Since, $m \leq k$, this explains the additive $k^{3/2}$ term in the complexity.

4   Proof of Theorem 2

In this section, we will present a low space algorithm for finding all the realisable sets for $k -$ SSSUM. Our low space algorithms build upon a fundamental number-theoretic identity [16], and efficient sparse multivariate polynomial reconstruction [13].

Proof of Theorem 2. Here are some notations that we will follow throughout the proof.

▶ **Basic Notations.** Let us assume that there are exactly m ($m \leq k$) many realisable sets $S_1, \ldots, S_m$, each $S_i \subseteq [n]$. We remark that for our algorithm we do not need to apriori calculate m.

▶ The Multivariate Polynomial. For our purpose, we will be working with the following $(n+1)$-variate polynomial:

$$f(x, y_1, \ldots, y_n) := \prod_{i \in [n]} (1 + y_i x^{a_i}) \ .$$

Since, we have a $k - \mathsf{SSSUM}$ instance $(a_1, \ldots, a_n, t)$, $\mathrm{coef}_{x^t}(f)$ has the following properties.

1. It is an n-variate polynomial $p_t(y_1, \ldots, y_n)$ with sparsity *exactly m*.
2. p_t is a multilinear polynomial in $y_1, \ldots, y_n$, i.e. individual degree of y_i is at most 1.
3. The total degree of p_t is at most n.
4. if $S \subseteq [n]$ is a realisable set, then $\mathbf{y}_S := \prod_{i \in S} y_i$, is a monomial in p_t.

In particular, the following is an immediate but important observation.

Observation 2. $p_t(y_1, \ldots, y_n) = \sum_{i \in [m]} \mathbf{y}_{S_i}$.

Therefore, it suffices to know the polynomial p_t. However, we cannot treat y_i as new variables and try to find the coefficient of x^t since the trivial multiplication algorithm (involving $n + 1$ variables) takes $\exp(n)$-time. This is because, $f(x, y_1, \ldots, y_n) \mod x^{t+1}$ can have $2^n \cdot t$ many monomials as coefficient of x^i, for any $i \leq t$ can have 2^n many multilinear monomials.

However, if we substitute $y_i = c_i \in \mathbb{F}_q$, for some prime q, we claim that we can figure out the value $p_t(c_1, \ldots, c_n)$ from the coefficient of x^t in $f(x, c_1, \ldots, c_n)$ efficiently (see Claim 3). Once we have figured out, we can simply interpolate using the following theorem to reconstruct the polynomial p_t. Before going into the technical details, we state the sparse interpolation theorem below; for simplicity we consider multilinearity (though [13] holds for general polynomials as well).

Theorem 6 ([13]). *Given a black box access to a multilinear polynomial $g(x_1, \ldots, x_n)$ of degree d and sparsity at most s over a finite field $\mathbb{F}$ with $|\mathbb{F}| \geq (nd)^6$, there is a poly(snd)-time and $O(\log(snd))$-space algorithm that outputs all the monomials of g.*

Remark. We represent one monomial in terms of indices (to make it consistent with the notion of realisable set), i.e. for a monomial $x_1 x_5 x_9$, the corresponding indices set is $\{1, 5, 9\}$. Also, we do not include the indices in the space complexity, as mentioned earlier.

▶ Brief Analysis on the Space Complexity of [13]. Klivans and Spielman [13], did not explicitly mention the space complexity. However, it is not hard to show that the required space is indeed $O(\log(snd))$. [13] shows that substituting $x_i = y^{k^{i-1} \mod p}$, for some $k \in [2s^2 n]$ and $p > 2s^2 n$, makes the exponents of the new univaraite polynomial (in y) *distinct* (see [13, Lemma 3]); the algorithm actually tries for all k and find the correct k. Note that the degree becomes $O(s^2 nd)$.

Then, it tries to first find out the coefficients by simple univariate interpolation [13, Section 6.3]. Since we have blackbox access to $g(a_1, \ldots, a_n)$, finding out a single coefficient, by univariate interpolation (which basically sets up linear equations and solve) takes $O(\log(snd))$ space and $\mathsf{poly}(snd)$ time only. In the last step, to find one coefficient, we can use the standard univariate interpolation algorithm which uses the Vandermonde matrices and one entry of the inverse of the Vandermonde is log-space computable[2].

At this stage, we know the coefficients (one by one), but we do not know which monomials the coefficients belong. However, it suffices to substitute $x_i = 2y^{k^{i-1} \bmod p}$. Using this, we can find the correct value of the first exponent in the monomial. For e.g. if after the correct substitution, y^{10} appears with coefficient say 5, next step, when we change just x_1, if it does not affect the coefficient 5, y_1 is not there in the monomial corresponding to the monomial which has coefficient 5, otherwise it is there (here we also use that it is multilinear and hence the change in the coefficient must be reflected). This step again requires univariate interpolation, and one has to repeat this experiment wrt each variable to know the monomial exactly corresponding to the coefficient we are working with. We can reuse the space for interpolation and after one round of checking with every variable, it outputs one exponent at this stage. This requires $O(\log(snd))$-space and $\mathsf{poly}(snd)$ time.

With a more careful analysis, one can further improve the field requirement to $|\mathbb{F}| \geq (nd)^6$ only (and not dependent on s); for details see [13, Thm. 5 & 11].

Now we come back to our subset sum problem. Since we want to reconstruct an n-variate m sparse polynomial p_t which has degree at most n, it suffices to work with $|\mathbb{F}| \geq n^{12}$. However, we also want to use Kane's identity (Lemma 2), which requires $q > \deg(f(x, c_1, \ldots, c_n)) + 2$, and $\deg(f(x, c_1, \ldots, c_n)) \leq nt$. Denote $M := \max(nt+3, n^{12})$. Thus, it suffices to we work with $\mathbb{F} = \mathbb{F}_q$ where $q \in [M, (6/5) \cdot M]$, such prime exists (Theorem 5) and easy to find deterministically in $\mathsf{poly}(nt)$ time and $O(\log(nt))$ space using [29]. In particular, we will substitute $y_i = c_i \in [0, q-1]$.

Claim 3. Fix $c_i \in [0, q-1]$, where $q \in [M, (6/5) \cdot M]$. Then, there is a $\mathsf{poly}(nt)$-time and $O(\log(nt))$ space algorithm which computes $p_t(c_1, \ldots, c_n)$ over $\mathbb{F}_q$.

Proof. Note that, we can evaluate each $1 + c_i x^{a_i}$, at some $x = \alpha \in \mathbb{F}_q$, in $\tilde{O}(\log nt)$ time and $O(\log(nt))$ space. Multiplying n of them takes $\tilde{O}(n \log(nt))$-time and $O(\log(nt))$ space.

Once we have computed $f(\alpha, c_1, \ldots, c_n)$ over $\mathbb{F}_q$, using Kane's identity (Lemma 2), we can compute $p_t(c_1, \ldots, c_n)$, since

$$p_t(c_1, \ldots, c_n) = -\sum_{\alpha \in \mathbb{F}_q^*} \alpha^{q-1-t} f(\alpha, c_1, \ldots, c_n) .$$

---

[2] In fact Vandermonde determinant and inverse computations are in $\mathsf{TC}^0 \subset \mathsf{LOGSPACE}$, see [32].

As each evaluation $f(\alpha, c_1, \ldots, c_n)$ takes $\tilde{O}(n \log(nt))$ time, and we need $q - 1$ many additions, multiplications and modular exponentiations, total time to compute is $\mathsf{poly}(nt)$. The required space still remains $O(\log(nt))$. $\qquad\square$

Once, we have calculated $p_t(c_1, \ldots, c_n)$ efficiently, now we try different values of $(c_1, \ldots, c_n)$ to reconstruct p_t using Theorem 6. Since, p_t is a n-variate at most k sparse polynomial with degree at most n, it still takes $\mathsf{poly}(knt)$ time and $O(\log(knt))$ space. This finishes the proof. $\qquad\square$

5 Conclusion

This work introduces some interesting search versions of variants of SSUM problem and gives efficient algorithms for each of them. This opens a variety of questions which require further rigorous investigations.

1. Can we improve the time complexity of Theorem 2? Because of using Theorem 6, the complexity for interpolation is already cubic. Whether some other algebraic (non-algebraic) techniques can improve the time complexity, while keeping it low space, is not at all clear.
2. Can we use these algebraic-number-theoretic techniques, to give a *deterministic* $\tilde{O}(n + t)$ time algorithm for decision version of SSUM?
3. Can we improve Remark 1 to find both the hamming weights w_i as well as the multiplicities λ_i, in $\tilde{O}(k(n + t))$ time?

References

1. Lewis, H.R.: Computers and Intractability. A Guide to the Theory of NP-Completeness (1983)
2. Lyubashevsky, V., Palacio, A., Segev, G.: Public-key cryptographic primitives provably as secure as subset sum. In: Micciancio, D. (ed.) TCC 2010. LNCS, vol. 5978, pp. 382–400. Springer, Heidelberg (2010). https://doi.org/10.1007/978-3-642-11799-2_23
3. Faust, S., Masny, D., Venturi, D.: Chosen-ciphertext security from subset sum. In: Cheng, C.-M., Chung, K.-M., Persiano, G., Yang, B.-Y. (eds.) PKC 2016. LNCS, vol. 9614, pp. 35–46. Springer, Heidelberg (2016). https://doi.org/10.1007/978-3-662-49384-7_2
4. Bringmann, K., Wellnitz, P.: On near-linear-time algorithms for dense subset sum. In: Proceedings of the 2021 ACM-SIAM Symposium on Discrete Algorithms (SODA), pp. 1777–1796. SIAM (2021)
5. Bringmann, K.: A near-linear pseudopolynomial time algorithm for subset sum. In: Proceedings of the Twenty-Eighth Annual ACM-SIAM Symposium on Discrete Algorithms, pp. 1073–1084. SIAM (2017)
6. Jin, C., Wu, H.: A simple near-linear pseudopolynomial time randomized algorithm for subset sum. arXiv preprint arXiv:1807.11597 (2018)
7. Jin, C., Vyas, N., Williams, R.: Fast low-space algorithms for subset sum. In: Proceedings of the 2021 ACM-SIAM Symposium on Discrete Algorithms (SODA), pp. 1757–1776. SIAM (2021)

250   P. Dutta and M. S. Rajasree

8. Esser, A., May, A.: Low weight discrete logarithm and subset sum in 20. 65n with polynomial memory. Memory **1**, 2 (2020)
9. Bernstein, D.J., Jeffery, S., Lange, T., Meurer, A.: Quantum algorithms for the subset-sum problem. In: Gaborit, P. (ed.) PQCrypto 2013. LNCS, vol. 7932, pp. 16–33. Springer, Heidelberg (2013). https://doi.org/10.1007/978-3-642-38616-9_2
10. Helm, A., May, A.: Subset sum quantumly in $1.17^{\wedge}$ n. In: 13th Conference on the Theory of Quantum Computation, Communication and Cryptography (TQC 2018). Schloss Dagstuhl-Leibniz-Zentrum fuer Informatik (2018)
11. Li, Y., Li, H.: Improved quantum algorithm for the random subset sum problem. arXiv preprint arXiv:1912.09264 (2019)
12. Bellman, R.E.: Dynamic Programming (1957)
13. Klivans, A.R., Spielman, D.: Randomness efficient identity testing of multivariate polynomials. In: Proceedings of the Thirty-Third Annual ACM Symposium on Theory of Computing, pp. 216–223 (2001)
14. Abboud, A., Bringmann, K., Hermelin, D., Shabtay, D.: Seth-based lower bounds for subset sum and bicriteria path. In: Proceedings of the Thirtieth Annual ACM-SIAM Symposium on Discrete Algorithms, pp. 41–57. SIAM (2019)
15. McKay, D.M., Williams, R.R.: Quadratic time-space lower bounds for computing natural functions with a random oracle. In: 10th Innovations in Theoretical Computer Science Conference (ITCS 2019). Schloss Dagstuhl-Leibniz-Zentrum fuer Informatik (2018)
16. Kane, D.M.: Unary subset-sum is in logspace. arXiv preprint arXiv:1012.1336 (2010)
17. Pisinger, D.: Linear time algorithms for knapsack problems with bounded weights. J. Algorithms **33**(1), 1–14 (1999)
18. Koiliaris, K., Chao, X.: Faster pseudopolynomial time algorithms for subset sum. ACM Trans. Algorithms (TALG) **15**(3), 1–20 (2019)
19. Koiliaris, K., Xu, C.: Subset sum made simple. arXiv preprint arXiv:1807.08248 (2018)
20. Lokshtanov, D., Nederlof, J.: Saving space by Algebraization. In: Proceedings of the Forty-Second ACM Symposium on Theory of Computing, pp. 321–330 (2010)
21. Mulmuley, K., Vazirani, U.V., Vazirani, V.V.: Matching is as easy as matrix inversion. In: Proceedings of the Nineteenth Annual ACM Symposium on Theory of Computing, pp. 345–354 (1987)
22. Von Zur Gathen, J., Gerhard, J.: Modern Computer Algebra. Cambridge University Press, Cambridge (2013)
23. Moree, P.: Artin's primitive root conjecture-a survey. Integers **12**(6), 1305–1416 (2012)
24. Gupta, R., Ram Murty, M.: A remark on Artin's conjecture. Inventiones Math. **78**(1), 127–130 (1984)
25. Erdös, P., Ram Murty, M.: On the order of a (mod p). In: CRM Proceedings and Lecture Notes, vol. 19, pp. 87–97 (1999)
26. Chinen, K., Murata, L.: On a distribution property of the residual order of a (mod p). J. Number Theory **105**(1), 60–81 (2004)
27. Shparlinski, I.: On finding primitive roots in finite fields. Theoret. Comput. Sci. **157**(2), 273–275 (1996)
28. Nagura, J.: On the interval containing at least one prime number. Proc. Jpn. Acad. **28**(4), 177–181 (1952)
29. Agrawal, M., Kayal, N., Saxena, N.: Primes is in p. Ann. Math. **160**, 781–793 (2004)

30. Shoup, V.: On the deterministic complexity of factoring polynomials over finite fields. Inf. Process. Lett. **33**(5), 261–267 (1990)
31. Bourgain, J., Konyagin, S., Shparlinski, I.: Character sums and deterministic polynomial root finding in finite fields. Math. Comput. **84**(296), 2969–2977 (2015)
32. Maciel, A., Therien, D.: Threshold circuits of small majority-depth. Inf. Comput. **146**(1), 55–83 (1998)

Hardness and Approximation Results for Some Variants of Stable Marriage Problem

B. S. Panda and Sachin[✉]

Department of Mathematics, Indian Institute of Technology Delhi,
New Delhi 110016, India
{bspanda,maz198086}@maths.iitd.ac.in

Abstract. We study several key variants of SMTI - Stable Marriage problem in which the preference lists may contain ties and may be incomplete. A matching is called *weakly stable* unless there is a man and a woman such that they are currently not matched with each other but if they get matched with each other, then both of them become better off. The COM SMTI problem is to decide whether there exists a complete (in which all men and women are matched) weakly stable matching in an SMTI instance. It is known that the COM SMTI problem is NP-complete. We strengthen this result by proving that this problem remains NP-complete even for the instance SMTI-C, instance where members in each preference list are consecutive with respect to some orderings of the set of men and set of women. On the positive side, we give a polynomial time algorithm for COM SMTI problem for the instance SMTI-STEP, where the preference lists admit *step-property*, that is, preference list of every man m_i is the set of all women w_j such that $j \leq i$ for some ordering of men and some ordering of women. Further, DECIDE_MAX SMTI (resp. DECIDE_MIN SMTI) is the decision version of MAX SMTI (resp. MIN SMTI), the problem of finding a weakly stable matching of maximum (resp. minimum) cardinality in an SMTI instance. Both DECIDE_MAX SMTI and DECIDE_MIN SMTI problems are known to be NP-complete. We improve these results by showing that DECIDE_MAX SMTI and DECIDE_MIN SMTI problems remain NP-complete even for the case where the preference lists admit inclusion ordering and even for the case where the preference lists admit step-property, respectively. Finally, we present a 3/2-approximation algorithm for the MIN SMTI problem with inclusion ordering.

Keywords: Stable matching · Polynomial time algorithm · NP-complete · Approximation algorithm

1 Introduction

An instance of the Stable Marriage problem (SM) consists of n men, n women, and their *preference lists*. A *preference list* of a man (resp. woman) is a set containing all women (resp. men) in strict order of preference. We denote preference

© Springer Nature Switzerland AG 2022
N. Balachandran and R. Inkulu (Eds.): CALDAM 2022, LNCS 13179, pp. 252–264, 2022.
https://doi.org/10.1007/978-3-030-95018-7_20

list of a member a by $P(a)$. The task is to pair the men and women together such that there are no two individuals of the opposite sex who would both prefer each other over their current partners. When there are no such pairs, the set of marriages is said to be *stable*. Furthermore, a *tie* is a set of individuals which are preferred equally by some person. We use SMT to denote the variant of SM that may contain ties in the preference lists. Note that the preference lists are considered to be complete in SM as well as in SMT. Further, we use SMI to denote the variant of SM where preference lists may be incomplete, whereas SMTI stands for the variant of SM where the preference lists may be incomplete as well as may contain ties. An ordering of men (resp. women) is said to be *consecutive ordering* if members in each woman's (resp. man's) preference list are consecutive. An SMTI instance having consecutive ordering of men as well as of women is said to be *consecutive*, denoted by SMTI-C. An SMTI instance I is said to be *inclusive*, denoted by SMTI-INC, if the members of one of the two sets, say men, can be linearly ordered, i.e., $m_1, m_2, \ldots, m_n$ such that, for i, $j = 1$ to n, $P(m_i) \subseteq P(m_j)$ if $i < j$. Furthermore, an SMTI instance satisfying the *step property*, that is, for all $m_i \in M$, $P(m_i) = \{w_j \in W | j \leq i\}$ for some ordering of men and some ordering of women, is denoted by SMTI-STEP.

   Three notions of stability namely *weak*, *strong*, and *super* are established in the literature [4] when ties are allowed in the preference lists. A matching is called *weakly stable* if there is no man and woman such that they are currently not matched with each other but if they get matched together, then both of them would strictly improve. A matching is called *strongly stable* if there is no man and woman such that they are currently not matched with each other but if they get matched together, then one of them is better off and the other is not worse off. A matching is called *super stable* if there is no man and woman such that they are currently not matched with each other but if they get matched together, then both of them are not worse off. We define these notions formally in the next section. However, of three notions of stability in the literature, weak stability has received most attention till now. In this paper, we are solely concerned with weakly stable matching. Henceforth for the rest of the paper, in presence of ties, the terms *stability* and *stable matching* will be considered as weak stability and weakly stable matching, respectively, unless stated otherwise. Based on these variants, the following decision problems are identified in the literature.

COM SMTI
Instance: An SMTI instance, i.e., n men, n women, and their preference lists.
Question: Does there exists a stable matching in which all men and women are matched?

DECIDE_MAX (resp. DECIDE_MIN) SMTI
Instance: n men, n women, their preference lists, and an integer $k \in \mathbb{Z}^+$.
Question: Does there exists a stable matching of cardinality *atleast* (resp. *atmost*) k in the given instance?

   MAX SMTI and MIN SMTI, the optimisation versions of DECIDE_MAX SMTI and DECIDE_MIN SMTI, respectively, are known to be NP-hard [6,8].

These problems remain NP-hard even if the ties are present at the end of preference lists and on one side only, each tie is of size (length) 2, and there is at
most one tie per list [8]. Furthermore, COM SMTI is known to be NP-complete
[6,8]. Also, COM SMTI remains NP-complete for the case when each preference
list is of size at most 3 and ties occur on one side only [5]. It also implies the
NP-hardness of MAX SMTI for this restricted case. Regarding the approximability results, 3/2-approximation algorithm is known for MAX SMTI [7,9], but
for MIN SMTI, no constant factor approximation has been identified in the literature. Halldórsson et al. [2] proposed a $(1 + \frac{t(I)}{OPT(I)})$-approximation algorithm
for MIN SMTI, where t(I) is the number of preference lists that contain ties in
the instance I of SMTI and OPT(I) is the optimal solution size of I, i.e., the
cardinality of minimum size stable matching.

In this paper, we present the first ever study of the Stable Marriage problem involving ties and incomplete lists solely based on analyzing the pattern of
preference lists. The following list summarizes our key contributions.

1. We strengthen the NP-completeness result of COM SMTI problem by establishing that this problem remains NP-complete for SMTI-C instance.
2. We present $O(n^2)$ time algorithm for COM SMTI-STEP problem.
3. We improve the NP-completeness result of DECIDE_MAX SMTI problem by
   showing that DECIDE_MAX SMTI-INC is NP-complete.
4. Further, we prove that DECIDE_MIN SMTI-STEP is NP-complete, strength-
   ening the NP-completeness of DECIDE_MIN SMTI problem.
5. Finally, we propose a 3/2-approximation algorithm for the MIN SMTI-INC
   problem.

2   Preliminaries

We define *stable matching* formally. Let $M = \{m_1, m_2, m_3, \ldots, m_n\}$ and $W = \{w_1, w_2, w_3, \ldots, w_n\}$ be two sets, each of cardinality n, consisting of men and
women, respectively. Each member of M and W has a *preference list* in which
he/she ranks the members of opposite set in a decreasing order of preference.
We say that a pair (m, w) is *admissible* if w is present in m's preference list
and m is present in w's preference list. A *matching* M' is a subset of $M \times W$
such that $|M'(m_i)| \leq 1$ for all $m_i \in M$ and $|M'(w_j)| \leq 1$ for all $w_j \in G$, where
$M'(m_i)$ denotes the set of women matched with m_i and $M'(w_j)$ denotes the
set of men matched with w_j in M'. Note that $|M'(a)|$ can be either 0 or 1. If
$|M'(a)| = 1$, i.e., $M'(a) = \{b\}$ where b is a person of opposite sex, then we say
that a is matched with b in M'. Otherwise, if $|M'(a)| = 0$, then we say that
a is unmatched in M'. A *complete matching* is a matching in which all men
and women are matched. A *blocking pair* of a matching M' is an admissible
man-woman pair $(m_i, w_j) \in (M \times W) \setminus M'$ such that m_i is unmatched or prefers
w_j to his current partner, i.e., $M'(m_i)$ and w_j is unmatched or prefers m_i to
her current partner, i.e., $M'(w_j)$. A matching M' is said to be *stable* if it has no
blocking pair. The existence of a stable matching in SM is implied by the classical

Gale-Shapley algorithm [1] given in 1962. The above definition can be extended to the case where preference lists may contain ties. A *tie* of an individual m's preference list is a set of individuals whom m prefers equally, and the preference list of m is a strict order of ties. We say that m *strictly prefers* w_1 to w_2 (denoted by $w_1 >_m w_2$), if w_1 is in tie T_1 and w_2 is in tie T_2 in m's preference list, and m ranks T_1 before T_2. We write $w_1 =_m w_2$, if w_1 and w_2 are present in the same tie or if w_1 and w_2 are the same person. We say that m *weakly prefers* w_1 to w_2, if $w_1 =_m w_2$ or $w_1 >_m w_2$ holds, and write $w_1 \geq_m w_2$. When ties are involved, three notions of stability named *weak*, *strong*, and *super* are identified in the literature [4].

A *weak blocking pair* for a matching M' is an admissible pair $(m, w) \notin M'$ such that $w \geq_m M'(m)$ and $m \geq_w M'(w)$. A *super stable* matching is a matching that admits no weak blocking pair.

A *strong blocking pair* for a matching M' is an admissible pair $(m, w) \notin M'$ such that either $w \geq_m M'(m)$ and $m >_w M'(w)$ or $w >_m M'(m)$ and $m \geq_w M'(w)$. A *strongly stable* matching is a matching that admits no strong blocking pair.

A *super blocking pair* for a matching M' is an admissible pair $(m, w) \notin M'$ such that $w >_m M'(m)$ and $m >_w M'(w)$. A *weakly stable* matching is a matching that admits no super blocking pair. Since we are concerned with weakly stable matching in presence of ties, henceforth, for the rest of the paper, by a *blocking pair*, we mean a *super blocking pair* unless stated otherwise.

A *consecutive ordering* of men is an ordering $\alpha = < m_1, m_2, \ldots, m_n >$ of the members of M such that for all $w \in W$, the men in the preference list of w are consecutive. Consecutivity of women can be defined analogously. An SMTI instance is said to be *consecutive*, denoted by *SMTI-C*, if it admits a consecutive ordering of men as well as of women. Next, I is said to be *inclusive* if the members of one set, say men, can be linearly ordered, that is, $m_1, m_2, \ldots, m_n$ such that $P(m_1) \subseteq P(m_2) \subseteq \cdots \subseteq P(m_n)$. An ordering $< m_1, m_2, \ldots, m_n, w_1, w_2, \ldots, w_n >$ of $M \cup W$ is called an *inclusion ordering* if $P(m_1) \subseteq P(m_2) \subseteq \cdots \subseteq P(m_n)$ and $P(w_1) \supseteq P(w_2) \supseteq \cdots \supseteq P(w_n)$. Further, an SMTI instance I is said to possess *step property* if for all $m_i \subset M$, $P(m_i) = \{w_j \in W | j \leq i\}$ for some ordering of men and some ordering of women. An SMTI instance satisfying the step property is denoted by SMTI-STEP.

3   Complete Stable Matching in SMTI-C and SMTI-STEP

In this section, we show that the problem of finding a complete stable matching in an SMTI-C instance is NP-complete, whereas the same problem is $O(n^2)$ time solvable for an SMTI-STEP instance.

3.1   COM SMTI-C Problem

We prove that COM SMTI-C is NP-complete by giving a polynomial reduction from the EXACT-MM problem for bipartite graphs which asks whether, given a graph G and a positive integer k, there exists a maximal matching of size exactly k in G. The NP-completeness of EXACT-MM problem for bipartite graphs follows from MIN MM-D which is known to be NP-complete for bipartite graphs [10], where MIN MM-D is the decision version of MIN MM, the problem of finding a minimum cardinality maximal matching in a graph. Note that one can obtain a polynomial time reduction from the MIN MM-D problem to EXACT-MM problem by making use of the fact that maximal matchings satisfy the interpolation property, i.e., G has a maximal matching of size s, for $m_{min} \leq s \leq m_{max}$, where m_{min} and m_{max} are the sizes of minimum maximal matching and maximum matching in G, respectively.

Theorem 1. *The COM SMTI-C problem is NP-complete.*

Proof. Given a matching M of an SMTI-C instance, it can be easily verified in polynomial time whether M is complete and stable or not. Hence COM SMTI-C problem is in NP. We give a polynomial reduction from the EXACT-MM problem which remains NP-complete for bipartite graphs.

Let $I_E = (G,k)$, where $G = (X \cup Y, E)$ is a bipartite graph with $X = \{x_1, x_2, x_3, \ldots, x_p\}$ and $Y = \{y_1, y_2, y_3, \ldots, y_q\}$, be an EXACT-MM instance. If $k > \min\{p, q\}$, then the EXACT-MM instance would not admit any maximal matching of size exactly k. Therefore, we assume that $k \leq \min\{p, q\}$.

We construct an instance I_C of COM SMTI-C problem by using following steps.

1. Let $X \cup Z \cup \{m\}$ and $Y \cup W \cup \{w\}$ be the set of men and women, respectively, where $Z = \{z_1, z_2, z_3, \ldots, z_{q-k}\}$ and $W = \{w_1, w_2, w_3, \ldots, w_{p-k}\}$.
2. Let Y_i (resp. X_j) be the set of vertices in Y (resp. X) which are adjacent to $x_i \in X$ (resp. $y_j \in Y$). Create preference list for each person as follows:

   Men: $m : w$

   $(1 \leq i \leq q - k)$ $z_i : (Y)$

   $(1 \leq i \leq p)$ $x_i : (Y_i)\,(W)\,w\,(Y \backslash Y_i)$

   Women: $w : (X)\,m$

   $(1 \leq j \leq q)$ $y_j : (X_j)\,(Z)\,(X \backslash X_j)$

   $(1 \leq j \leq p - k)$ $w_j : (X)$

In a preference list, the symbol (T) denotes a tie consisting of all members of T. Clearly, this constrution can be completed in polynomial time. Further since, $< m, x_1, x_2, \ldots, x_p,\ z_1, z_2, \ldots, z_{q-k} >$ is a consecutive ordering of men and $< w, y_1, y_2, \ldots, y_q,\ w_1, w_2, \ldots, w_{p-k} >$ is a consecutive ordering of women in I_C. Therefore, I_C is an instance of COM SMTI-C problem. An illustration of the construction of instance I_C from a bipartite graph G is shown in Fig. 1.

Claim. G has a maximal matching of size exactly k iff I_C has a complete stable matching.

Proof. Let M be a maximal matching of size exactly k in G. Define $M_c = $ M $\cup \{(m,w)\} \cup \{(x_{a_i}, w_i) \mid 1 \leq i \leq p - k\} \cup \{(z_j, y_{b_j}) \mid 1 \leq j \leq q - k\}$ where x_{a_i}'s (resp. y_{b_j}'s) are the men (resp. women) who are unmatched w.r.t. M with

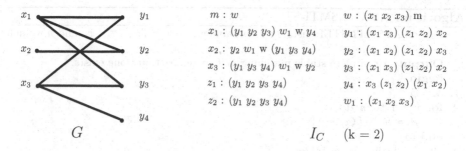

$$G \qquad\qquad\qquad I_C \quad (\text{k} = 2)$$

Fig. 1. An illustration of the construction of instance I_C from a bipartite graph G.

$a_1 < a_2 < ... < a_{p-k}$ (resp. $b_1 < b_2 < ... < b_{q-k}$). We show that M_c is complete stable matching in I_C. Since all men and women are matched in M_c, so M_c is complete. Further, since m, $z_j(1 \leq j \leq q - k)$, and $x_i \in M$ are matched to their first choice woman, so none of them can form a blocking pair in M_c. So, let (x_u, y_v) be a blocking pair of M_c for some x_u which is unmatched with respect to M and $y_v \in Y_u$. But such a $y_v \in Y_u$ is already matched with her first preference, as M is maximal. Therefore, such a blocking pair is not possible. Hence no man can participate in any blocking pair of M_c. Therefore, M_c is a stable matching.

Conversely, suppose M_c is a complete stable matching in I. Note that m must be matched with w in M_c as M_c is complete. Also, note that no $x_i(1 \leq i \leq p)$ can match with any woman in $Y \backslash Y_i$ because if it is so, then (x_i, w) blocks M_c. Define $M = M_c \backslash M_1$ where $M_1 = \{(z_j, y_{g_j}) \mid 1 \leq j \leq q - k\} \cup \{(x_{h_i}, w_i) \mid 1 \leq i \leq p - k\} \cup \{(m,w)\}$, where y_{g_j} (resp. x_{h_i}) is the woman (resp. man) who is matched with man z_j (resp. woman w_i) in M_c. Note that $|M| = |M_c| - |M_1| = (p+q-k+1) - ((q-k) + (p-k) + 1) = k$. Further, we show that the matching M is maximal in G. Assume M is not maximal in G. Then $M \cup \{(x_r, y_s)\}$ is a matching in G for some $(x_r, y_s) \in E$ with $x_r \in X$ and $y_s \in Y$. Hence $(z_\beta, y_s) \in M_c$ for some $z_\beta \in Z$ and $(x_r, w_\alpha) \in M_c$ for some $w_\alpha \in W$ because M_c is complete. But this implies (x_r, y_s) is a blocking pair of M_c, a contradiction. Hence M is maximal in G. □

Hence, the theorem is proved. □

3.2 COM SMTI-STEP Problem

We present a polynomial time algorithm to find a complete stable matching in an SMTI-STEP instance, if it exists.

Theorem 2. *Algorithm 1 gives a complete stable matching in an SMTI-STEP instance or reports that no such matching exists in $O(n^2)$ time.*

Proof. First note that the orderings $< x_1, ..., x_n >$ and $< y_1, ..., y_n >$ can be found by arranging the men (resp. women) in decreasing (resp. increasing) order of the size of their preference lists. This can be done by using bucket sort

---

Algorithm 1. COM SMTI-STEP

Input: An SMTI-STEP instance I containing n men $x_1, x_2, \ldots, x_n$ and n women $y_1, y_2, \ldots, y_n$.
Output: A complete stable matching in I *or* report that none exists.
begin
1: M$= \phi$;
2: **for** i = 1 **to** n **do**
    $M = M \cup \{(x_i, y_i)\}$;
3: **end for**
4: Check whether M is stable;
    If yes, output M.
    Else, output *no complete stable matching*.
end

---

algorithm in atmost $O(n^2)$ time. Further let the instance I has a complete stable matching M_1. Note that M_1 must be unique and $M_1 = \{(x_i, y_i)|1 \leq i \leq n\}$. Because if x_i is matched with some woman other than y_i in M_1, then there exists atleast one pair of man and woman who are unmatched in M_1 and are not admissible to each other. Hence M_1 will not be complete.

Claim. Algorithm 1 outputs M_1.

Proof. After completing step 2, the algorithm constructs matching M which is same as M_1, and since M_1 is stable, so is M. Therefore, following step 3, the algorithm outputs M which is same as M_1.   □

Now, suppose I has no complete stable matching. This implies $\{(x_i, y_i)| 1 \leq i \leq n\}$ must not be stable as this is the only complete matching in I. Therefore, step 4 of algorithm reports that no complete weakly stable matching exists.

Note that the algorithm clearly terminates. Also, step 1 and 2 take $O(1)$ and $O(n)$ time, respectively. Further we can check whether a matching is stable or not in $O(n^2)$ time. Therefore, overall time complexity for the algorithm is $O(n^2)$.   □

4   Maximum Stable Matching in SMTI-INC Problem

Theorem 3. *The DECIDE_MAX SMTI-INC problem is NP-complete.*

Proof. Given a matching S of an SMTI-INC instance and a positive integer k_1, it can be easily verified in polynomial time whether S is stable and $|S| \geq k_1$. Hence DECIDE_MAX SMTI-INC is in NP. To show NP-hardness, we give a polynomial reduction from the MIN MM-D problem which is known to be NP-complete for subdivision graphs of cubic graphs [3]. Note that the subdivision graph of a graph H is a graph G obtained by replacing each edge of H with a 2-length path. Let $I_{min} = (G, k)$, where G is a subdivision graph (and hence a bipartite graph) of some cubic graph H, be a MIN MM-D instance. Let $G = (X \cup Y, E)$, where

$X = \{x_1, x_2, x_3, ..., x_p\}$, and $Y = \{y_1, y_2, y_3, ..., y_q\}$. Without loss of generality, suppose $k \leq min\{p, q\}$.

We construct an instance I_{max} of DECIDE_MAX SMTI-INC by using following steps.

1. Let $X' = \{x_{p+1}, x_{p+2}, ..., x_{p+q}\}$ and $Y' = \{y_{q+1}, y_{q+2}, ..., y_{q+p}\}$. Suppose the set of men and women in I_{max} be $X \cup X'$ and $Y \cup Y'$, respectively.
2. In G, let Y_i be the set of vertices in Y that are adjacent to $x_i \in X$ and let X_j be the set of vertices in X that are adjacent to $y_j \in Y$. Create preference list for each person as follows:

Men: $(1 \leq i \leq p)$ $x_i : (Y_i)\ y_{q+i}\ y_{q+i-1}\ y_{q+i-2}\ \cdots\ y_{q+1}\ (Y \backslash Y_i)$
   $(p+1 \leq i \leq p+q)$ $x_i : y_{i-p}\ (Y \backslash \{y_{i-p}\})$

Women: $(1 \leq j \leq q)$ $y_j : (X_j)\ x_{p+j}\ (X' \backslash \{x_{p+j}\})\ (X \backslash X_j)$
   $(q+1 \leq j \leq q+p)$ $y_j : x_{j-q}\ x_{j-q+1}\ x_{j-q+2}\ \cdots\ x_p$

Clearly, this construction can be completed in polynomial time. Since, $N(x_{p+1}) \subseteq N(x_{p+2}) \subseteq \cdots \subseteq N(x_{p+q}) \subseteq N(x_1) \subseteq N(x_2) \subseteq \cdots \subseteq N(x_p)$ and $N(y_1) \supseteq N(y_2) \supseteq \cdots \supseteq N(y_q) \supseteq N(y_{q+1}) \supseteq N(y_{q+2}) \supseteq \cdots \supseteq N(y_{q+p})$. Therefore, this is an instance of DECIDE_MAX SMTI-INC problem, with parameter $k_1 = p+q-k$ (let). An illustration of the construction of instance I_{max} from the bipartite graph G which is the subdivision graph of a cubic graph H is shown in Fig. 2.

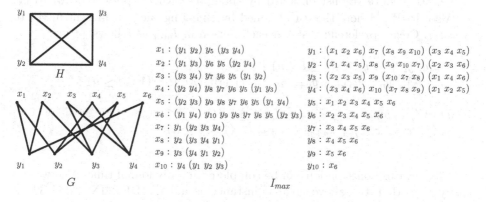

Fig. 2. An illustration of the construction of instance I_{max} from the bipartite graph G which is the subdivision graph of a cubic graph H.

Claim. G has a maximal matching of size atmost k iff I_{max} has a stable matching of size atleast k_1.

Proof. The proof is omitted due to space constraint.

Hence, the theorem is proved.   □

5 Minimum Stable Matching in SMTI-STEP Instance

Theorem 4. *The DECIDE_MIN SMTI-STEP problem is NP-complete.*

Proof. Given a matching S of an SMTI-STEP instance and a positive integer k_1, we can easily verify in polynomial time whether S is stable and $|S| \leq k_1$. Hence DECIDE_MIN SMTI-STEP problem is in NP. To show hardness, we give a polynomial reduction from DECIDE_MAX SMTI-INC problem which we have shown to be NP-complete. Let $I_{max} = (M_e \cup W_o, k)$ be a DECIDE_MAX SMTI-INC instance with parameter k, that consists of n men, n women, and their preference lists. So, let $M_e = \{m_1, m_2, m_3, ..., m_n\}$, $W_o = \{w_1, w_2, w_3, ..., w_n\}$ and $< m_1, m_2, ..., m_n, w_1, w_2, ..., w_n >$ be the inclusion ordering. Without loss of generality, we can assume that $k \leq n$, otherwise the DECIDE_MAX SMTI-INC instance would result in a 'no' instance.

We construct an instance I_{min} of DECIDE_MIN SMTI-STEP as follows.

1. Corresponding to each man $m_i \in M_e$ and each woman $w_j \in W_o$ in I_{max}, we create a man x_{n+i} and a woman y_j, respectively, in I_{min}. So, let $X \cup M_e'$ and $W_o' \cup Y$ be the set of men and women, respectively, in I_{min}, where $X = \{x_1, x_2, x_3, ..., x_n\}$, $M_e' = \{x_{n+1}, x_{n+2}, ..., x_{2n}\}$, $W_o' = \{y_1, y_2, y_3, ..., y_n\}$, and $Y = \{y_{n+1}, y_{n+2}, ..., y_{2n}\}$. Note that M_e' and W_o' are created from M_e and W_o, respectively.

2. Let M_i (resp. W_j) be the preference list of man m_i (resp. woman w_j) in I_{max}. Let M_i' denote the list obtained by changing each w_k present in M_i to y_k. Also, let W_j' denote the list obtained by changing each m_l present in W_j to x_{n+l}. Create preference lists for each person in I_{min} as follows:

$$x_i : y_i \ y_{i-1} \ \cdots \ y_1 \quad (1 \leq i \leq n)$$
$$x_{n+i} : M_i' \ y_{n+i} \ y_{n+i-1} \ y_{n+i-2} \ \cdots \ y_{n+1} \ (W_o' \backslash M_i') \quad (1 \leq i \leq n)$$

$$y_j : W_j' \ x_j \ x_{j+1} \ x_{j+2} \ \cdots \ x_n \ (M_e' \backslash W_j') \quad (1 \leq j \leq n)$$
$$y_{n+j} : x_{n+j} \ x_{n+j+1} \ x_{n+j+2} \ \cdots \ x_{2n} \quad (1 \leq j \leq n)$$

Clearly, this construction can be completed in polynomial time. Also we can easily note that the above created instance is a DECIDE_MIN SMTI-STEP instance, with parameter $k_1 = 2n - k$ (let). An illustration of the construction of instance I_{min} from the instance I_{max} is shown in Fig. 3.

Claim. I_{max} has a stable matching of size atleast k iff I_{min} has a stable matching of size atmost k_1.

Proof. The proof is omitted due to space constraint.

Hence, the theorem is proved.   □

$$
\begin{array}{lll}
m_1 : w_1 & x_1 : y_1 & y_1 : x_6\ x_5\ x_7\ x_8 \quad x_1\ x_2\ x_3\ x_4 \\
m_2 : w_1\ w_2\ (w_3\ w_4) & x_2 : y_2\ y_1 & y_2 : x_7\ x_8\ x_6 \quad x_2\ x_3\ x_4 \quad x_5 \\
m_3 : w_1\ (w_2\ w_3)\ w_4 & x_3 : y_3\ y_2\ y_1 & y_3 : x_7\ x_8\ x_6 \quad x_3\ x_4 \quad x_5 \\
m_4 : (w_1\ w_2\ w_3\ w_4) & x_4 : y_4\ y_3\ y_2\ y_1 & y_4 : x_8\ x_7\ x_6 \quad x_4 \quad x_5 \\
& x_5 : y_1 \quad y_5 \quad (y_2\ y_3\ y_4) & y_5 : x_5\ x_6\ x_7\ x_8 \\
w_1 : m_2\ m_1\ m_3\ m_4 & x_6 : y_1\ y_2\ (y_3\ y_4) \quad y_6\ y_5 & y_6 : x_6\ x_7\ x_8 \\
w_2 : m_3\ m_4\ m_2 & x_7 : y_1\ (y_2\ y_3)\ y_4 \quad y_7\ y_6\ y_5 & y_7 : x_7\ x_8 \\
w_3 : m_3\ m_4\ m_2 & x_8 : (y_1\ y_2\ y_3\ y_4) \quad y_8\ y_7\ y_6\ y_5 & y_8 : x_8 \\
w_4 : m_4\ m_3\ m_2
\end{array}
$$

$$I_{max} \hspace{7cm} I_{min}$$

Fig. 3. An illustration of the construction of instance I_{min} from the instance I_{max} in the proof of Theorem 4.

6   Approximation Algorithm for MIN SMTI-INC

In this section, we present a 3/2-approximation algorithm for the MIN SMTI-INC problem, as the NP-hardness of this problem clearly follows from the MIN SMTI-STEP problem.

Algorithm 2. MIN SMTI-INC APPROX

Input: An SMTI-INC instance I containing n men $x_1, x_2, \ldots, x_n$ and n women $y_1, y_2, \ldots, y_n$.

Output: A stable matching of size atmost 3/2 times the cardinality of minimum size stable matching in I.

begin

1: **Find an inclusion ordering** $< m_1, \ldots, m_n, w_1, \ldots, w_n >$ **in I.**
   Let I' be the instance obtained after finding the inclusion ordering of men and women in I and changing their preference lists accordingly.

2: **If** $\exists$ **a tie** $T = (w_{\alpha_1}, w_{\alpha_2}, \ldots, w_{\alpha_p})$ **in any man** m_i**'s list in** I'**, then break the tie in the following way:** m_i **prefers** w_{α_g} **to** w_{α_h} **if** $\alpha_g < \alpha_h$.
   If $\exists$ **a tie** $T' = (m_{\beta_1}, m_{\beta_2}, \ldots, m_{\beta_q})$ **in any woman** w_j**'s list in** I', **then break the tie in the following way:** w_j **prefers** m_{β_u} **to** m_{β_v} **if** $\beta_u > \beta_v$.
   Let I'' be the modified instance resulted from above breaking of ties in I'.

3: **Apply Gale-Shapley algorithm in** I'' **to find a stable matching, say** M_1.

4: **In** M_1**, change the corresponding** m**'s to** x**'s and** w**'s to** y**'s as in the original instance I, which were changed in step 1.**
   Name the matching containing x's and y's obtained by this transformation of m's to x's and w's to y's as M.

5: **return M**
   end

Theorem 5. *Algorithm 2 outputs a stable matching of size at most* $\frac{3}{2} \cdot |M_{OPT}|$ *in an SMTI-INC instance in* $O(n^2)$ *time, where* M_{OPT} *is the minimum size stable matching.*

Proof. Since the Gale-Shapley algorithm always returns a stable matching in an SMI instance in polynomial time, the matching M returned by algorithm 2 is stable.

Let M_{OPT} be an optimal matching, that is, minimum size stable matching of the instance I. Let $G = (V, E)$ be the corresponding bipartite graph of I''. Now each component of $G[M \triangle M_{OPT}]$ is either a path or an even cycle (having equal number of edges from M and M_{OPT}). We denote any M_{OPT}-augmenting path of length 3 as *3AP*. Note that any *3AP* contains two edges of M and one edge of M_{OPT}. We prove that $G[M \triangle M_{OPT}]$ has no *3AP*.

Claim. $G[M \triangle M_{OPT}]$ has no 3AP.

Proof. Note that 3AP can be of four types as shown in Fig. 4, where the black edges and blue edges are from M and M_{OPT}, respectively as M_{OPT} is a minimum size stable matching.

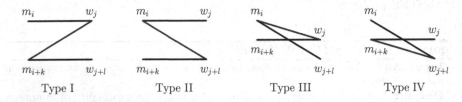

Type I   Type II   Type III   Type IV

Fig. 4. Types of M_{OPT}-augmenting paths of length 3.

We will show that none of these four types of *3AP* are possible.

First assume that $G[M \triangle M_{OPT}]$ has a Type I 3AP. Then w_j does not prefer m_i over m_{i+k} else M_{OPT} will not be stable. Due to similar reason, m_{i+k} does not prefer w_{j+l} over w_j, else (m_{i+k}, w_{j+l}) will be a blocking pair for M_{OPT}, a contradiction. Also, if both w_j and m_{i+k} prefer each other over m_i and w_{j+l}, respectively, then the matching M reported by algorithm will not be stable. Hence either (i) w_j prefers m_{i+k} over m_i, and m_{i+k} is tied between w_j and w_{j+l} or (ii) w_j is tied between m_i and m_{i+k}, and m_{i+k} prefers w_j over w_{j+l}, or (iii) w_j is tied between m_i and m_{i+k}, and m_{i+k} is tied between w_j and w_{j+l}.
Case I: w_j prefers m_{i+k} over m_i, and m_{i+k} is tied between w_j and w_{j+l}.

After step 2 of algorithm, m_{i+k} prefers w_j over w_{j+l}, as $j < j + l$. This implies (m_{i+k}, w_j) is a blocking pair of M, a contradiction.
Case II: w_j is tied between m_i and m_{i+k}, and m_{i+k} prefers w_j and w_{j+l}.

After step 2 of algorithm, w_j prefers m_{i+k} over m_i, as $i + k > i$. This implies (m_{i+k}, w_j) is a blocking pair of M, a contradiction.
Case III: w_j is tied between m_i and m_{i+k}, and m_{i+k} is tied between w_j and w_{j+l}.

After step 2 of algorithm, w_j prefers m_{i+k} over m_i, and m_{i+k} prefers w_j over w_{j+l}. This implies (m_{i+k}, w_j) is a blocking pair of M, a contradiction.
Therefore, Type I 3AP is not possible.

Next assume that Type II 3AP is possible. Since $P(m_i) \subseteq P(m_{i+k})$, so (m_{i+k}, w_j) edge must be in G and hence this edge blocks M_{OPT}, a contradiction.

Assume that Type III 3AP is possible. Since $P(m_i) \subseteq P(m_{i+k})$, so (m_{i+k}, w_{j+l}) blocks M_{OPT}, a contradiction.

Assume that Type IV 3AP is possible. Since $P(w_{j+l}) \subseteq P(w_j)$, so (m_i, w_j) edge must be in G and hence this edge blocks M_{OPT}, a contradiction.

Hence $G[M \triangle M_{OPT}]$ has no $3AP$.   □

Also, any other M_{OPT}-augmenting path in $M \triangle M_{OPT}$ will have length at least 5 (containing two edges of M_{OPT} and three edges of M).

Claim. $|M| \leq \frac{3}{2} \cdot |M_{OPT}|$.

Proof. Let $d_1, d_2, \ldots, d_r$ be the components of $G[M \triangle M_{OPT}]$ which are M_{OPT}-augmenting paths. Let m_i $(1 \leq i \leq r)$ be the number of edges of M_{OPT} in d_i. This implies number of edges of M in d_i are $m_i + 1$. Further, let $c_1, c_2, \ldots, c_s$ be other components of $M \triangle M_{OPT}$ which contribute equally to both M and M_{OPT}. Let n_j $(1 \leq j \leq s)$ be the number of edges of M_{OPT}, and hence of M, in c_j. By above claim, $m_i \geq 2$ $\forall i = 1$ to r, so

$$\frac{|M|}{|M_{OPT}|} = \frac{\sum_{i=1}^{r}(m_i + 1) + \sum_{j=1}^{s} n_j}{\sum_{i=1}^{r} m_i + \sum_{j=1}^{s} n_j} = 1 + \frac{r}{\sum_{i=1}^{r} m_i + \sum_{j=1}^{s} n_j} \leq 1 + \frac{1}{2}.$$

□

Furthermore, each of step of the algorithm takes atmost $O(n^2)$ time. Therefore, the running time of the algorithm is $O(n^2)$. Hence, the theorem is proved.   □

7   Conclusion

We have proposed a first ever study of SMTI problem by analysing the pattern of the involved preference lists. We have strengthened the NP-completeness result of COM SMTI problem by showing that this problem remains NP-complete for SMTI-C instance. Also, we have improved the NP-completeness result of DECIDE_MAX SMTI problem by establishing that DECIDE_MAX SMTI-INC problem is NP-complete. Furthermore, for an SMTI-STEP instance, we have shown that the problem of finding minimum size stable matching is NP-hard whereas the problem of finding a complete stable matching is polynomial time solvable. Finally, we have proposed the first constant factor approximation algorithm linked with MIN SMTI problem by presenting a 3/2-approximation algorithm for the MIN SMTI-INC problem. It remains open to further improve the approximability bounds for the MIN SMTI problem.

References

1. Gale, D., Shapley, L.S.: College admissions and the stability of marriage. Am. Math. Mon. **69**(1), 9–15 (1962)
2. Halldórsson, M.M., et al.: Approximability results for stable marriage problems with ties. Theoret. Comput. Sci. **306**(1–3), 431–447 (2003)
3. Horton, J.D., Kilakos, K.: Minimum edge dominating sets. SIAM J. Discret. Math. **6**(3), 375–387 (1993)
4. Irving, R.W.: Stable marriage and indifference. Discret. Appl. Math. **48**(3), 261–272 (1994)
5. Irving, R.W., Manlove, D.F., O'Malley, G.: Stable marriage with ties and bounded length preference lists. J. Discrete Algorithms **7**(2), 213–219 (2009)
6. Iwama, K., Miyazaki, S., Morita, Y., Manlove, D.: Stable marriage with incomplete lists and ties. In: Wiedermann, J., van Emde Boas, P., Nielsen, M. (eds.) ICALP 1999. LNCS, vol. 1644, pp. 443–452. Springer, Heidelberg (1999). https://doi.org/10.1007/3-540-48523-6_41
7. Király, Z.: Linear time local approximation algorithm for maximum stable marriage. Algorithms **6**(3), 471–484 (2013)
8. Manlove, D.F., Irving, R.W., Iwama, K., Miyazaki, S., Morita, Y.: Hard variants of stable marriage. Theoret. Comput. Sci. **276**(1–2), 261–279 (2002)
9. Paluch, K.: Faster and simpler approximation of stable matchings. Algorithms **7**(2), 189–202 (2014)
10. Yannakakis, M., Gavril, F.: Edge dominating sets in graphs. SIAM J. Appl. Math. **38**(3), 364–372 (1980)

On Fair Division with Binary Valuations Respecting Social Networks

Neeldhara Misra[(✉)] and Debanuj Nayak

Indian Institute of Technology, Gandhinagar, India
neeldhara.m@iitgn.ac.in, debanuj.nayak@alumni.iitgn.ac.in

Abstract. We study the computational complexity of finding fair allocations of indivisible goods in the setting where a social network on the agents is given. Notions of fairness in this context are "localized", that is, agents are only concerned about the bundles allocated to their neighbors, rather than every other agent in the system. We comprehensively address the computational complexity of finding locally envy-free and Pareto efficient allocations in the setting where the agents have binary valuations for the goods and the underlying social network is modeled by an undirected graph. We study the problem in the framework of parameterized complexity.

We show that the problem is computationally intractable even in fairly restricted scenarios, for instance, even when the underlying graph is a path. We show NP-hardness for settings where the graph has only two distinct valuations among the agents. We demonstrate W-hardness with respect to the number of goods or the size of the vertex cover of the underlying graph. We also consider notions of proportionality that respect the structure of the underlying graph.

Keywords: Fair division · Social networks · Envy-freeness · Parameterized complexity

1 Introduction

The problem of fairly allocating resources among a set of agents with (possibly distinct) interests in said resources is a fundamental problem with important and varied practical applications. We focus on the problem of allocating indivisible items: in this setting, we have n agents and m resources, and every agent expresses their utilities for the resources, either as a ranking over the resources or by specifying a valuation function. The goal is to determine an *allocation* of the items to the agents that respects some notion of "fairness" and "efficiency". We use the term *bundle* to refer to the set of items that an agent receives in an allocation.

Envy-freeness is one of the most widely used notions of fairness. Given an allocation, an agent envies another if it perceives the bundle of the other agent to be more valuable than her own. An allocation is *envy-free* if no agent envies

© Springer Nature Switzerland AG 2022
N. Balachandran and R. Inkulu (Eds.): CALDAM 2022, LNCS 13179, pp. 265–278, 2022.
https://doi.org/10.1007/978-3-030-95018-7_21

another. Note that the trivial allocation that leaves every agent empty-handed is always envy-free. Therefore, one is typically interested in fair allocations that also satisfy some criteria of economic *efficiency*, such as completeness (every good should be allocated to some agent), non-wastefulness (no agent receives a piece of cake that is worth nothing to her and worth something to another agent), or Pareto-efficiency (there is no other feasible agreement that would make at least one agent strictly better off while not making any of the others worse off). We remark here that just as there are trivial allocations that are fair, it is also possible to trivially achieve efficiency if we had no fairness considerations involved: for instance, the allocation that gives all goods to a single agent is Pareto-efficient assuming that the agent has a strictly monotonic utility function over the items.

The question of finding allocations that respect fairness and efficiency demands simultaneously is non-trivial: in particular, such allocations may not exist (if there are two agents and one good, and both agents have positive utility for this single resource), and can be computationally hard to find (for instance, the problem of finding a complete envy-free allocation between even two agents who hold identical valuations over m goods is equivalent to the PARTITION problem).

The focus of this work is the notion of local envy-freeness. In this setting, the agents are related by a graph, which might be thought of as modeling a social network over the agents, and we explore notions of fairness that account for the structure of this network. For instance, the notion of envy is now restricted: it only manifests between agents who are friends in the network. This is a compelling model of fairness, since agents are likely to not envy agents about whom they have little or no information. We note that the problem of fair division respecting a social network generalizes the classical notion, which can be captured by considering a complete graph on the agents. Thus, the problem of finding allocations that are "locally fair" is a generalization of the classical allocation problem.

1.1   Related Work

The model of local envy-freeness has been proposed and considered in several recent lines of work. Some of the earliest considerations for incorporating a graph structure on the agents were made in the context of the *cake-cutting* problem, which is the closely related setting of allocating a divisible resource among agents [1,4]. Abebe et al. [1] consider both directed and undirected graphs and focus on characterizing the structure of graphs that admit algorithms with certain bounds. They also consider the issue of the *price of envy-freeness* in this setting, which compares the total utility of an optimal allocation to the best utility of an allocation that is envy-free. Bei et al. [4], on the other hand, propose a moving-knife algorithm that outputs an envy-free allocation on trees and an algorithm for computing a proportional allocation on descendant graphs.

We now turn to the literature in the context of indivisible items. Beynier et al. [5] study the fair division problem in the setting of "house allocation":

here agents have (strict) preferences over items, and each agent must receive exactly one item. An agent envies another in this setting if she prefers the item received by the other agent over her own. In the case of a complete network, for an allocation to be envy-free, each agent must get her top object, and this assignment is automatically Pareto-efficient as well. This motivates the setting of local envy-freeness with respect to a graph on the agents. The authors consider the case when the underlying graph is undirected, and they also consider a variant of the problem where agents themselves can be located on the network by the central authority. These problems turn out to be computationally intractable even on very simple graph structures.

Bredereck et al. [10] consider the problem of graph-based envy-freeness in the context of *directed* graphs and for various classes of valuations: including binary, identical, additive, and even valuations that are both identical and binary. They also consider the complexity of the allocation problem in the framework of parameterized complexity.[1] Somewhat surprisingly, it turns out that finding complete envy-free allocations in the setting of a graph is NP-hard even when the valuations are binary *and* identical. Note that in this setting, every agent in every strongly connected component must get the same number of items: thus, the allocation problem is trivial for directed graphs that are strongly connected, but NP-hard for general directed graphs. Also, it turns out that for general binary preferences, the problem of finding a complete envy-free allocation is NP-hard even when the graph is strongly connected. The problem is also tractable for DAGs: indeed, allocating all resources to a single source agent (corresponding to a vertex with no incoming arcs) is both complete and locally envy-free since nobody can envy a source agent, and empty-handed agents have no envy for each other.

More recently, Eiben et al. [13] consider the problem of finding locally envy-free allocations and envy-free allocations that are additionally proportional in the setting of directed graphs in the framework of parameterized complexity, and specifically considering parameters such as treewidth, cliquewidth, and vertex cover — all of these reflect the structure of the underlying network. It turns out that the problem of finding fair and efficient allocations is tractable for networks that have bounded values for these parameters with some additional assumptions that bound the number of item types or the size of the largest bundle received by an agent. The authors also show hardness results in both the parameterized and classical settings. For instance, the authors show that finding a locally envy-free allocation is NP-hard even when the underlying network is a star, but we note that this is in the setting of general utilities.

The work of Bredereck et al. [8,9] demonstrates that the problem of finding fair and efficient allocations in various settings (including graph-based constraints) is fixed-parameter tractable in the combined parameter "number of agents" and "number of item types" for general utilities. In contrast, our work here focuses on smaller parameters for the special case of binary utilities.

In [11], Chevaleyre, Endriss, and Maudet consider *distributed* mechanisms for allocating indivisible goods, in which agents can locally agree on deals to

---

[1] The terminology relevant to this framework is introduced in the next section.

exchange some of the goods in their possession. This study focuses on convergence properties for such distribution mechanisms both in the context of the classical setting and the setting involving social constraints coming from an underlying undirected graph. Here, the notions of fairness localized according to the graph, and the network also constraints the exchanges that can take place — agents can engage in an exchange only if they are friends in the network. There are also some lines of work that suggest eliminating envy by some mechanism for *hiding* information [14].

1.2   Our Contributions

Our focus in this paper is on the setting when agents have binary valuations over the goods and the underlying social network is modeled by an undirected graph. Our focus is on exploring the computational complexity of finding locally envy-free allocations that are also Pareto efficient (EEF) in the framework of parameterized complexity, building most closely on the works of [6, 10, 13].

Bounded Agent Types. We begin by noting that the setting of undirected graphs can be significantly different from their directed counterparts: indeed, recall that finding a complete and locally envy-free allocation was NP-hard for even identical binary valuations for directed graphs, but the analogous question is easily seen to be tractable for undirected graphs (indeed, observe that the notions of strong connectivity and connectivity coincide). This motivates the question of whether the problem of finding locally EEF allocations is easier for undirected graphs with a bounded number of agent types. We answer this question in the negative by showing that the problem of determining locally envy-free allocations is NP-hard even when there are only two distinct binary valuations among the agents by a reduction from a graph separation problem called CUTTING ℓ VERTICES (Theorem 1).

Sparse and Dense Graphs. In contrast with the result for DAGs, we show that finding locally envy-free allocations that are Pareto efficient (EEF) is NP-hard even when the underlying graph is a path (Theorem 4 and Corollary 1). Although Beynier et al. [5] also show hardness results for very sparse graphs, we note that our methods are significantly different since the models for the valuations are different and additionally, the allocations we seek need not give every agent exactly one item. Moving away from sparsity, we recall that finding complete envy-free allocations for binary valuations is known to be NP-hard even for complete graphs [3, 14], which justifies the need for using additional parameters[2] in the XP[3] algorithm for finding locally envy-free allocations shown by [13, Theorem 10].

---

[2] The algorithm referred to is XP in the cliquewidth of the underlying graph, the number of agent types and item types.

[3] XP is the class of parameterized problems that can be solved in time $n^{f(k)}$ for some computable function f.

Structural Parameters I: Treewidth and Cliquewidth. Informally speaking, the parameters treewidth and cliquewidth of graphs quantitatively capture the sparsity and density of the graph by measuring their "likeness" to trees and complete graphs. The results we have already for sparse and dense graphs demonstrate that these parameters being bounded alone is not enough to obtain tractable algorithms. On the other hand, the results of [13] imply that the problem of finding complete and locally envy-free allocations admits XP algorithms when parameterized by either the treewidth or cliquewidth of the underlying graph jointly with the number of item types and agent types. Since their model allows for bidirectional edges, these results apply to the setting of undirected graphs as well. We note that the algorithms described in [13] focus on complete allocations, but can be adapted to account for Pareto efficiency as well.

Structural Parameters II: Vertex Cover and Twin Cover. In the setting of directed graphs and general utilities, we note that the problem of finding a complete and locally envy-free allocation is NP-hard even when the underlying graph is a star. In particular, this demonstrates hardness on graphs with a constant-sized vertex cover.[4] It is not clear if this is the case for undirected graphs and binary utilities. We show that the problem of finding locally EEF allocations is W[1]-hard when parameterized by the vertex cover number (Theorem 3). We remark that a stronger hardness result can be observed for the closely related parameter of twin cover[5] —indeed, the known NP-hardness of finding envy-free allocations for binary valuations on complete graphs [3,14] implies hardness for graphs that have a twin cover of size zero.

Few Resources or Agents. We also consider the cases where the number of goods or the number of agents are relatively small. When considering these parameters, the work of Bliem et al. [6] shows that the computation of EEF allocations is FPT when parameterized by the number of goods or the number of agents for additive 0/1 valuations. In contrast, we show that finding EEF allocations respecting the structure of an underlying undirected graph is W[1]-hard when parameterized by the number of goods (Theorem 2). On the other hand, the FPT algorithm when parameterized by the number of agents can be extended to account for the graph constraints (noted in Observation 1 in the full version [15]).

Other Notions of Fairness. Finally, we also consider notions of proportionality in the context of graphs — we refer to these as local and quasi-global proportionality concepts, representing the extent to which the definitions account for the underlying graph. We demonstrate that computing a locally proportional allocation is NP-hard (Theorem 5), while computing a proportional allocation that is quasi-global is tractable (Theorem 6). Notions of local proportionality

---

[4] A vertex cover of a graph is a subset of vertices that contains at least one endpoint of every edge. A graph with a bounded vertex cover also has bounded treewidth.

[5] A twin cover of a graph is a subset of vertices S such that $G \setminus S$ is a disjoint union of cliques, and further, every pair of vertices u, v in any clique of $G \setminus S$ are "twins", that is, $N[v] = N[u]$.

have been proposed and studied in several of the papers that were summarized in the previous section.

2 Preliminaries

We use standard terminology from graph theory and fair division. Unless mentioned otherwise, the graphs we consider are simple and undirected. For a graph $G = (V, E)$, consisting of a set V of vertices and a set E of edges, by $N(v)$ we denote the *neighborhood* of vertex $v \in V$, i.e., the set $W \subset V$ of vertices such that for each vertex $w \in W$ there exists an edge $e = (v, w) \in E$. The *closed neighborhood* of a vertex v is $N(v) \cup \{v\}$ and is denoted $N[v]$. The *degree* of a vertex v, denoted $d(v)$, is $|N(v)|$. A *clique* is a subset of vertices which are pairwise adjacent. An *independent set* is a subset of vertices, no two of which are adjacent. For $X \subseteq V$, the *induced subgraph* $G[X]$ denotes the subgraph whose vertex set is X and the edge set consists of all edges whose both end points are in X.

An instance of fair division for indivisible goods consists of n *agents* $A = \{1, \ldots, n\}$ and m *goods* (also called *items* or *resources*), $R = \{o_1, \ldots, o_m\}$. Further, we are also given *valuations* (also called *preference functions* or *utilities*) $v_\ell : 2^R \rightarrow \mathbb{Z}$ for every agent $\ell \in A$. We will assume throughout that the valuation functions are *additive*, i.e., for each agent $\ell \in A$ and any set of goods $S \subseteq R$, $v_\ell(S) := \sum_{o \in S} v_\ell(\{o\})$. A 0/1 *valuation* is a function that takes values in $\{0, 1\}$, while valuations are said to be *identical* if every agent has the same preference function. In the context of 0/1 valuations, we say that an agent *values* or *approves* a good if her utility for the good is 1. We will use $\mathcal{V}$ to denote the valuations of the agents A over R. When considering fair division in the context of social networks, we are also given an undirected graph G over the agents A.

Every subset $S \subseteq R$ is called a *bundle*. An allocation is a function $\pi : A \rightarrow 2^R$ mapping each agent to the bundle she receives, such that $\pi(i) \cap \pi(j) = \emptyset$ when $i \neq j$ because the items cannot be shared. When $\bigcup_{a \in A} \pi(a) = R$, the allocation π is said to be *complete*, otherwise it is *partial*. An allocation is *non-wasteful* if every good is allocated to an agent that assigns positive utility to it.

An allocation π' *dominates* π if for all $\ell \in A$ it holds that $v_\ell(\pi(\ell)) \leqslant v_\ell(\pi'(\ell))$ and for some $a_j \in A$ it holds that $v_{a_j}(\pi(a_j)) < v_{a_j}(\pi'(a_j))$. An allocation π is *Pareto-efficient* if there exists no allocation π' that dominates π. In the case of 0/1 preferences, we note that an allocation is Pareto-efficient if and only if it is complete and non-wasteful, assuming that each resource provides a value of 1 to at least one agent.

Given an instance of fair division $(A, R, G = (A, E), \mathcal{V})$ as described above, we now introduce the following fairness notions:

▷ **Graph Envy-Freeness (GEF).** We call allocation π graph-envy-free if for each pair of (distinct) agents $i, j \in A$ such that $j \in N(i)$, it holds that $v_i(\pi(i)) \geqslant v_i(\pi(j))$.

▷ **Quasi-Global Proportionality (QP).** We say that an allocation π achieves quasi-global proportionality if for each agent $\ell \in A$, $v_i(\pi(i)) \geqslant \frac{1}{d(\ell)+1} v_i(R)$.

▷ **Local Proportionality (LP).** We say that an allocation π achieves local proportionality if for each agent $\ell \in A$, $v_i(\pi(i)) \geqslant \frac{1}{d(\ell)+1} \sum_{j \in N[i]} v_i(\pi(j))$.

Note that the graph versions of variants of envy-freeness (such as EF1 or EFX) can be defined analogously in a straightforward manner. It is easy to see that any graph envy-free allocation is also locally proportional and that if the underlying graph is complete, then local proportionality coincides with the standard notion of proportionality. For the problems we consider, we are typically given an instance of fair division on a graph, and the goal is to determine if there exists an allocation that satisfies some notion of fairness and efficiency. For instance, consider the following problems:

GRAPH ENVY-FREE ALLOCATION ($\mathcal{E}$-GEFA)
Input: An instance of fair division on a graph
$(A, R, G = (A, E), \mathcal{V})$.
Question: Does there exist an envy-free, Pareto-efficient allocation?

LOCALLY PROPORTIONAL ALLOCATION ($\mathcal{E}$-LPA)
Input: An instance of fair division on a graph
$(A, R, G = (A, E), \mathcal{V})$.
Question: Does there exist a Pareto-efficient allocation that achieves local proportionality?

For any efficiency concept (X) and fairness notion (Y), the X-YA problem is defined in a similar fashion. Although our questions are posed as decision versions, we note that most of our algorithms can be easily adapted to handle the natural "search" version of these problems. We refer the reader to the books [7, 16] and the article [11] for additional background on fair division.

A problem *parameterized* by k is fixed-parameter tractable if it is solvable in $f(k)|I|^{O(1)}$ time for some computable function f and the input size $|I|$ according to the problem's encoding. Informally, W-hard problems are presumably not fixed-parameter tractable. The problem of finding a clique on at least k vertices is W[1]-hard when parameterized by k. We call a problem para-NP-hard if it is NP-hard even for a constant value of the parameter. For a comprehensive introduction to the paradigm of parametrized complexity and algorithms, we refer the reader to the book [12].

3 Envy-Freeness

3.1 NP-Hardness for Two Agent Types

In this section, we show that finding $\mathcal{E}$-GEFA allocations is NP-hard even in the setting of near-identical binary valuations: in particular, when all agents have one of two possible utilities over the items. Note that in the setting of identical binary valuations when the graph G is connected, it is easy to see that all agents must value all goods without loss of generality, and that desirable allocations are the ones that allocates the same number of goods to each agent, where the goods themselves may be arbitrarily chosen. Indeed, it is clear that an allocation with equal bundle sizes is $\mathcal{E}$-GEFA. On the other hand, consider a $\mathcal{E}$-GEFA allocation that does not allocate bundles of equal size to all agents. Let a_i and a_j be two agents that receive bundles of different size. We can always find two adjacent agents on a path from a_i to a_j who have received bundles of different sizes, contradicting envy-freeness.

We now show that even a slightly more general situation is computationally intractable — in particular, if all agents have one of two valuations over the goods, the problem of identifying $\mathcal{E}$-GEFA allocations is NP-hard. Due to lack of space, the proof is deferred to the full version [15].

Theorem 1. *The $\mathcal{E}$-GEFA problem is NP-complete even when there are two agent types, and further, agents have 0/1 valuations over the goods.*

3.2 W-Hardness Parameterized by Goods

In this section, we demonstrate the hardness of finding $\mathcal{E}$-GEFA allocations even when the number of goods is bounded by showing that the problem is $W[1]$-hard when parameterized by the number of goods.

Theorem 2. *The $\mathcal{E}$-GEFA problem is $W[1]$-hard when parameterized by the number of goods, even when agents have 0/1 valuations over the goods.*

We describe a reduction from the $W[1]$-hard problem CLIQUE, given a graph G and an integer k, does there exist a clique on k vertices in G. Let $\mathcal{I} = (G, k)$ be an instance of clique, where $G = (V, E)$ and further, $V = \{v_1, \ldots, v_n\}$ and $E = \{e_1, \ldots, e_m\}$. We assume, without loss of generality, that $m \geqslant \binom{k}{2}$, since we can always return a trivial NO-instance when this is not the case. We begin by describing the construction of the reduced instance $\mathcal{J}_\mathcal{I} := (A, R, H = (A, F), \mathcal{V})$. We define the set of goods R as follows:

$$R = \{q_1, \ldots, q_\ell, p_1, \ldots, p_k, d_1, \ldots, d_{\ell+1}\},$$

where $\ell = \binom{k}{2}$. For ease of discussion, we call the first ℓ goods *popular* and the next k goods *specialized*. The remaining are *dummy* items. We now define the set of agents as $A = V \cup E \cup S \cup W$, where:

$$S := \{s_1, \ldots, s_{\ell+1}\} \text{ and } W := \{w_{ij} \mid i \in [n], j \in [\ell+1]\}.$$

We indulge in a mild abuse of notation and use v_i to refer to both an element of V from the clique instance and an agent of A in the reduced instance (similarly for edges). The edges of H are as follows:

▷ $e = (u, v) \in E$ is adjacent to all vertices of S and u, v.
▷ $v_i \in V$ is adjacent to all vertices w_{ij} for $j \in [\ell+1]$.
▷ For each $1 \leqslant i \leqslant n$, $H[\cup_{j=1}^{\ell+1} w_{ij}]$ induces a clique.

The preferences of the agents are as follows:

▷ All agents have an utility of 1 for the popular goods.
▷ All agents in V have an utility of 1 for the specialized goods.
▷ The agent $s_i \in S$ has an utility of 1 for d_i, for all $i \in [\ell + 1]$.

This completes the construction of the instance (Fig. 1). Note that the number of goods is a function of k alone. We now turn to the argument for equivalence, a clique $X \subseteq V$, of size k exists in G *iff*, there is an GEF allocation for the instance constructed $\mathcal{J}_\mathcal{J} := (A, R, H = (A, F), \mathcal{V})$.

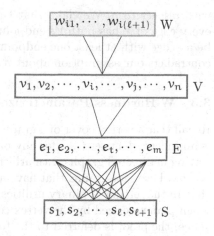

Fig. 1. A sketch of the reduced instance based on an instance $G = (V, E), k$ of CLIQUE. Recall that ℓ denotes $\binom{k}{2}$, and only some vertices of W are shown for clarity. The shaded vertices induce a complete subgraph. The edge $e_t = (v_i, v_j)$ is adjacent only to v_i and v_j among vertices in V.

Proof. In the forward direction, let $X \subseteq V$ be a clique in G and let $Y := G[X] \subseteq G$. Consider now the following allocation π. We let each agent corresponding to Y receive one popular item, each agent corresponding to X receive one specialized item, and finally allocate the item d_i to s_i for all $i \in [\ell + 1]$. It is straightforward to verify that the allocation π is Pareto-efficient and envy-free with respect to H.

This concludes our description of a fair and complete allocation strategy given a clique in G. We now turn to the reverse direction, where we are given an allocation π that is Pareto-efficient and envy-free with respect to H. It is useful to make the following observation about π to begin with.

Claim. Let H be defined as above, and let π be an allocation that is Pareto-efficient and envy-free with respect to H. Then, any popular good is assigned by π to an agent from E. Further, no agent in E can receive more than one popular good in the allocation π.

Since π is non-wasteful, the specialized goods must be distributed among agents corresponding to V. The following is easy to see.

Claim. Let H be defined as above, and let π be an allocation that is Pareto-efficient and envy-free with respect to H. No agent in V can receive more than one specialized good in the allocation π.

Let $X \subseteq V$ be the subset of k agents that receive at least one specialized item and let $Y \subseteq E$ be the subset of ℓ agents that receive at least one popular item with respect to π. We claim that $G[X]$ is a clique. In particular, we claim that every edge of Y has both its endpoints in X. Indeed, suppose not, and let $e \in Y$ be an edge with at least one endpoint (say v) outside X. Then, v envies e, which contradicts our assumption about π being envy-free with respect to H.   □

3.3   W-Hardness Parameterized by Vertex Cover

Recall that a vertex cover of a graph $G = (V, E)$ is a subset $S \subseteq V$ such that $G\backslash S$ is an independent set (i.e., for any pair of vertices $u, v \in G\backslash S$, $(u, v) \notin E$). In the setting of directed graphs with arbitrary utilities, finding $\mathcal{E}$-GEFA allocations is NP-hard even for graphs that have a constant-sized vertex cover. Here, we show that in the setting of binary utilities, finding a $\mathcal{E}$-GEFA allocation is W[1]-hard when parameterized by the vertex cover of the underlying graph. Due to lack of space, the proof is deferred to the full version [15].

Theorem 3 ($\star$).   *The $\mathcal{E}$-GEFA problem is W[1]-hard when parameterized by the vertex cover of the underlying graph, even when agents have 0/1 valuations over the goods.*

3.4   NP-Hardness on Paths

To show the hardness of $\mathcal{E}$-GEFA even when the underlying graph is a path, we reduce from a variant of SAT called LINEAR SAT (abbreviated LSAT). In an LSAT instance, each clause has at most three literals, and further the literals of the formula can be sorted such that every clause corresponds to at most three consecutive literals in the sorted list, and each clause shares at most one of its literals with another clause, in which case this literal is extreme in both clauses. The hardness of LSAT was shown in [2]. In fact, by studying the reduced instance, one may assume that a "hard" instance of LSAT has the following structure: the first $2q$ clauses have two literals each and are of the following form:

$$A_i = \{s_i, \ell_i\}, B_i = \{\ell_i, t_i\}; 1 \leqslant i \leqslant q,$$

where s_i, ℓ_i, and t_i denote literals, while the remaining p clauses have three literals each and are mutually disjoint from each other as well as the first $2q$ clauses. For ease of description, we will assume that the LSAT formula that we reduce from has this particular structure. We are now ready to describe our reduction — in the interest of simplicity, our proof is designed to address the case when the graph is a disjoint union of paths, although it is easy to "stitch" these components into a single, longer path, as we will explain later.

Theorem 4.   *The $\mathcal{E}$-GEFA problem is NP-complete even when the graph induced by the agents is a disjoint union of paths, and further, agents have 0/1 valuations over the goods.*

Membership in NP is straightforward to check. We focus here on the reduction demonstrating hardness. Let ϕ be an instance of LSAT over variables $\hat{X} := \{x_1, \ldots, x_n\}$ and clauses:

$$C := \{A_1, B_1, \ldots, A_q, B_q, C_1, \ldots, C_p\},$$

as described above. We refer to the first $2q$ clauses as the *coupled* clauses and the remaining as *isolated* clauses. We now turn to the construction of the reduced instance $\mathcal{J}_\phi := (V, R, H, \mathcal{V})$. We define the set of goods R as $R_{\hat{X}} \cup R_C$, where:

$$R_{\hat{X}} = \{y_1, \ldots, y_n, x_1, \ldots, x_n, \bar{x}_1, \ldots, \bar{x}_n, \},$$

and:

$$R_C = \{g_1, \ldots, g_q, d_1, \ldots, d_p\}.$$

The set of agents V is given by $X \cup C \cup Y \cup G \cup D$, where C is denoted in the same way as in the LSAT instance, and further:

$$X = \{X_1, \ldots, X_n\}, Y = \{Y_1, \ldots, Y_n\},$$

$$G = \{G_1, \ldots, G_q\} \text{ and } D = \{D_1, \ldots, D_p\}.$$

We now simultaneously describe the structure of the graph H and the preferences of the agents.

Assignment Gadgets. For each $1 \leqslant i \leqslant n$, add an edge between X_i and Y_i. The agent X_i values $\{x_i, \bar{x}_i, y_i\}$, while Y_i values the good y_i (and nothing else).

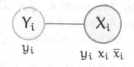

Isolated Clause Gadgets. For each $1 \leqslant i \leqslant p$, we add an edge between agents C_i and D_i. The agent C_i values the literal ℓ if and only if $\ell \in C_i$ along with d_i, while D_i values the good d_i (and nothing else).

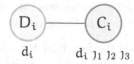

Coupled Clause Gadgets. For each $1 \leqslant i \leqslant q$, we add an edge between agents A_i and B_i, and also an edge between G_i and A_i. The agent G_i values the good g_i (and nothing else). Agents A_i and B_i value, respectively, the goods $\{g_i, s_i, t_i, \ell_i\}$ and $\{s_i, t_i\}$.

This completes the description of the construction (Fig.2). We defer the proof of

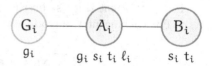

Fig. 2. A schematic of the reduced instance in the proof of Theorem 4, which is a disjoint union of paths. $\jmath_1$ $\jmath_2$ $\jmath_3$ are the literals of clause C_i

equivalence to the full version [15] due to lack of space.

We remark that it is possible to combine the connected components in the reduced instance above by simply introducing "dummy connector agents" that each value a corresponding dummy item and nothing else, leading to the following consequence.

Corollary 1. *The $\mathcal{E}$-GEFA problem is* NP-*complete even when the graph induced by the agents is a path, and further, agents have 0/1 valuations over the goods.*

4 Proportionality for Graphs

4.1 Local Proportionality: NP-hardness

Theorem 5. *The $\mathcal{E}$-LPA problem is* NP-*complete on undirected graphs, even when all agents have 0/1 valuations over the resources.*

Proof. We defer the proof to the full version [15] due to lack of space.   □

4.2 Quasi-Global Proportionality: Efficient Algorithms

To obtain efficient algorithms for finding Pareto-efficient allocations that respect quasi-global proportionality, We model the problem of finding Pareto-efficient allocations respecting quasi-global proportionality using an integer linear program (ILP) with a structured constraint matrix. In particular, it is well-known that if the constraint matrix of an ILP is totally unimodular,[6] then the corresponding instance can be solved in polynomial time. We turn to an explanation of our encoding.

Theorem 6. *The problem of finding a Pareto-efficient allocation that is quasi-globally proportional with respect to an underlying undirected graph on the agents can be solved in polynomial time if all agents have 0/1 valuations.*

Proof. Let us assume there are n agents and m goods. We will introduce a variable x_{ij} which indicates whether agent i gets good j, and a_{ij} indicates whether agent i likes good j. These constraints are as follows.

▷ We encode the fact that the allocation defined by x is well-defined by introducing the following constraint for each good j:
▷ For each agent i, let s_i be the number of items that have utility 1 for agent i. For each agent i, introduce the following proportionality constraint:

$$\forall j \sum_{i=1}^{n} x_{ij} \leqslant 1 \text{ and } \forall i \sum_{j=1}^{m} a_{ij}x_{ij} \geqslant \frac{s_i}{(d_i + 1)}$$

We let the objective function be $\sum_{i=1}^{n} a_{ij}x_{ij}$. Note that any assignment for which this function achieves a value of m is complete and non-wasteful, and also respects quasi-global proportionality. It is straightforward to verify that the constraint matrix for the ILP described above is totally unimodular for any underlying graph H.   □

---

[6] A unimodular Matrix is a square integer matrix having determinant $+1$ or -1. A totally unimodular matrix is a matrix for which every square non-singular submatrix is unimodular.

We remark that the problem of assigning goods in a proportional fashion (for any of the notions of proportionality that we have introduced) beyond 0/1 valuations is NP-hard even when there are only two agents with identical valuations, by a standard reduction from PARTITION, with the graph being a singe edge on two agents.

Acknowledgments. The first author is grateful to SERB for their support of this project through the ECR grant ECR/2018/002967.

References

1. Abebe, R., Kleinberg, J.M., Parkes, D.C.: Fair division via social comparison. In: Proceedings of the 16th Conference on Autonomous Agents and MultiAgent Systems, AAMAS, pp. 281–289 (2017)
2. Arkin, E.M., et al.: Choice is hard. In: Elbassioni, K., Makino, K. (eds.) ISAAC 2015. LNCS, vol. 9472, pp. 318–328. Springer, Heidelberg (2015). https://doi.org/10.1007/978-3-662-48971-0_28
3. Aziz, H., Gaspers, S., Mackenzie, S., Walsh, T.: Fair assignment of indivisible objects under ordinal preferences. Artif. Intell. **227**, 71–92 (2015)
4. Bei, X., Qiao, Y., Zhang, S.: Networked fairness in cake cutting. In: Proceedings of the Twenty-Sixth International Joint Conference on Artificial Intelligence, IJCAI, pp. 3632–3638 (2017)
5. Beynier, A., Chevaleyre, Y., Gourvès, L., Lesca, J., Maudet, N., Wilczynski, A.: Local envy-freeness in house allocation problems. In: Proceedings of the 17th International Conference on Autonomous Agents and MultiAgent Systems, AAMAS, pp. 292–300 (2018)
6. Bliem, B., Bredereck, R., Niedermeier, R.: Complexity of efficient and envy-free resource allocation: few agents, resources, or utility levels. In: Proceedings of the Twenty-Fifth International Joint Conference on Artificial Intelligence, IJCAI, pp. 102–108 (2016)
7. Brams, S.J., Taylor, A.D.: Fair Division - from Cake-Cutting to Dispute Resolution. Cambridge University Press, Cambridge (1996)
8. Bredereck, R., Kaczmarczyk, A., Knop, D., Niedermeier, R.: High-multiplicity fair allocation using parametric integer linear programming. CoRR arXiv:2005.04907 (2020)
9. Bredereck, R., Kaczmarczyk, A., Knop, D., Niedermeier, R.: High-multiplicity fair allocation: Lenstra empowered by n-fold integer programming. In: Proceedings of the 2019 ACM Conference on Economics and Computation. Association for Computing Machinery (2019)
10. Bredereck, R., Kaczmarczyk, A., Niedermeier, R.: Envy-free allocations respecting social networks. In: Proceedings of the 17th International Conference on Autonomous Agents and MultiAgent Systems, AAMAS, 2018, pp. 283–291 (2018)
11. Chevaleyre, Y., Endriss, U., Maudet, N.: Distributed fair allocation of indivisible goods. Artif. Intell. **242**, 1–22 (2017)
12. Cygan, M., et al.: Parameterized Algorithms. Springer, Cham (2015). https://doi.org/10.1007/978-3-319-21275-3
13. Eiben, E., Ganian, R., Hamm, T., Ordyniak, S.: Parameterized complexity of envy-free resource allocation in social networks. In: Proceedings of the Thirty-Fourth AAAI Conference on Artificial Intelligence, AAAI, pp. 7135–7142. AAAI Press (2020)

14. Hosseini, H., Sikdar, S., Vaish, R., Wang, H., Xia, L.: Fair division through information withholding. In: Proceedings of the Thirty-Fourth AAAI Conference on Artificial Intelligence, AAAI, pp. 2014–2021 (2020)
15. Misra, N., Nayak, D.: On fair division with binary valuations respecting social networks - full version. CoRR arXiv:2111.11528 (2021)
16. Robertson, J.M., Webb, W.A.: Cake-Cutting Algorithms - Be Fair If You Can. A K Peters, Natick (1998)

Parameterized Intractability of Defensive Alliance Problem

Ajinkya Gaikwad, Soumen Maity$^{(\boxtimes)}$, and Shuvam Kant Tripathi

Indian Institute of Science Education and Research, Pune, India
{ajinkya.gaikwad,tripathi.shuvamkant}@students.iiserpune.ac.in,
soumen@iiserpune.ac.in

Abstract. The DEFENSIVE ALLIANCE problem has been studied extensively during the last twenty years. A set R of vertices of a graph is a defensive alliance if, for each element of R, the majority of its neighbours are in R. The problem of finding a defensive alliance of minimum size in a given graph is NP-hard. Fixed-parameter tractability results have been obtained for the solution size and some structural parameters such as the vertex cover number and neighbourhood diversity. For the parameter treewidth the problem is W[1]-hard. However, for the parameters pathwidth and feedback vertex set, the question of whether the problem is FPT has remained open. In this work we prove that (1) the DEFENSIVE ALLIANCE problem is W[1]-hard when parameterized by the pathwidth of the input graph, (2) the EXACT DEFENSIVE ALLIANCE problem is W[1]-hard parameterized by a wide range of fairly restrictive structural parameters such as the feedback vertex set number, pathwidth, treewidth and treedepth and (3) a generalization of the DEFENSIVE ALLIANCE problem is W[1]-hard parameterized by the size of a vertex deletion set into trees of height at most 6.

Keywords: Defensive alliance · Parameterized complexity · FPT · W[1]-hard · Treewidth

1 Introduction

Alliances in graphs were introduced first in 2000 by Kristiansen, Hedetniemi, and Hedetniemi [12]. The purpose is to form coalitions of vertices able to defend each other from attacks of other vertices (in the case of defensive alliances) or able to collaborate to attack non-allied vertices (in the case of offensive alliances). Alliances can be formed between nations in a security context, between companies in a business context, or between people wishing to gather by affinity. The alliance problems have been studied extensively during last fifteen

---

The first author gratefully acknowledges support from the Ministry of Human Resource Development, Government of India, under Prime Minister's Research Fellowship Scheme (No. MRF-192002-211).

The second author's research was supported in part by the Science and Engineering Research Board (SERB), Govt. of India, under Sanction Order No. MTR/2018/001025.

N. Balachandran and R. Inkulu (Eds.): CALDAM 2022, LNCS 13179, pp. 279–291, 2022.
https://doi.org/10.1007/978-3-030-95018-7_22

years [2,7,15,17,18], and generalizations called r-alliances are also studied [16]. Throughout this article, $G = (V, E)$ denotes a finite, simple and undirected graph of order $|V| = n$. The subgraph induced by $S \subseteq V$ is denoted by $G[S]$. For a vertex $v \in V$, we use $N_G(v) = \{u : (u, v) \in E(G)\}$ to denote the (open) neighbourhood of vertex v in G, and $N_G[v] = N_G(v) \cup \{v\}$ to denote the closed neighbourhood of v. The degree $d_G(v)$ of a vertex $v \in V(G)$ is $|N_G(v)|$. For a subset $S \subseteq V(G)$, we define its closed neighbourhood as $N_G[S] = \bigcup_{v \in S} N_G[v]$ and its open neighbourhood as $N_G(S) = N_G[S] \setminus S$. For a non-empty subset $S \subseteq V$ and a vertex $v \in V(G)$, $N_S(v)$ denotes the set of neighbours of v in S, that is, $N_S(v) = \{u \in S : (u, v) \in E(G)\}$. We use $d_S(v) = |N_S(v)|$ to denote the degree of vertex v in $G[S]$. The complement of the vertex set S in V is denoted by S^c.

Definition 1. A non-empty set $R \subseteq V$ is a defensive alliance in $G = (V, E)$ if $d_R(v) + 1 \geq d_{R^c}(v)$ for all $v \in R$.

A vertex $v \in R$ is said to be protected if $d_R(v) + 1 \geq d_{R^c}(v)$. A set $R \subseteq V$ is a defensive alliance if every vertex in R is protected. In this paper, we consider DEFENSIVE ALLIANCE and EXACT DEFENSIVE ALLIANCE under structural parameters. We define the problems as follows:

DEFENSIVE ALLIANCE
Input: An undirected graph $G = (V, E)$ and an integer $r \geq 1$.
Question: Is there a defensive alliance $R \subseteq V$ such that $|R| \leq r$?

EXACT DEFENSIVE ALLIANCE
Input: An undirected graph $G = (V, E)$ and an integer $r \geq 1$.
Question: Is there a defensive alliance $R \subseteq V$ such that $|R| = r$?

For standard notations and definitions in graph theory and parameterized complexity, we refer to West [19] and Cygan et al. [3], respectively. The graph parameters we explicitly use in this paper are feedback vertex set number, pathwidth, treewidth and treedepth.

Definition 2. For a graph $G = (V, E)$, the parameter *feedback vertex set* is the cardinality of the smallest set $S \subseteq V(G)$ such that the graph $G - S$ is a forest and it is denoted by $fvs(G)$.

We now review the concept of a tree decomposition, introduced by Robertson and Seymour in [14]. Treewidth is a measure of how "tree-like" the graph is.

Definition 3. [4] A *tree decomposition* of a graph $G = (V, E)$ is a tree T together with a collection of subsets X_t (called bags) of V labeled by the vertices t of T such that $\bigcup_{t \in T} X_t = V$ and (1) and (2) below hold:

1. For every edge $uv \in E(G)$, there is some t such that $\{u, v\} \subseteq X_t$.
2. (Interpolation Property) If t is a vertex on the unique path in T from t_1 to t_2, then $X_{t_1} \cap X_{t_2} \subseteq X_t$.

Definition 4. [4] The *width* of a tree decomposition is the maximum value of $|X_t| - 1$ taken over all the vertices t of the tree T of the decomposition. The treewidth $tw(G)$ of a graph G is the minimum width among all possible tree decomposition of G.

Definition 5. If the tree T of a tree decomposition is a path, then we say that the tree decomposition is a *path decomposition*, and use *pathwidth* in place of treewidth.

A rooted forest is a disjoint union of rooted trees. Given a rooted forest F, its *transitive closure* is a graph H in which $V(H)$ contains all the nodes of the rooted forest, and $E(H)$ contain an edge between two vertices only if those two vertices form an ancestor-descendant pair in the forest F.

Definition 6. The *treedepth* of a graph G is the minimum height of a rooted forest F whose transitive closure contains the graph G. It is denoted by $td(G)$.

For the standard concepts in parameterized complexity, see the recent textbook by Cygan et al. [3].

1.1 Our Main Results

The goal of this paper is to provide new insight into the complexity of DEFENSIVE ALLIANCE parameterized by the structure of the input graph. In this paper, we prove the following results:

- the DEFENSIVE ALLIANCE problem is W[1]-hard parameterized by the path-width of the input graph.
- the EXACT DEFENSIVE ALLIANCE problem is W[1]-hard parameterized by any of the following parameters: the feedback vertex set number, treedepth and pathwidth of the input graph.
- a generalization of the DEFENSIVE ALLIANCE problem is W[1]-hard parameterized by the size of a vertex deletion set into trees of height at most 6.

1.2 Known Results

Fernau and Raible showed in [6] that the defensive, offensive and powerful alliance problems and their global variants are fixed parameter tractable when parameterized by solution size k. Kiyomi and Otachi showed in [10], the problems of finding smallest alliances of all kinds are fixed-parameter tractable when parameteried by the vertex cover number. The problems of finding smallest defensive and offensive alliances are also fixed-parameter tractable when parameteried by the neighbourhood diversity [8]. Enciso [5] proved that finding defensive and global defensive alliances is fixed parameter tractable when parameterized by domino treewidth. Bliem and Woltran [1] proved that deciding if a graph contains a defensive alliance of size at most k is W[1]-hard when parameterized by treewidth of the input graph. This puts it among the few problems that are FPT when parameterized by solution size but not when parameterized by treewidth (unless FPT = W[1]).

2 Hardness Results of Defensive Alliance

In this section we prove the following theorems:

Theorem 1. The DEFENSIVE ALLIANCE problem is W[1]-hard parameterized by the pathwidth of the input graph.

Theorem 2. The EXACT DEFENSIVE ALLIANCE problem is W[1]-hard parameterized by any of the following parameters: the feedback vertex set number, treedepth and pathwidth of the input graph.

We introduce several variants of DEFENSIVE ALLIANCE that we require in our proofs. The problem DEFENSIVE ALLIANCE^F generalizes DEFENSIVE ALLIANCE where some vertices are forced to be outside the solution; these vertices are called "forbidden" vertices. This variant can be formalized as follows:

DEFENSIVE ALLIANCE^F
Input: An undirected graph $G = (V, E)$, an integer r and a set $V_\square \subseteq V$.
Question: Is there a defensive alliance $R \subseteq V$ such that (i) $|R| \leq r$, and (ii) $R \cap V_\square = \emptyset$?

DEFENSIVE ALLIANCE^{FN} is a further generalization that, in addition, requires some "necessary" vertices to be in R. This variant can be formalized as follows:

DEFENSIVE ALLIANCE^{FN}
Input: An undirected graph $G = (V, E)$, an integer r, a set $V_\triangle \subseteq V$, and a set $V_\square \subseteq V(G)$.
Question: Is there a defensive alliance $R \subseteq V$ such that (i) $|R| \leq r$, (ii) $R \cap V_\square = \emptyset$, and (iii) $V_\triangle \subseteq R$?

While the DEFENSIVE ALLIANCE problem asks for defensive alliance of size at most r, we also consider the EXACT DEFENSIVE ALLIANCE problem that concerns defensive alliance of size exactly r. Analogously, we also define exact versions of the two generalizations of DEFENSIVE ALLIANCE presented above. To show W[1]-hardness of DEFENSIVE ALLIANCE, we consider the MULTIDIMENSIONAL SUBSET SUM (MSS) problem.

MULTIDIMENSIONAL SUBSET SUM (MSS)
Input: An integer k, a set $S = \{s_1, \ldots, s_n\}$ of vectors with $s_i \in \mathbb{N}^k$ for every i with $1 \leq i \leq n$ and a target vector $t \in \mathbb{N}^k$.
Parameter: k
Question: Is there a subset $S' \subseteq S$ such that $\sum_{s \in S'} s = t$?

We introduce two variants of MSS that we require in our proofs. In the MULTI-DIMENSIONAL RELAXED SUBSET SUM (MRSS) problem, an additional integer k' is given (which will be part of the parameter) and we ask whether there is a subset $S' \subseteq S$ with $|S'| \leq k'$ such that $\sum_{s \in S'} s \geq t$. It is known that MRSS is W[1]-hard when parameterized by the combined parameter $k + k'$, even if all integers in the input are given in unary [9]. For EXACT MRSS problem, both the input as well as the parameters are the same as in the case of MRSS however one now asks whether there is a subset $S' \subseteq S$ with $|S'| = k'$ such that $\sum_{s \in S'} s \geq t$. This variant can be formalized as

---

EXACT MULTIDIMENSIONAL RELAXED SUBSET SUM (EXACT MRSS)
Input: An integer k, a set $S = \{s_1, \ldots, s_n\}$ of vectors with $s_i \in \mathbb{N}^k$ for every i with $1 \leq i \leq n$ and a target vector $t \in \mathbb{N}^k$.
Parameter: k, k'
Question: Is there a subset $S' \subseteq S$ with $|S'| = k'$ such that $\sum_{s \in S'} s \geq t$?

Lemma 1. EXACT MRSS is W[1]-hard when parameterized by the combined parameter $k + k'$, even if all integers in the input are given in unary.

This follows from the fact that the MRSS problem is W[1]-hard even if all integers in the input are given in unary. We now show that the DEFENSIVE ALLIANCE^{FN} problem is W[1]-hard parameterized by a vertex deletion set to trees of height at most four, i.e., a subset D of the vertices of the graph such that every component in the graph, after removing D, is a tree of height at most four.

Lemma 2. The DEFENSIVE ALLIANCE^{FN} problem is W[1]-hard parameterized by the size of a vertex deletion set into trees of height at most 4.

Proof. To prove this we reduce from MRSS, which is known to be W[1]-hard when parameterized by the combined parameter $k + k'$ [9]. Let $I = (k, k', S, t)$ be an instance of MRSS. We construct an instance $I' = (G, r, V_\triangle, V_\square)$ of DEFENSIVE ALLIANCE^{FN} the following way. See Fig. 1 for an illustration. First, we introduce a set of k new vertices $U = \{u_1, u_2, \ldots, u_k\}$. For every $u_i \in U$, we create a set $V_{u_i \square}$ of $\sum_{s \in S} s(i)$ one degree forbidden vertices and a set $V_{u_i \triangle}$ of $2\left(\sum_{s \in S} s(i) - t(i) \right)$ one degree necessary vertices; and make u_i adjacent to every vertex of $V_{u_i \square} \cup V_{u_i \triangle}$. For each vector $s = (s(1), s(2), \ldots, s(k)) \in S$, we introduce a tree T_s into G. Define $\max(s) = \max_i \{s(i)\}$. We introduce two vertices x_s and y_s, and introduce two sets of new vertices $A_s = \{a_1^s, \ldots, a_{\max(s)}^s\}$ and $B_s = \{b_1^s, \ldots, b_{\max(s)}^s\}$. For each $i \in \{1, 2, \ldots, k\}$ and for each $s \in S$, we make u_i adjacent to exactly $s(i)$ many vertices of A_s in an arbitrary manner. Next, for every vertex $a^s \in A_s$, we add a set $V_{a^s}^\square$ of $|N_U(a^s)| + 3$ many one degree forbidden vertices adjacent to a^s. The vertex set of tree T_s is defined as follows:

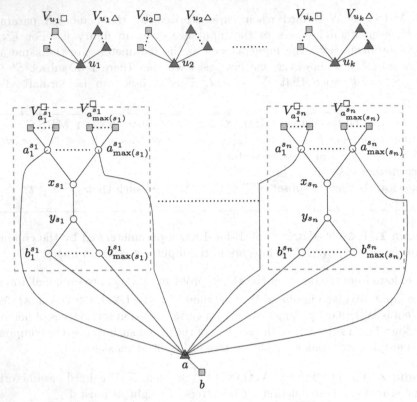

Fig. 1. The reduction from MRSS to DEFENSIVE ALLIANCE[FN] in Lemma 2. Note that the edges between the vertices in U and A_s are not shown. Gadgets in the green square correspond to vectors in set S.

$V(T_s) = A_s \cup B_s \bigcup_{a^s \in A_s} V_{a^s}^{\square} \bigcup \{x_s, y_s\}$. We now create the edge set of T_s,

$$E(T_s) = \Big\{ (x_s, \alpha), (y_s, \beta), (x_s, y_s) \mid \alpha \in A_s, \beta \in B_s \Big\}$$

$$\bigcup_{a^s \in A_s} \Big\{ (a^s, \alpha) \mid \alpha \in V_{a^s}^{\square} \Big\}.$$

Next, two vertices a, b are introduced into G. Make a adjacent to all the vertices of $\bigcup_{s \in S} A_s \cup B_s$ and also make a adjacent to b. We define $V_\triangle = \bigcup_{i=1}^{k} V_{u_i \triangle} \bigcup \{a\} \cup U$, $V_\square = \{b\} \bigcup_{s \in S} \bigcup_{a^s \in A_s} V_{a^s}^{\square} \bigcup_{i=1}^{k} V_{u_i \square}$, and set $r = \sum_{i=1}^{k} 2 \Big(\sum_{s \in S} s(i) - t(i) \Big) + \sum_{i=1}^{n} \max(s_i) + k + k' + 1$. Observe that if we remove the set $U \cup \{a\}$ of $k + 1$ vertices from G, each connected component of the resulting graph is a tree with height at most 4.

Formally, we claim that I is a yes-instance if and only if I' is a yes-instance. Let $S' \subseteq S$ be such that $|S'| \leq k'$ and $\sum_{s \in S'} s \geq t$. Then we claim that the set

$$R = V_\triangle \cup \bigcup_{s \in S'} A_s \cup \{x_s\} \cup \bigcup_{s \in S \setminus S'} B_s = \bigcup_{i=1}^{k} V_{u_i \triangle} \cup \{a\} \cup U \cup \bigcup_{s \in S'} A_s \cup \{x_s\} \cup \bigcup_{s \in S \setminus S'} B_s \text{ is}$$

a defensive alliance in G such that $|R| \leq r$, $V_\triangle \subseteq R$, and $V_\square \cap R = \emptyset$. Let x be an arbitrary element of R.

Case 1: If $x = u_i \in U$, then

$$d_R(u_i) = \sum_{s \in S'} s(i) + |V_{u_i \triangle}| = \sum_{s \in S'} s(i) + 2 \sum_{s \in S} s(i) - 2t(i)$$

$$= \left(\sum_{s \in S'} s(i) - t(i) \right) + \left(\sum_{s \in S} s(i) - t(i) \right) + \sum_{s \in S} s(i)$$

$$\geq \left(\sum_{s \in S} s(i) - t(i) \right) + \sum_{s \in S} s(i)$$

$$= \sum_{s \in S \setminus S'} s(i) + \left(\sum_{s \in S'} s(i) - t(i) \right) + \sum_{s \in S} s(i)$$

$$\geq \sum_{s \in S \setminus S'} s(i) + \sum_{s \in S} s(i) = \sum_{s \in S \setminus S'} s(i) + |V_{u_i \square}|$$

$$= d_{R^c}(u_i)$$

Therefore, we have $d_R(u_i) + 1 \geq d_{R^c}(u_i)$, and hence u_i is protected.

Case 2: If $x = a^s \in A_s$, then $d_R(a^s) = |N_U(a^s)| + |\{a, x_s\}| = |N_U(a^s)| + 2$ and $d_{R^c}(a^s) = |V_{a^s}^\square| = |N_U(a^s)| + 3$. Therefore, we have $d_R(a^s) + 1 \geq d_{R^c}(a^s)$.

Case 3: If $x = a$, then $N_R(a) = \bigcup_{s \in S'} A_s \cup \bigcup_{s \in S \setminus S'} B_s$ and $N_{R^c}(a) = \bigcup_{s \in S'} B_s \cup \bigcup_{s \in S \setminus S'} A_s \cup \{b\}$. As $|A_s| = |B_s|$, we have $d_R(a) + 1 \geq d_{R^c}(a)$.

For the rest of the vertices in R, it is easy to see that they have more neighbours in R than in R^c. Therefore, I' is a yes-instance.

For the reverse direction, suppose that G has a defensive alliance R of size at most r such that $V_\triangle \subseteq R$ and $V_\square \cap R = \emptyset$. From the definition of $V_\triangle$, we have $U \subseteq R$. We know $V_\triangle$ contains $1 + k + \sum_{i=1}^{k} 2 \left(\sum_{s \in S} s(i) - t(i) \right)$ vertices; thus besides the vertices of $V_\triangle$, there are at most $\sum_{i=1}^{n} \max(s_i) + k'$ vertices in R. Since $a \in R$, it must have at least $\sum_{i=1}^{n} \max(s_i)$ many neighbours in R from the set $\bigcup_{s \in S} A_s \cup B_s$. We also observe that if a vertex a^s from the set A_s is in the solution then x_s also lie in the solution for the protection of a^s. This shows that at most k' many sets of the form A_s contribute to the solution as otherwise the size of solution exceeds r. Therefore, any arbitrary defensive alliance R of size at most r can be transformed to another defensive alliance R' of size at most r as follows:

$$R' = V_\triangle \cup \bigcup_{x_s \in R} A_s \cup \{x_s\} \cup \bigcup_{x_s \in V(G) \setminus R} B_s.$$

We define a subset $S' = \left\{ s \in S \mid x_s \in R' \right\}$. Clearly, $|S'| \leq k'$. We claim that $\sum_{s \in S'} s(i) \geq t(i)$ for all $1 \leq i \leq k$. Assume for the sake of contradiction that $\sum_{s \in S'} s(i) < t(i)$ for some $i \in \{1, 2, \ldots, k\}$. Then, we have

$$
\begin{aligned}
d_{R'}(u_i) &= \sum_{s \in S'} s(i) + |V_{u_i \triangle}| = \sum_{s \in S'} s(i) + 2 \sum_{s \in S} s(i) - 2t(i) \\
&= \sum_{s \in S'} s(i) - t(i) + \sum_{s \in S} s(i) - t(i) + \sum_{s \in S} s(i) \\
&< \sum_{s \in S} s(i) - t(i) + \sum_{s \in S} s(i) \\
&= \sum_{s \in S \setminus S'} s(i) + \left(\sum_{s \in S'} s(i) - t(i) \right) + \sum_{s \in S} s(i) \\
&< \sum_{s \in S \setminus S'} s(i) + \sum_{s \in S} s(i) = \sum_{s \in S \setminus S'} s(i) + |V_{u_i \square}| \\
&= d_{R'^c}(u_i)
\end{aligned}
$$

and we also know that $u_i \in R'$, which is a contradiction to the fact that R' is a defensive alliance. This shows that I is a yes-instance. $\square$

Corollary 1. The DEFENSIVE ALLIANCE^{FN} problem is W[1]-hard parameterized by the size of a vertex deletion set into trees of height at most 5, even when $|V_\triangle| = 1$.

Proof. Given an instance $I = (G, r, V_\triangle, V_\square)$ of DEFENSIVE ALLIANCE^{FN}, we construct an equivalent instance $I' = (G', r', V'_\triangle, V'_\square)$ of DEFENSIVE ALLIANCE^{FN} where $|V'_\triangle| = 1$. See Fig. 2 for an illustration. Let $v_1, v_2, \ldots, v_\ell$ be vertices of $V_\triangle$. We introduce a necessary vertex x and make x adjacent to all the vertices in $V_\triangle$. We introduce a set $V_{x\square}$ of $\ell + 1$ one degree forbidden vertices adjacent to x. For every $v_i \in V_\triangle$, add a degree one forbidden vertex $v_i^\square$ adjacent to v_i. We define $V'_\triangle = \{x\}$ and $V'_\square = V_{x\square} \cup V_\square \bigcup_{v_i \in V_\triangle} v_i^\square$. We define G' as follows:

$$
V(G') = V(G) \cup \{x\} \cup V_{x\square} \bigcup_{v_i \in V_\triangle} v_i^\square
$$

and

$$
E(G') = E(G) \bigcup \{(x, \alpha), (x, v), (v, v^\square) \mid \alpha \in V_{x\square}, v \in V_\triangle\}.
$$

Let H be a set with at most k vertices in G such that $G - H$ is a forest with trees of height at most 4. Clearly, the set $H \cup \{x\}$ is of size at most $k + 1$ and $G' - (H \cup \{x\})$ is a forest with trees of height at most 5. It is easy to see that I and I' are equivalent instances. $\square$

We can get an analogous result for the exact variant.

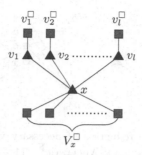

Fig. 2. An illustration of the gadget to reduce the number of necessary vertices to one.

Corollary 2. The EXACT DEFENSIVE ALLIANCE^{FN} problem is W[1]-hard when parameterized by the size of a vertex deletion set into trees of height at most 5, even when $|V_\triangle| = 1$.

Corollary 3. The DEFENSIVE ALLIANCE^{FN} problem is W[1]-hard when parameterized by pathwidth and treedepth of the input graph.

Proof. To prove this we reduce from MRSS, which is known to be W[1]-hard when parameterized by the combined parameter $k + k'$ [9]. Let $I = (k, k', S, t)$ be an instance of MRSS. We construct an instance $I' = (G, r, V_\triangle, V_\square)$ of DEFEN-SIVE ALLIANCE^{FN} as in Lemma 2. Note that the pathwidth and treedepth of a tree are at most its height. We claim that as G has a $k + 1$ size vertex deletion set $D = U \cup \{a\}$ into trees of height at most 4, then G has a path decomposition with pathwidth at most $k + 4$. First, we get a path decomposition of trees of height at most 4, it has pathwidth at most 3; then add $k + 1$ vertices of D to all the bags to get a path decomposition of G. This implies that the pathwidth of G is bounded by $k + 4$. To compute treedepth of G, note that $G \setminus D$ is a rooted forest where the trees are of height at most 4. We add paths of length at most $k + 1$ at the roots, covering the vertices of D, such that the resulting forest is a transitive closure of G. The height of the resulting forest is at most $k + 5$. This implies that the treedepth of G is bounded by $k + 5$. $\square$

Lemma 3. The EXACT DEFENSIVE ALLIANCE^F problem is W[1]-hard parameterized by the size of a vertex deletion set into trees of height at most 5.

Proof. To prove this we reduce from EXACT DEFENSIVE ALLIANCE^{FN} when $|V_\triangle| = 1$, which is W[1]-hard when parameterized by the size of a vertex deletion set into trees of height at most 5. See Corollary 1. Let $I = (G, r, V_\triangle, V_\square)$ with $|V_\triangle| = 1$ be an instance of EXACT DEFENSIVE ALLIANCE^{FN}. Let $n = |V(G)|$, $r \le n$ and $V_\triangle = \{x\}$. We construct an instance $I' = (G', r', V'_\square)$ of EXACT DEFENSIVE ALLIANCE^F problem the following way. See Fig. 3 for an illustration.

We introduce a set of vertices $H = \{h_1, h_2, \ldots, h_{2n}\}$ and a vertex x'. We also add a set H^x of one degree forbidden vertices adjacent to x. Similarly, we

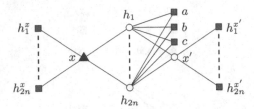

Fig. 3. An illustration of the reduction of necessary vertices in EXACT DEFENSIVE ALLIANCE^{FN} to EXACT DEFENSIVE ALLIANCE^F. The vertex x may have additional neighbours in G.

add a set $H^{x'}$ of one degree forbidden vertices adjacent to x'. We add three new forbidden vertices a, b and c. All the vertices in set H are adjacent to x, x', a, b and c. We define G' as follows:

$$V(G') = V(G) \cup H \cup H^x \cup H^{x'} \cup \{a, b, c, x'\}$$

and

$$E(G') = E(G) \bigcup_{h \in H} \{(x, h), (x', h), (a, h), (b, h), (c, h)\} \bigcup_{h^x \in H^x} (x, h^x) \bigcup_{h^{x'} \in H^{x'}} (x', h^{x'}).$$

We define $V'_{\square} = V_{\square} \cup H^x \cup H^{x'} \cup \{a, b, c\}$ and set $r' = r + 2n + 1$. Suppose for G the size of a vertex deletion set into trees of height at most 4, is k. Deleting that set along with vertices $\{x, x', a, b, c\}$, we get a vertex deletion set into trees of height at most 4. The size is clearly bounded by $k + 5$.

We claim that I is a yes-instance if and only if I' is a yes-instance. Suppose there is a defensive alliance R of size exactly r in G such that $x \in R$ and $V_{\square} \cap R = \emptyset$. It is easy to check that $R' = R \cup H \cup \{x'\}$ is a defensive alliance of size $r + 2n + 1$ such that $V'_{\square} \cap R' = \emptyset$. This implies that I' is a yes-instance.

To prove the reverse direction of the equivalence, suppose there is a defensive alliance R' of size r' such that $V'_{\square} \cap R' = \emptyset$. We claim that $x \in R'$. For the sake of contradiction, suppose $x \notin R'$. Then no vertex from the set H is part of R' as $d_{R'}(h) \leq 2$ and $d_{R'^c}(h) \geq 4$ for each $h \in H$. This implies that $|R'| \leq n < r'$, a contradiction. Therefore $x \in R'$. Observe that at least one vertex from H must be part of R' for protection of x in R'. Without loss of generality assume that $h_1 \in R'$. We see that the protection of h_1 requires x' to be inside the solution. Therefore, x' is in R'. Now, the protection of x' requires $2n$ many vertices which can only be contributed by H. It implies that $H \subseteq R'$. We claim that $R = R' \setminus \{H, x\}$ forms a defensive alliance of size exactly r in G. Since $R' \cap V(G) = \{x\}$, we only need to show that x is protected R. This is true since x looses $2n$ neighbours from inside and outside the solution in G'. This shows that I is a yes-instance. $\square$

In [1], Bliem and Woltran proved that DEFENSIVE ALLIANCE^F problem is W[1]-hard when parameterized by the treewidth of the input graph by giving a reduction from DEFENSIVE ALLIANCE^{FN} problem which again they showed to be W[1]-hard when parameterized by the treewidth of the graph. Since we have a stronger result that the DEFENSIVE ALLIANCE^{FN} is W[1]-hard when parameterized by the pathwidth of the input graph, we now prove that the DEFENSIVE ALLIANCE^F problem is W[1]-hard when parameterized by the pathwidth of the input graph. We obtain the following hardness results using the reduction of Lemma 5 given in [1] and see that the resulting graph has bounded pathwidth.

Lemma 4. The DEFENSIVE ALLIANCE^F problem is W[1]-hard when parameterized by the pathwidth of the graph.

Now we give an FPT reduction to get rid of forbidden vertices. The same reduction holds for the exact version of the problem as well.

2.1   Proof of Theorem 1

Proof. To prove Theorem 1 we reduce from the DEFENSIVE ALLIANCE^F problem, which is W[1]-hard when parameterized by the pathwidth of the input graph. See Lemma 4. Let $I = (G, r, V_\square)$ be an instance of the DEFENSIVE ALLIANCE^F problem. We construct an instance $I' = (G', r')$ of DEFENSIVE ALLIANCE problem the following way. We set $r' = r$. For every vertex $u \in V_\square$, we introduce a set $V_u = \{u_1, u_2, \ldots, u_{2r+2}\}$ of $2r + 2$ many vertices adjacent to u. We also add a new vertex t which is adjacent to all the vertices in $\bigcup_{u \in V_\square} V_u$.
Clearly, we can see that the pathwidth of G' is at most the pathwidth of G plus two. We claim that I is a yes-instance if and only if I' is a yes-instance. Suppose there is a defensive alliance R of size at most r in G such that $V_\square \cap R = \emptyset$. Clearly $R' = R$ is also a defensive alliance of size at most $r' = r$ in G'. This implies that I' is a yes-instance.

To prove the reverse direction of the equivalence, suppose there is a defensive alliance R' in G' of size at most $r' = r$. We observe that any defensive alliance in G' containing a vertex from the set $V_\square \bigcup_{u \in V_\square} V_u \cup \{t\}$ is of size at least $r + 1$. Since $|R'| \leq r' = r$, this implies that $R' \cap (V_\square \bigcup_{u \in V_\square} V_u \cup \{t\}) = \emptyset$. We see that $R = R'$ is a defensive alliance of size at most r such that $R \cap V_\square = \emptyset$. This shows that I is a yes-instance.   □

2.2   Proof of Theorem 2

Proof. To prove Theorem 2 we reduce from EXACT DEFENSIVE ALLIANCE^F problem, which is W[1]-hard when parameterized by the vertex deletion set into trees of height at most 4 of the input graph. See Lemma 3. The reduction here is the same as in the proof of Theorem 1. Therefore, the EXACT DEFENSIVE ALLIANCE problem is W[1]-hard when parameterized by the vertex deletion set into trees of height at most 6. Clearly trees of height at most 6 are trivially acyclic. Moreover, it is easy to verify that such trees have pathwidth [11] treedepth [13] at most 6, which implies the EXACT DEFENSIVE ALLIANCE problem is W[1]-hard when parameterized by any of the following parameters: the feedback vertex set, pathwidth, treewidth and treedepth of the input graph.   □

Corollary 4. The DEFENSIVE ALLIANCE^N problem with exactly one necessary vertex is W[1]-hard when parameterized by the size of a vertex deletion set into trees of height at most 6.

Proof. To prove this we reduce from DEFENSIVE ALLIANCE^{FN} problem with $|V_\triangle| = 1$, which is W[1]-hard when parameterized by the size of a vertex deletion set into trees of height at most 5. See Corollary 1. The reduction here is the same as in the proof of Theorem 1.   □

3   Conclusions

In this paper, we have proved that the DEFENSIVE ALLIANCE problem is W[1]-hard when parameterized by the pathwidth of the input graph and the EXACT DEFENSIVE ALLIANCE problem is W[1]-hard parameterized by a wide range of fairly restrictive structural parameters such as the feedback vertex set number, pathwidth, treewidth and treedepth of the input graph. The parameterized complexity of the DEFENSIVE ALLIANCE problem remains unsettled when parameterized by the feedback vertex set number, pathwidth and treedepth of the input graph. It would also be interesting to consider the parameterized complexity with respect to twin cover and modular width.

References

1. Bliem, B., Woltran, S.: Defensive alliances in graphs of bounded treewidth. Discret. Appl. Math. **251**, 334–339 (2018)
2. Chellali, M., Haynes, T.W.: Global alliances and independence in trees. Discuss. Math. Graph Theory **27**(1), 19–27 (2007)
3. Cygan, M., et al.: Parameterized Algorithms. Springer, Cham (2015). https://doi.org/10.1007/978-3-319-21275-3
4. Downey, R.G., Fellows, M.R.: Parameterized Complexity. Springer, New York (2012)
5. Enciso, R.: Alliances in graphs: parameterized algorithms and on partitioning series -parallel graphs. Ph.D. thesis, USA (2009)
6. Fernau, H., Raible, D.: Alliances in graphs: a complexity-theoretic study. In: Proceeding Volume II of the 33rd International Conference on Current Trends in Theory and Practice of Computer Science (2007)
7. Fricke, G., Lawson, L., Haynes, T., Hedetniemi, M., Hedetniemi, S.: A note on defensive alliances in graphs. Bull. Inst. Combin. Appl. **38**, 37–41 (2003)
8. Gaikwad, A., Maity, S., Tripathi, S.K.: Parameterized complexity of defensive and offensive alliances in graphs. In: Goswami, D., Hoang, T.A. (eds.) ICDCIT 2021. LNCS, vol. 12582, pp. 175–187. Springer, Cham (2021). https://doi.org/10.1007/978-3-030-65621-8_11

9. Ganian, R., Klute, F., Ordyniak, S.: On structural parameterizations of the bounded-degree vertex deletion problem. Algorithmica **83**, 297–336 (2020)
10. Kiyomi, M., Otachi, Y.: Alliances in graphs of bounded clique-width. Discret. Appl. Math. **223**, 91–97 (2017)
11. Kloks, T. (ed.): Treewidth. LNCS, vol. 842. Springer, Heidelberg (1994). https://doi.org/10.1007/BFb0045375
12. Kristiansen, P., Hedetniemi, M., Hedetniemi, S.: Alliances in graphs. J. Comb. Math. Comb. Comput. **48**, 157–177 (2004)
13. Nešetřil, J., Ossona de Mendez, P.: Sparsity. AC, vol. 28. Springer, Heidelberg (2012). https://doi.org/10.1007/978-3-642-27875-4
14. Robertson, N., Seymour, P.: Graph minors. III. planar tree-width. J. Comb. Theor. Ser. B **36**(1), 49–64 (1984)
15. Rodríguez-Velázquez, J., Sigarreta, J.: Global offensive alliances in graphs. Electron. Notes Discrete Math. **25**, 157–164 (2006)
16. Sigarreta, J., Bermudo, S., Fernau, H.: On the complement graph and defensive k-alliances. Discret. Appl. Math. **157**(8), 1687–1695 (2009)
17. Sigarreta, J., Rodríguez, J.: On defensive alliances and line graphs. Appl. Math. Lett. **19**(12), 1345–1350 (2006)
18. Sigarreta, J., Rodríguez, J.: On the global offensive alliance number of a graph. Discret. Appl. Math. **157**(2), 219–226 (2009)
19. West, D.B.: Introduction to Graph Theory. Prentice Hall (2000)

On the Approximability of Path and Cycle Problems in Arc-Dependent Networks

Piotr Wojciechowski[1], K. Subramani[1(⊠)], Alvaro Velasquez[2], and Matthew Williamson[3]

[1] LDCSEE, West Virginia University, Morgantown, WV, USA
{pwojciec,k.subramani}@mail.wvu.edu
[2] Information Directorate, Air-Force Research Laboratory, Rome, NY, USA
alvaro.velasquez.1@us.af.mil
[3] Marietta College, Marietta, OH, USA
williamm@marietta.edu

Abstract. In the field of transportation planning, it is often insufficient to model transportation networks by using networks with fixed arc costs. There may be additional factors that modify the time or cost of a single trip. These include turn prohibitions, fare rebates, and transfer times. Each of these factors causes the cost of a portion of the trip to depend directly on the previous portion of the trip. This dependence can be modeled using arc-dependent networks. In an arc-dependent network, the cost of an arc a depends upon the arc used to enter a. In this paper, we study the approximability of a number of negative cost cycle problems in arc-dependent networks. In a general network, the cost of an arc is a fixed constant and part of the input. Arc-dependent networks can be used to model several real-world problems, including the turn-penalty shortest path problem. Previous literature established that corresponding path problems in these networks are **NP-hard**. We extend that research by providing inapproximability results for several of these problems. In [7], it was established that a more general form of the shortest path problem in arc-dependent networks, known as the quadratic shortest path problem, cannot be approximated to within a constant factor. In this paper, we strengthen that result by showing **NPO PB-completeness**.

1 Introduction

This paper studies the inapproximability of several problems associated with negative cost cycles in arc-dependent networks. Recall that in a traditional network $\mathbf{G} = \langle \mathbf{V}, \mathbf{E} \rangle$, we have a set of vertices $\mathbf{V} = \{v_1, v_2, \ldots, v_n\}$, a set of directed arcs $\mathbf{E} = \{e_1, e_2, \ldots, e_m\}$, and a cost function $\mathbf{c} : \mathbf{E} \to \mathbb{R}$. The cost of an arc is a fixed constant. This is in contrast to arc-dependent networks in which the cost of an arc depends upon the arc taken to enter it. The inapproximability results obtained in this paper are stronger than similar results obtained for the quadratic shortest path problem [7].

© Springer Nature Switzerland AG 2022
N. Balachandran and R. Inkulu (Eds.): CALDAM 2022, LNCS 13179, pp. 292–304, 2022.
https://doi.org/10.1007/978-3-030-95018-7_23

In addition, the shortest paths do not necessarily need to be simple. Recall that a simple path is a path without repeated vertices or repeated arcs. In fact, in some cases, non-simple paths are actually shorter than simple paths depending on the path taken. In [13], an $O(m^2 \cdot n^{2 \cdot k - 1})$ algorithm is provided for solving the path-dependent shortest path problem, where k is the number of predecessors used to determine the arc cost. Our work, however, focuses exclusively on simple paths. Our problem is **NP-complete** even when the cost of an arc depends on only one predecessor [14].

Our problem is a restricted version of the quadratic shortest paths problem (QSPP) [7,8]. QSPP cannot be approximated to within any constant factor unless $\mathbf{P} = \mathbf{NP}$ [7]. This result applies even if an arc's cost only depends on adjacent arcs. However, in this paper, we are able to obtain stronger inapproximability results.

This paper is also concerned with the detection of negative cost cycles in arc-dependent networks. In this problem, the goal is to find a simple cycle NC such that the cost of NC is negative. Note that the cost of e_1, the first arc in NC, depends on e_k, the last arc in NC. Thus, the negative cost of the cycle depends only on the cycle itself and not on how the cycle was reached initially. The negative cost cycle (NCC) problem in general is one of the more widely-studied problems in theoretical computer science and operations research due to its wide applicability in a number of domains.

The study of arc-dependent networks is motivated by several applications including highway engineering. It is desirable to find optimal routes between points in a city or even a freeway network. These optimal routes are often measured in terms of the time or distance to travel from one point to one or more other points. There already exist efficient algorithms for finding such optimal routes [2,5]. However, these algorithms assume that there are no restrictions or delays at intersections. If there were delays at intersections, it could alter the intended optimal routes [16]. This introduces the notion of "turn penalties" at each intersection [3]. These penalties can increase the time and/or distance of a route based on which turn is taken at intersections. Turn penalties can be modeled in arc-dependent networks by having the cost of an arc depend on the turn taken from an intersection.

Arc-dependent networks also have applications in public transportation systems. It is common for travelers to receive rebates when transferring from one service line to another [12,13]. These fare rebates introduce an additional layer of complexity in public transportation. Some routes may result in a discount, while other routes might not have a discount. If we use standard shortest path algorithms, the optimal route might not be optimal when fare discounts are included.

The principal contributions of this paper are as follows:

1. Establishing that the shortest negative cost cycle problem is **NPO PB-complete** (see Sect. 3).
2. Establishing that the longest negative cost cycle problem is **NPO PB-complete** (see Sect. 3).

3. A fixed-parameter tractable algorithm for the shortest path problem in terms of the number of arcs in the path (see Sect. 4).
4. Showing that the shortest path problem does not admit a kernel whose size is polynomial in the number of arcs in the path (see Sect. 5).

2   Statement of Problems

In this section, we describe the structure of arc-dependent networks. We also present the network optimization problems that are studied in this paper.

Let $\mathbf{G} = \langle \mathbf{V}, \mathbf{E}, \mathbf{C} \rangle$ denote a directed network. $\mathbf{V}$ is the vertex set with n vertices, and $\mathbf{E} = \{e_1, e_2, \ldots, e_m\}$ is the set of arcs. Let $s \in \mathbf{V}$ be the source vertex. The cost structure is represented by the matrix $\mathbf{C}$, where entry $\mathbf{C}[e_i, e_j]$ stores the cost of arc e_j assuming that e_j was entered through arc e_i. The matrix $\mathbf{C}$ has $(m+1)$ rows and m columns. The $(m+1)^{th}$ row of $\mathbf{C}$ contains the cost of arcs that do not have any incoming arcs. We use the phantom arc e_0 entering s to account for these costs. We refer to $\mathbf{G}$ as an arc-dependent network.

Let P_i denote a path $(e_1 - e_2 - e_3 - \cdots - e_k)$, where e_1 is an arc leaving some vertex s, e_k is an arc entering some vertex t, and i is a positive integer to help label the unique path from s to t using arcs e_1 to e_k. It should be noted that there can be more than one path from s to t. Note that $\mathbf{C}[e_j, e_k]$ only matters when the head of arc e_j is the tail of arc e_k. Otherwise, e_j cannot be the predecessor of e_k. In cases where the head of e_j is not the tail of e_k, we define $\mathbf{C}[e_j, e_k] = 0$. As a result, we note that $\mathbf{C}$ does not represent the connectivity of $\mathbf{G}$. The cost of a path P_i between two vertices is given by: $cost(P_i) = \mathbf{C}[e_0, e_1] + \mathbf{C}[e_1, e_2] + \cdots + \mathbf{C}[e_{k-1}, e_k]$.

In this paper, we explore the following optimization problem:

Definition 1. *Shortest Path (SP) problem: Given an arc-dependent network $\mathbf{G}$, source vertex s, and target vertex t, what is the arc-dependent simple path from s to t in $\mathbf{G}$ with the least cost?*

We also explore the problem of finding simple negative cost cycles in arc-dependent networks. We call this problem the negative cost cycle problem, and it is defined as follows:

Definition 2. *Negative Cost Cycle (NCC) problem: Given an arc-dependent network $\mathbf{G}$, does $\mathbf{G}$ contain a simple cycle NC consisting of arcs e_1 through e_k such that:*

$$\mathbf{C}[e_k, e_1] + \sum_{i=2}^{k} \mathbf{C}[e_{i-1}, e_i] < 0?$$

Note that the cost of e_1, the first arc in NC, depends on e_k, the last arc in NC. Thus, the negative cost of the cycle depends only on the cycle itself and not on how the cycle was reached initially.

In this paper, we study several variants of the NCC problem. These are the following:

1. Shortest Negative Cost Cycle (SNCC) problem: Given an arc-dependent network $\mathbf{G}$, what is the simple arc-dependent negative cycle in $\mathbf{G}$ with the fewest arcs?
2. Longest Negative Cost Cycle (LoNCC) problem: Given an arc-dependent network $\mathbf{G}$, what is the simple arc-dependent negative cycle in $\mathbf{G}$ with the most arcs?

3 Computational Complexity of SNCC and LoNCC

In this section, we show that the shortest negative cost cycle (SNCC) and longest negative cost cycle (LoNCC) problems in arc-dependent networks are **NPO PB-complete** [10]. This is done by a reduction from the Minimum Ones and Maximum Ones problems respectively.

Definition 3. *Minimum Ones: Given a 3 CNF formula Φ, what is the minimum number of variables assigned to* **true** *in any satisfying assignment to Φ?*

Definition 4. *Maximum Ones: Given a 3 CNF formula Φ, what is the maximum number of variables assigned to* **true** *in any satisfying assignment to Φ?*

Both the Minimum Ones and Maximum Ones problems are known to be **NPO PB-complete** [9]. Let Φ be a 3CNF formula with n variables and m clauses. From Φ, we create an arc-dependent network $\mathbf{G}$ as follows:

1. Create the vertices x_0 and z_0.
2. For each variable x_i in Φ:
   (a) Create the vertex x_i, the vertices $y_{i,1}^+$ through $y_{i,m+(n+2)\cdot(m+1)}^+$, and the vertices $y_{i,1}^-$ through $y_{i,m}^-$.
   (b) Create the arcs $(x_{i-1}, y_{i,1}^+)$ and $(x_{i-1}, y_{i,1}^-)$ with cost 0.
   (c) Create the arc $(y_{i,1}^+, y_{i,2}^+)$ with cost 0 if the preceding arc is $(x_{i-1}, y_{i,1}^+)$ and cost 1 otherwise, and the arc $(y_{i,1}^-, y_{i,2}^-)$ with cost 0 if the preceding arc is $(x_{i-1}, y_{i,1}^-)$ and cost 1 otherwise.
   (d) For each $j = 3 \ldots m$, create the arc $(y_{i,j-1}^-, y_{i,j}^-)$ with cost 0 if the preceding arc is $(y_{i,j-2}^-, y_{i,j-1}^-)$ and cost 1 otherwise.
   (e) Create the arc $(y_{i,m}^-, x_i)$ with cost 0 if the preceding arc is $(y_{i,m-1}^-, y_{i,m}^-)$ and cost 1 otherwise.
   (f) For each $j = 3 \ldots (m + (n + 2) \cdot (m + 1))$, create the arc $(y_{i,j-1}^+, y_{i,j}^+)$ with cost 0 if the preceding arc is $(y_{i,j-2}^+, y_{i,j-1}^+)$ and cost 1 otherwise.
   (g) Create the arc $(y_{i,m+(n+2)\cdot(m+1)}^+, x_i)$ with cost 0 if the preceding arc is $(y_{i,m+(n+2)\cdot(m+1)-1}^+, y_{i,m+(n+2)\cdot(m+1)}^+)$ and cost 1 otherwise.
3. Create the arc (x_n, z_0) with cost 0.
4. For each clause ϕ_j in Φ:
   (a) Create the vertex z_j.

(b) For each literal x_i in ϕ_j, create the arc $(z_{j-1}, y_{i,j}^-)$ with cost 0 and the arc $(y_{i,j}^-, z_j)$ with cost 0 if the preceding arc is $(z_{j-1}, y_{i,j}^-)$ and cost 1 otherwise.

(c) For each literal $\neg x_i$ in ϕ_j, create the arc $(z_{j-1}, y_{i,j}^+)$ with cost 0 and the arc $(y_{i,j}^+, z_j)$ with cost 0 if the preceding arc is $(z_{j-1}, y_{i,j}^+)$ and cost 1 otherwise.

5. Create the arc (z_m, x_0) with cost -1.

We construct **G** so that the only arc with negative cost is (z_m, x_0). Thus, a negative cost cycle must consist of the arc (z_m, x_0) and a 0-cost path from x_0 to z_m. The arc costs in a negative cost cycle are defined in a manner such that the only possible path uses the vertices x_1 through x_n followed by the vertices z_0 through z_m, with additional intermediate vertices.

Between each pair of vertices x_{i-1} and x_i, there are two possible 0-cost paths. The first path uses $(m + 1 + (n + 2) \cdot (m + 1))$ arcs and corresponds to setting the variable x_i to **true**. The second path uses $(m + 1)$ arcs and corresponds to setting the variable x_i to **false**. Thus, the path from x_0 to x_n uses $(n \cdot (m+1) + k \cdot (n + 2) \cdot (m + 1))$ arcs, where k is the number of **true** variables.

Between each pair of vertices z_{j-1} and z_j, there are up to three possible 0-cost paths each with two arcs. These paths are constructed such that only paths corresponding to **true** literals can be used in a simple cycle. Thus, the path from z_0 to z_m uses $(2 \cdot m)$ arcs. With the additional arcs (x_n, z_0) and (z_m, x_0), the cycle will use a total of $(k + 1) \cdot (n + 2) \cdot (m + 1)$ arcs.

We construct **G** such that any negative cost cycle in **G** corresponds to choosing a truth assignment to each variable in Φ and ensuring that each clause has a **true** literal. We show that for any k, Φ has a satisfying assignment in which k variables set to **true** if and only if there is a simple negative cycle in **G** with $(k + 1) \cdot (n + 2) \cdot (m + 1)$ arcs.

Lemma 1. *Let Φ be a 3CNF formula with n variables and m clauses. Let **G** be the corresponding network for Φ. For any k, Φ has a satisfying assignment with k variables set to **true** if and only if there is a simple negative cycle in **G** with $(k + 1) \cdot (n + 2) \cdot (m + 1)$ arcs.*

Proof. First assume that Φ has a satisfying assignment in which k variables are set to **true**. Let **x** be such a satisfying truth assignment to Φ. From **x**, we construct a cycle NC in **G** as follows:

We examine each variable $x_i \in \mathbf{x}$. If $x_i = \textbf{true}$, add the arcs $(x_{i-1}, y_{i,1}^+)$, $(y_{i,1}^+, y_{i,2}^+)$, ..., $(y_{i,m+(n+2)\cdot(m+1)}^+, x_i)$ to NC. This adds $(m+1+(n+2)\cdot(m+1))$ arcs to NC. If $x_i = \textbf{false}$, add the arcs $(x_{i-1}, y_{i,1}^-)$, $(y_{i,1}^-, y_{i,2}^-)$, ..., $(y_{i,m-1}^-, y_{i,m}^-)$, $(y_{i,m}^-, x_i)$ to NC. This adds $(m + 1)$ arcs to NC. NC now contains a total of $(n \cdot (m + 1) + k \cdot (n + 2) \cdot (m + 1))$ arcs. We then add the arc (x_n, z_0) to NC.

We next examine each clause $\phi_j \in \Phi$. Since **x** is a satisfying assignment, at least one literal in ϕ_j is set to **true**. If this **true** literal is x_i, then add the arcs $(z_{j-1}, y_{i,j}^-)$ and $(y_{i,j}^-, z_j)$ to NC. Since x_i is **true**, the vertex $y_{i,j}^-$ is not already on NC. If this

true literal is $\neg x_i$, then add the arcs $(z_{j-1}, y_{i,j}^+)$ and $(y_{i,j}^+, z_j)$ to NC. Since x_i is **false**, the vertex $y_{i,j}^+$ is not already on NC. We then add the arc (z_m, x_0) to NC. This means NC now has a total of $(k+1) \cdot (n+2) \cdot (m+1)$ arcs.

By construction of $\mathbf{G}$ and NC, the only arc in NC with non-zero cost is (z_m, x_0) with cost -1. Thus, NC is a negative cycle with $(k+1) \cdot (n+2) \cdot (m+1)$ arcs.

Now assume that $\mathbf{G}$ has a simple negative cycle NC. From NC, we construct an assignment $\mathbf{x}$ to Φ as follows: *For each variable x_i, if the arc $(x_{i-1}, y_{i,1}^+)$ is in NC, then set $x_i =$ **true**. Otherwise, set $x_i =$ **false***. By construction, the only negative cost arc in $\mathbf{G}$ is (z_m, x_0). Since (z_m, x_0) has cost -1, NC cannot include any arcs with cost 1.

Since NC contains an arc entering x_0, NC must contain an arc leaving x_0. For each $i = 1, \ldots, n$, the only arcs leaving x_{i-1} are $(x_{i-1}, y_{i,1}^+)$ and $(x_{i-1}, y_{i,1}^-)$. If NC uses the arc $(x_{i-1}, y_{i,1}^+)$, then the only way to avoid arcs of cost 1 is to use the arcs $(x_{i-1}, y_{i,1}^+), (y_{i,1}^+, y_{i,2}^+), \ldots, (y_{i,m+(n+2)\cdot(m+1)}^+, x_i)$. Similarly, if NC uses the arc $(x_{i-1}, y_{i,1}^-)$, then the only way to avoid arcs with cost 1 is to use the arcs $(x_{i-1}, y_{i,1}^-)$, $(y_{i,1}^-, y_{i,2}^-), \ldots, (y_{i,m-1}^-, y_{i,m}^-), (y_{i,m}^-, x_i)$. Thus, for each $i = 1, \ldots, n$, the cycle NC uses the vertex x_i. In particular, NC must contain an arc leaving x_n.

The only arc leaving x_n is (x_n, z_0). This arc must be on NC. Since NC contains an arc entering z_0, NC must contain an arc leaving z_0. For each $j = 1, \ldots, m$, the only arcs leaving z_{j-1} are $(z_{j-1}, y_{i,j}^-)$ for literals x_i in ϕ_j and $(z_{j-1}, y_{i,j}^+)$ for literals $\neg x_i$ in ϕ_j.

If NC uses the arc $(z_{j-1}, y_{i,j}^+)$, then NC could not have used the arc $(x_{i-1}, y_{i,1}^+)$ previously. Thus, $x_i =$ **false** and the clause ϕ_j is satisfied. Additionally, the only way to leave $y_{i,j}^+$ without using an arc with cost 1 is to use the arc $(y_{i,j}^+, z_j)$.

If NC uses the arc $(z_{j-1}, y_{i,j}^-)$, then NC must have used the arc $(x_{i-1}, y_{i,1}^+)$ previously. Thus, $x_i =$ **true** and the clause ϕ_j is satisfied. Additionally, the only way to leave $y_{i,j}^-$ without using an arc with cost 1 is to use the arc $(y_{i,j}^-, z_j)$.

Therefore, $\mathbf{x}$ is a satisfying assignment to Φ. As previously shown, the number of arcs in NC is $(k+1) \cdot (n+2) \cdot (m+1)$, where k is the number of variables set to **true** by $\mathbf{x}$. □

Using Lemma 1, we can show that both the SNCC and LoNCC problems for arc-dependent networks are **NPO PB-complete**.

Theorem 1. *The SNCC problem for arc-dependent networks is* **NPO PB-complete**.

Proof. First, we establish that a **PTAS** reduction [11] exists from the minimum ones problem to the SNCC problem for arc-dependent networks. This will be done by establishing the existence of the functions f, g, and α.

1. The function f: Earlier in this section, we provided a method for constructing an arc-dependent network $\mathbf{G}$ from a 3CNF formula Φ in polynomial time. This forms the function f required for the **PTAS** reduction.

2. The function g: In the proof of Lemma 1, we provided a method to take a simple negative cost cycle NC in $\mathbf{G}$ and construct a satisfying assignment to Φ. This forms the function g required for the **PTAS** reduction.

3. The function α: Let k^* be minimum number of **true** variables in any satisfying assignment to Φ. From Lemma 1, $\mathbf{G}$ has a negative cost simple cycle with $(k^* + 1) \cdot (n + 2) \cdot (m + 1)$ arcs. Additionally, if $\mathbf{G}$ has a simple negative cost cycle with fewer arcs, then Φ would have a satisfying assignment with fewer **true** variables. Thus, the SNCC of $\mathbf{G}$ has $(k^* + 1) \cdot (n + 2) \cdot (m + 1)$ arcs. Let $\alpha(\epsilon) = \frac{\epsilon - 1}{2}$.

Let NC be a simple negative cost cycle in $\mathbf{G}$ with $(k + 1) \cdot (n + 2) \cdot (m + 1)$ arcs. The function g produces a satisfying assignment to Φ with k **true** variables. Since we can determine if Φ is satisfied by the all **false** assignment in polynomial time, we can assume without loss of generality that $k^* \geq 1$. If $\frac{(k+1) \cdot (n+2) \cdot (m+1)}{(k^*+1) \cdot (n+2) \cdot (m+1)} \leq 1 + \alpha(\epsilon) = \frac{\epsilon + 1}{2}$, then:

$$\frac{k}{k^*} = \frac{2 \cdot k}{2 \cdot k^*} \leq \frac{2 \cdot (k+1)}{k^* + 1} = \frac{2 \cdot (k+1) \cdot (n+2) \cdot (m+1)}{(k^*+1) \cdot (n+2) \cdot (m+1)} \leq \frac{2 \cdot (\epsilon + 1)}{2} = 1 + \epsilon.$$

Thus, we have a **PTAS** reduction from the minimum ones problem to the SNCC problem for arc-dependent networks.

Since the minimum ones problem is **NPO PB-complete** [9], the SNCC problem for arc-dependent networks is **NPO PB-hard**. As argued previously, the SNCC problem for arc-dependent networks is in **NPO PB**. Thus, the SNCC problem for arc-dependent networks is **NPO PB-complete**.   □

Theorem 2. *The LoNCC problem for arc-dependent networks is* **NPO PB-complete**.

The proof of Theorem 2 is similar to the proof of Theorem 1. Note that the SNCC problem is a restricted version of QSPP in which the cost of each arc depends only on the previous arc and the path must go from a vertex to itself. Thus, the inapproximability result for the SNCC problem also applies to QSPP. As a result, we have the following corollary:

Corollary 1. *QSPP is* **NPO PB-hard**.

4   Fixed-Parameter Algorithm for SP

In this section, we present a Fixed-Parameter Tractable (FPT) algorithm for finding a shortest path p in an arc-dependent network $\mathbf{G}$. This algorithm uses k, the number of arcs in p, as its parameter.

4.1   Intuition

Observe that if k is small, then it is easy to enumerate all possible paths and then return the shortest path found. Thus, the length of the path is a natural parameter for this problem. Let **G** be an arc-dependent network with initial vertex s and target vertex t. Our approach for solving this problem proceeds as follows: First, we design a randomized algorithm (Algorithm 4.1) that randomly partitions the vertices of **G** into $(k - 1)$ sets. Then we find the shortest path from s to t that uses at most one intermediate vertex from each set.

Note that if the shortest path has k edges, then it has $(k - 1)$ intermediate vertices. Thus, it is possible to construct a partition that assigns each intermediate vertex to a different set. In Sect. 4.3, we prove that this happens with probability at least $\frac{1}{e^{k-1}}$. We then derandomize the algorithm by proving that only a limited number of partitionings need to be tested. This results in the desired **FPT** algorithm.

4.2 Randomized Algorithm

Let **G** be an arc-dependent network with initial vertex s and target vertex t. The algorithm proceeds by first partitioning the vertices of **G** into the sets $S_1, \ldots, S_{k-1}$. Then, the algorithm finds the shortest path p such that for each set of vertices $S \in \{S_1, \ldots, S_{k-1}\}$, at most one vertex from S is on p. We refer to such a path as a partitioned path. Note that every partitioned path is simple and that every simple path of length k can be made into a partitioned path by assigning every intermediate vertex in p to a different set. For a given vertex v_i in **G**, let $S(v_i) \in \{S_1, \ldots, S_{k-1}\}$ be the set containing the vertex v_i.

Let us consider a method that constructs the partitioned path backwards, starting from the destination t. Finding the shortest one-arc path to t can be done by simply looking at the costs of the arcs going into t. We can then look further back to consider all two-arc partitioned paths to t. As we continue backtracking, we need to keep track of the sets containing the intermediate vertices so that we do not use multiple vertices from the same set.

Suppose, for vertices v_i and v_j and $H \subseteq \{S_1, \ldots, S_{k-1}\} \setminus \{S(v_i), S(v_j)\}$, we know the shortest partitioned path p from v_i to t with predecessor v_j such that p uses at most one vertex from each set in H. Then, from the perspective of further backtracking, it does not matter what order the vertices are visited by p. We only need to know that the vertices in the sets in H cannot be used for further backtracking. Thus, we only need to know the shortest partitioned path for each starting vertex, predecessor vertex, and $H \subseteq \{S_1, \ldots, S_{k-1}\}$. This leads us to a dynamic programming based algorithm.

Let $P(v_i, v_j, H)$ be the least cost of any path from v_i to t with predecessor v_j and set of used partitions H. Note that the cost of any path from v_i to t depends only on the vertex v_j that precedes v_i. Thus, $P(v_i, v_j, H)$ is well defined. If there are no intermediate vertices, then the only possible path from v_i to t is the arc (v_i, t). Thus, $P(v_i, v_j, \emptyset) = \mathbf{C}[(v_j, v_i), (v_i, t)]$. We will now show that

$$P(v_i, v_j, H) = \min_{v_r : S(v_r) \in H} \left\{ \mathbf{C}[(v_j, v_i), (v_i, v_r)] + P(v_r, v_i, H \setminus \{S(v_r)\}) \right\}.$$

Theorem 3. *Let* $\mathbf{G} = \langle \mathbf{V}, \mathbf{E}, \mathbf{C} \rangle$ *be an arc-dependent network. For each* $v_i, v_j \in$ $\mathbf{V}$ *and each* $H \subseteq \{S_1, \ldots, S_{k-1}\} \setminus \{S(v_i), S(v_j)\}$,

$$P(v_i, v_j, H) = \min_{v_r : S(v_r) \in H} \left\{ \mathbf{C}[(v_j, v_i), (v_i, v_r)] + P(v_r, v_i, H \setminus \{S(v_r)\}) \right\}.$$

Proof. If $|H| = 1$, then there is only one set $S \in H$. By definition of P, the only intermediate vertex between v_i and t must belong to S. Thus, the partitioned path from v_i to v_t is of the form $(v_i, v_r) - (v_r, t)$, where $S(v_r) = S$. The cost of this path is

$$P(v_i, v_j, H) = \mathbf{C}[(v_i, v_r), (v_r, t)] + \mathbf{C}[(v_j, v_i), (v_i, v_r)]$$
$$= \mathbf{C}[(v_j, v_i), (v_i, v_r)] + P(v_r, v_i, \emptyset).$$

Now assume that this holds true for all sets H of size h. Let H' be a set of size $(h + 1)$. Let p be the shortest partitioned path from v_i to t with predecessor v_j and set of used partitions H'. We know that some v_r such that $S(v_r) \in H'$ immediately follows v_i on p. Thus, p can be broken up into the arc (v_i, v_r) and a path p' from v_r to t. Note that p' has $H' \setminus \{S(v_r)\}$ as its set of used partitions and has v_i as its predecessor. This means that the cost of p' is at least $P(v_r, v_i, H \setminus \{S(v_r)\})$.

If there is a partitioned path p^* from v_r to t with set of used partitions $H' \setminus \{S(v_r)\}$ and predecessor v_i that is shorter than p', then the path consisting of the arc (v_i, v_k) followed by p^* is a partitioned path from v_i to t with set used partitions H' that is shorter than p. Since this violates the optimality of p, p' must be the shortest path from v_k to t with set of used partitions $H' \setminus \{S(v_r)\}$ and predecessor v_i. Therefore, the cost of p' is $P(v_k, v_i, H \setminus \{S(v_r)\})$. This means that :

$$P(v_i, v_j, H') = \min_{v_r : S(v_r) \in H'} \left\{ \mathbf{C}[(v_j, v_i), (v_i, v_r)] + P(v_r, v_i, H' \setminus \{S(v_r)\}) \right\}.$$

$\square$

From Theorem 3, $P(v_i, v_j, H)$ can be found in $O(n)$ time once P is known for every pair of vertices and every subset of H. Note that we only need to find $P(v_i, v_j, H)$ when $(v_j, v_i) \in \mathbf{E}$. Thus, there are $O(m \cdot 2^k)$ possible inputs to P. This means that P can be computed in $O(m \cdot n \cdot 2^k)$ time using a dynamic program. Once P is computed, it is easy to see that the shortest partitioned path from s to t in $\mathbf{G}$ has cost $\min_{H \subseteq \{S_1, \ldots, S_{k-1}\}} P(s, \_, H)$, where $\_$ implies that s does not have a preceding arc in the shortest partitioned path from s to t.

We now provide a randomized algorithm for finding the shortest simple path in an arc-dependent network $\mathbf{G}$. This is represented by Algorithm 4.1. Algorithm 4.1 uses a similar color-coding technique to the one introduced in [1]. Recall that P can be computed in $O(m \cdot n \cdot 2^k)$ time. Thus, Algorithm 4.1 runs in $O(m \cdot n \cdot 2^k)$ time.

SP_RAND (arc-dependent network **G**, integer k)
1: Create the sets S_1 through S_{k-1}.
2: **for** (each vertex v_i in $\mathbf{V} \setminus \{s, t\}$) **do**
3:    Randomly assign v_i to a set $S(v_i) \in \{S_1, \ldots, S_{k-1}\}$.
4: Create the function $P(v_i, v_j, H)$ and define $P(v_i, v_j, \emptyset) = \mathbf{C}[(v_j, v_i), (v_i, t)]$ for each pair of vertices v_i, v_j.
5: **for** (each pair of vertices v_i, v_j and each $H \subseteq \{S_1, \ldots, S_{k-1}\} \setminus \{S(v_i), S(v_j)\}$) **do**
6:    $P(v_i, v_j, H) = \min_{v_r : S(v_r) \in H}\{\mathbf{C}[(v_j, v_i), (v_i, v_r)] + P(v_r, v_i, H \setminus \{S(v_r)\})\}$.
7: **return** $\min_{H \subseteq \{S_1, \ldots, S_{k-1}\}} P(s, \_, H)$.

Algorithm 4.1: Randomized SP Algorithm for Arc-Dependent Networks

4.3   Proof of Correctness

We now show that if Algorithm 4.1 returns L, then **G** has a simple path from s to t with total cost L using at most k arcs. We also show that if the shortest simple path from s to t in **G** with at most k arcs has total cost L, then Algorithm 4.1 returns L with probability at least $\frac{1}{e^{k-1}}$.

Theorem 4. *If Algorithm 4.1 returns L, then **G** has a simple path from s to t with total cost L using at most k arcs.*

Proof. Assume Algorithm 4.1 returns L. This means that there exist sets S_1 through S_{k-1} such that there exists a partitioned path p from s to t in **G** with total cost L. Note that p is a simple path. Additionally, p has at most $(k+1)$ vertices (s, t, and at most one vertex from each of the $(k-1)$ partitions). Thus, p is a simple path from s to t of total cost L with at most k arcs.   □

Theorem 5. *If a shortest simple path from s to t in **G** with at most k arcs has total cost L, then Algorithm 4.1 will return L with probability at least $\frac{1}{e^{k-1}}$.*

Proof. Let p be a shortest simple path from s to t in **G** with at most k arcs. Note that p has total cost L. Observe that if the sets S_1 through S_{k-1} are chosen so that p is a partitioned path, then by Theorem 3, Algorithm 4.1 will return L. We want to find the probability that p is a partitioned path. p is a partitioned path if each intermediate vertex v_i is assigned to a different set $S(v_i)$. Note that there are $(k-1)^{k-1}$ different ways to assign the $(k-1)$ intermediate vertices of p to the sets S_1 through S_{k-1}. Additionally, there are $(k-1)!$ ways to assign each vertex to a unique partition. Therefore, the probability of p being a partitioned path is $\frac{(k-1)!}{(k-1)^{k-1}} > \frac{1}{e^{k-1}}$.   □

4.4   Derandomization

To obtain an FPT algorithm for finding a shortest path, we will derandomize Algorithm 4.1 as described in [4]. This derandomization utilizes (m, k)-perfect hash families which are defined as follows:

Definition 5. *Let S be a set of size m. An (m, k)-perfect hash family is a family U of functions F that partition S into S_1 through S_k such that for any set $R \subseteq S$ of size k, there exists a function that assigns each element of R to a different partition.*

Let p be a shortest path in $\mathbf{G}$ that uses at most k arcs. Note that p has $(k-1)$ intermediate vertices. Let U be an $(n-2, k-1)$-perfect hash family of $\mathbf{V} \setminus \{s, t\}$. Then, for some $F \in U$, every intermediate vertex v_i used by p is assigned to a different set $S(v_i)$. Note that we can construct an $(n-2, k-1)$-perfect hash family for $\mathbf{V} \setminus \{s, t\}$ of size $e^{k-1} \cdot (k-1)^{O(\log k)} \cdot \log(n-2)$ in $O(e^k \cdot k^{O(\log k)} \cdot n \cdot \log n)$ time [4].

Thus, given P, finding the shortest path from s to t in $\mathbf{G}$ with at most k arcs can be done as follows:

1. Construct an $(n-2, k-1)$-perfect hash family U for $\mathbf{V} \setminus \{s, t\}$. Note that U contains $e^{k-1} \cdot (k-1)^{O(\log k)} \cdot \log(n-2)$ partitions of $\mathbf{V} \setminus \{s, t\}$. This can be done in $O(e^k \cdot k^{O(\log k)} \cdot n \cdot \log n)$ time.
2. For each $F \in U$, check if $\mathbf{G}$ has a partitioned path from s to t. Using the method described previously, this takes $O(m \cdot n \cdot 2^k)$ time for each of the $e^{k-1} \cdot (k-1)^{O(\log k)} \cdot \log(n-2)$ elements of U.

This algorithm runs in $O((2 \cdot e)^k \cdot k^{O(\log k)} \cdot m \cdot n \cdot \log n)$ time. Thus, this is an FPT algorithm for finding a shortest simple path in an arc-dependent network.

Note that Algorithm 4.1 can be easily modified to find partitioned negative cycles. Applying the same derandomization technique will result in FPT algorithms for the SNCC and LoNCC problems when parameterized by the number of arcs in the cycle.

5   Lower Bound on Kernel Size for the SP Problem

In this section, we show that the SP problem for arc-dependent networks does not have a kernel whose size is polynomial in k, where k is the number of arcs in the path. This is done through the use of an OR-distillation [6].

Definition 6. *Let P and Q be a pair of problems and let $t : \mathbb{N} \to \mathbb{N} \setminus \{0\}$ be a polynomially bounded function. A t-bounded OR-distillation from P into Q is an algorithm that for every s, given as $t(s)$ input strings $x_1, \ldots, x_{t(s)}$ with $|x_j| = s$ for all j:*

1. *Runs in polynomial time, and*
2. *Outputs a string y of length at most $t(s) \cdot \log s$ such that y is a **yes** instance of Q if and only if x_j is a **yes** instance of P for some $j \in \{1, \ldots, t(s)\}$.*

If any **NP-hard** problem has a t-bounded OR-distillation, then **coNP** $\subseteq$ **NP/poly** [6]. If **coNP** $\subseteq$ **NP/poly**, then $\mathbf{\Sigma_3^P} = \mathbf{\Pi_3^P}$ [15]. Thus, the polynomial hierarchy would collapse to the third level.

Theorem 6. *The SP problem for arc-dependent networks does not have a polynomial sized kernel unless* **coNP** $\subseteq$ **NP/poly**.

Proof. We will prove this by showing that if the SP problem for arc-dependent networks has a polynomial sized kernel, then there exists a t-bounded OR-distillation from the SP problem for arc-dependent networks into itself.

For each j, let $\mathbf{G}_j$ be an arc-dependent network with n vertices and m arcs such that, for pair of arcs (v_i, v_j) and (v_j, v_r), $|\mathbf{C}[(v_i, v_j), (v_j, v_r)]| \leq C_{max}$ for a fixed integer C_{max}. Note that $s = |\mathbf{G}_j| = m \cdot (m + \log C_{max})$.

Assume that for some constant c, the SP problem has a kernel of size k^c. Let $t(s) = s^c$. Note that $t(s)$ is a polynomial.

For each $j = 1 \ldots t(s)$, let $\mathbf{G}_j$ be an arc-dependent network with n vertices and m arcs such that $|\mathbf{G}_j| = s$. From these networks, we can create a new arc-dependent network $\mathbf{G}$ with $(t(s) \cdot (n-2) + 2)$ vertices and $t(s) \cdot m$ arcs such that $\mathbf{G}$ is a disjoint union of $\mathbf{G}_1, ..., \mathbf{G}_{t(S)}$ except the vertices s and t in $\mathbf{G}$ are used to represent the vertices s and t respectively in each $\mathbf{G}_j$.

Observe that no arc in $\mathbf{G}$ corresponding to an arc in $\mathbf{G}_j$ shares vertices with an arc in $\mathbf{G}$ corresponding to an arc in $\mathbf{G}_{j'}$, where $j' \neq j$. Thus, any path in $\mathbf{G}$ corresponds to a path in $\mathbf{G}_j$ for some $j \in \{1, \ldots, t(s)\}$. Consequently, $\mathbf{G}$ has a path from s to t with total cost L if and only if $\mathbf{G}_j$ has a has path from s to t with total cost L for some $j \in \{1, \ldots, t(s)\}$.

Let $\mathbf{G}'$ be a kernel of $\mathbf{G}$ such that $|\mathbf{G}'| \leq k^c$. Since $k \leq m \leq s$, $|\mathbf{G}'| \leq k^c \leq s^c = t(s)$. Additionally, $\mathbf{G}'$ has a path from s to t with total cost L if and only if $\mathbf{G}_j$ has a path from s to t with total cost L for some $j \in \{1, \ldots, t(s)\}$. Thus, we have a t-bounded OR-distillation from the SP problem for arc-dependent networks to itself. This cannot happen unless **coNP** $\subseteq$ **NP/poly**.   □

Note that the same t-bounded OR-distillation technique will work for the SNCC and LoNCC problems. This gives us the following result:

Theorem 7. *The SNCC and LoNCC problems for arc-dependent networks do not have polynomial sized kernels unless* **coNP** $\subseteq$ **NP/poly**.

6   Conclusion

In this paper, we presented inapproximability results for several negative cost cycle detection problems in arc-dependent networks. Specifically, we discussed detecting negative cost cycles in arc-dependent networks with the fewest arcs and the most arcs. We showed that the shortest and longest negative cost cycle problems are **NPO PB-complete**. We also designed a Fixed-Parameter Tractable algorithm for finding a shortest path in arc-dependent networks. Finally, we showed that the shortest path problem in arc-dependent networks is unlikely to admit a kernel whose size is polynomial in the number of arcs in the path. The inapproximability results in this paper strengthen the inapproximability results established for QSPP in [7].

References

1. Alon, N., Yuster, R., Zwick, U.: Color-coding. J. ACM **42**(4), 844–856 (1995)
2. Bellman, R.E.: Dynamic Programming. Princeton University Press, Princeton (1957)
3. Caldwell, T.: On finding minimum routes in a network with turn penalties. Commun. ACM **4**(2), 107–108 (1961)
4. Cygan, M., et al.: Parameterized Algorithms. Springer, Cham (2015). https://doi.org/10.1007/978-3-319-21275-3
5. Dijkstra, E.W.: A note on two problems in connexion with graphs. Numer. Math. **1**, 269–271 (1959)
6. Fomin, F.V., Lokshtanov, D., Saurabh, S., Zehavi, M.: Theory of Parameterized Preprocessing. Cambridge University Press, Kernelization (2019)
7. Hu, H., Sotirov, R.: Special cases of the quadratic shortest path problem. J. Comb. Optim. **35**(3), 754–777 (2017). https://doi.org/10.1007/s10878-017-0219-9
8. Hao, H., Sotirov, R.: On solving the quadratic shortest path problem. INFORMS J. Comput. **32**(2), 219–233 (2020)
9. Kann, V.: On the approximability of NP-complete Optimization Problems. Ph.D. thesis, Royal Institute of Technology Stockholm (1992)
10. Kann, V.: Polynomially bounded minimization problems that are hard to approximate. Nordic J. Comput. **1**(3), 317–331 (1994)
11. Orponen, P., Mannila, H.: On approximation preserving reductions: complete problems and robust measures. Technical report, Department of Computer Science, University of Helsinki (1987)
12. Schöbel, A., Urban, R.: Cheapest paths in public transport: properties and algorithms. In: Huisman, D., Zaroliagis, C.D. (eds.) 20th Symposium on Algorithmic Approaches for Transportation Modelling, Optimization, and Systems (ATMOS 2020), vol. 85. OpenAccess Series in Informatics (OASIcs), pp. 13:1–13:16. Schloss Dagstuhl-Leibniz-Zentrum für Informatik, Dagstuhl (2020)
13. Tan, J., Leong, H.W.: Least-cost path in public transportation systems with fare rebates that are path- and time-dependent. In: Proceedings, The 7th International IEEE Conference on Intelligent Transportation Systems (IEEE Cat. No. 04TH8749), pp. 1000–1005, October 2004
14. Wojciechowski, P., Williamson, M., Subramani, K.: On finding shortest paths in arc-dependent networks. In: Baïou, M., Gendron, B., Günlük, O., Mahjoub, A.R. (eds.) ISCO 2020. LNCS, vol. 12176, pp. 249–260. Springer, Cham (2020). https://doi.org/10.1007/978-3-030-53262-8_21
15. Yap, C.K.: Some consequences of non-uniform conditions on uniform classes. Theoret. Comput. Sci. **26**(3), 287–300 (1983)
16. Ziliaskopoulos, A.K., Mahmassani, H.S.: A note on least time path computation considering delays and prohibitions for intersection movements. Transp. Res. Part B Methodol. **30**(5), 359–367 (1996)

Approximation Algorithms in Graphs with Known Broadcast Time of the Base Graph

Puspal Bhabak and Hovhannes A. Harutyunyan[✉]

Department of Computer Science and Software Engineering, Concordia University,
Montreal, QC H3G 1M8, Canada
haruty@cs.concordia.ca

Abstract. *Broadcasting* is an information dissemination problem in a connected network in which one node, called the *originator*, must distribute a message to all other nodes of the network by placing a series of calls along the communication lines of the network. The broadcast time of a vertex is defined to be the minimum number of time units required to broadcast the message to all vertices of the graph (network) from that vertex. Finding the broadcast time of any vertex in an arbitrary graph is NP-complete. The polynomial time solvability is shown only for certain tree-like graphs. In this paper we study the broadcast problem in graph of trees where broadcast algorithms for the base graph is known. In such graphs we design a linear time constant approximation algorithm to determine the broadcast time of any originator in general case. In a particular case when the base graph is the hypercube or another minimum broadcast graph (graph with minimum possible broadcast time having the smallest number of edges) containing one tree we present a linear time exact algorithm to find the broadcast time of any originator vertex. When the base graph is the hypercube graph we improve the known result by presenting a 1.5-approximation algorithm to find the broadcast time of the whole graph which runs in linear time instead of known quadratic algorithm.

1 Introduction

In today's world, due to massive parallel processing, processors have become faster and more efficient. In recent years, a lot of work has been dedicated to studying properties of interconnection networks in order to find the best communication structures for parallel and distributed computing. One of the main problems of information dissemination investigated in this research area is broadcasting. The broadcast problem is one in which the knowledge of one processor must spread to all other processors in the network. For this problem we can view any interconnection network as a connected undirected graph $G = (V, E)$, where V is the set of vertices (or processors) and E is the set of edges (or communication lines) of the network. According to [13], the broadcast time problem was

© Springer Nature Switzerland AG 2022
N. Balachandran and R. Inkulu (Eds.): CALDAM 2022, LNCS 13179, pp. 305–316, 2022.
https://doi.org/10.1007/978-3-030-95018-7_24

introduced in 1977 by Slater, Cockayne and Hedetniemi. Large sources of information about broadcasting and related problems are survey articles [6, 13, 14], book [15] and book chapter [9].

Formally, *broadcasting* is the message dissemination problem in a connected network in which one informed node, called the *originator*, must distribute a message to all other nodes by placing a series of calls along the communication lines of the network. The informed nodes aid the originator in distributing the message. This is assumed to take place in discrete time units. The broadcasting is to be completed as quickly as possible subject to the following constraints:

- Each call requires one unit of time.
- A vertex can participate in only one call per unit of time.
- Each call involves only two adjacent vertices, a sender and a receiver.

Given a connected graph G and a message originator, vertex u, the natural question is to find the minimum number of time units required to complete broadcasting in graph G from vertex u. We define this number as the *broadcast time* of vertex u, denoted $b(u, G)$ or $b(u)$. The broadcast time $b(G)$ of the graph G is defined as $\max\{b(u)|u \in V\}$. It is easy to see that for any vertex u in a connected graph G with n vertices, $b(u) \geq \lceil \log n \rceil$ (all log's in the paper are base 2), since during each time unit the number of informed vertices can at most double. Determining $b(u)$ for an arbitrary originator u in an arbitrary graph G has been proved to be NP-complete in [22]. The problem remains NP-Complete even for 3-regular planar graphs [18] and for split graphs, the graphs whose vertex set can be partitioned into a clique and an independent set [16]. The best theoretical upper bound is obtained by the approximation algorithm in [4] which produces a broadcast scheme with $O(\frac{\log(|V|)}{\log\log(|V|)}b(G))$ rounds. Research in [21] has showed that the broadcast time cannot be approximated within a factor $\frac{57}{56} - \epsilon$. This result has been improved within a factor of $3 - \epsilon$ in [4]. As a result research has been made in the direction of finding approximation or heuristic algorithms to determine the broadcast time in arbitrary graphs (see [1, 2, 4, 5, 7, 8, 17, 19, 20]).

Since the broadcast problem in general is very difficult, another direction is to design polynomial algorithms for some classes of graphs. The first result in this direction was a linear algorithm to determine the broadcast time of any tree [22]. Recent research shows that there are polynomial time algorithms for the broadcast problem in tree-like graphs where two cycles do not intersect - unicyclic graphs, tree of cycles, or in graphs containing no intersecting cliques - fully connected trees and tree of cliques [10–12]. No other results are known in this area. The broadcasting problem becomes very difficult when graphs contain intersecting cycles. The exception is of course the complete graph, where all edges are available to broadcast optimally.

In this paper we consider broadcasting in so called graph of trees, where every vertex of the base graph is the root of a tree. Polynomial time broadcasting algorithms in such graphs becomes approachable when there is a polynomial time broadcast algorithm for the base graph. In Sect. 2 we generalize the existing result on hypercube of trees for any minimum broadcast graph with dimensional broadcast scheme, and present an exact algorithm which runs in linear time. In Sect. 3 we improve an earlier result on hypercube of trees by reducing the algorithm complexity from quadratic to linear, and also improve the approximation ratio from 2 to 1.5. In Sect. 4 we design a linear time constant approximation algorithm to determine the broadcast time of any originator for the graph of trees when the broadcast algorithm for the base graph is known.

2 Exact Algorithm When the Base Graph is a k-Regular mbgs with One Tree

In this section we generalize the result presented in [3] for Hypercube of trees with one tree, and present a linear algorithm for any minimum broadcast graph (mbg) on 2^k vertices with one tree. Our new algorithm and the proof of correctness are different from the algorithm and correctness proof presented in [3].

Let G_k be a *minimum broadcast graph (mbg)* on 2^k vertices. Recall that mbg is a graph on n vertices with broadcast time $\lceil \log_2 n \rceil$ containing minimum possible number of edges. There are three non-isomorphic mbgs known in the literature; hypercube $H(k)$, Knödel graph $W_{k,2^k}$ and regular circulant graph $C(4, 2^k)$ (see for example [9]). All the three graphs share some graph-theoretic and communication properties. In our algorithm below we will use some of these properties. All the three graphs are k-regular, with 2^k vertices, broadcast time equal to k and diameter upper bounded by k. Another property that all the three graphs share is the *dimensionality*, which will be used in our algorithm.

Property 1. [9]: In any minimum time broadcast scheme there are no idle vertices i.e. every vertex that receives the message at time i, $0 \leq i \leq k-1$ has to send the message to its neighbors during time units $i+1, ..., k$. Moreover if one informed vertex stays idle at time unit i, for any i, $2 \leq i \leq k$, then the other $2^k - 1$ vertices can finish broadcasting in $k + 1$ time units for any $2 \leq i \leq k$.

Let G be one of the three mbgs on 2^k vertices described above and denoted by H_k where r is the root of a tree T. The remaining $2^k - 1$ root vertices do not contain any tree. One can assume that these $2^k - 1$ trees are empty trees, containing only the root of the trees. Let us also assume that r has m neighbors in T, vertices $v_1, v_2, ..., v_m$. v_i is the root of the subtree T_i, $1 \leq i \leq m$. Let us consider $b(v_i, T_i) = t_i$ and without loss of generality we assume that $t_1 \geq t_2 \geq ... \geq t_m$. Then it follows from [22] that $b(r, T) = \max\{i + t_i\}$, where $1 \leq i \leq m$. Let $b(r, T) = \tau$ and $\tau \geq 1$ (see Fig. 1). Let us consider the largest index j such that $\tau = t_j + j$ for $1 \leq j \leq m$.

2.1 Broadcast Algorithm When Originator is r

Consider two cases depending on the relationship between τ and k, the dimension of the mbg in G (the broadcast time of mbg). Let all the root vertices will be informed by $\tau(r)$ time units. The algorithm A calls another algorithm Broadcast-mbg which returns $\tau(r)$. When a tree vertex is informed then it follows the well-known broadcast algorithm in trees from [22].

The algorithm below is given for Knödel graph $W_{k,2^k}$. Similar algorithm for regular circulant graph $C(4, 2^k)$ is easy present, but we will skip it because of the space limitation. For hypercube H_k the simple algorithm is presented in [3] (Fig. 2).

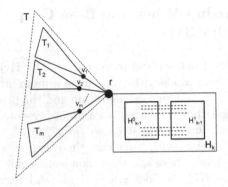

Fig. 1. G is one of three mbgs where r is the root of a tree T. In this case mbg is a k-dimensional hypercube

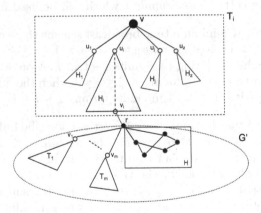

Fig. 2. G with originator v. The subtree T_i is separated from the rest of graph G'

Recall that Knödel graph on 2^k vertices is a k-regular graph with vertex set $V = \{0, 1, ..., 2^k - 1\}$ and set of edges $E = \{(i, j) | i + j = 2^p - 1 \bmod 2^k,$ for all $p = 1, 2, ..., k\}$.

Broadcast Algorithm A:

INPUT: $G = (V, E)$, originator r, $b(r, T) = \tau$, m, $t_1 \geq t_2 \geq ... \geq t_m$

OUTPUT: Broadcast time $b_A(r)$ and broadcast scheme for G

BROADCAST-SCHEME-A(G, r, τ, m, $t_1 \geq t_2 \geq ... \geq t_m$)

  0. $\tau = \max\{i + t_i\}$, where $1 \leq i \leq m$

  1. If $\tau \leq k$

      1.1. r informs another root vertex r_1 in the first time unit.

      1.2. $\tau(r) = $ BROADCAST-MBG(G, r_1, 1).

      1.3. For each time unit $i = 2$ to $m + 1$

         1.3.1. r informs tree vertex v_{i-1}.

  2. If $\tau > k$

      2.1. If $\tau \geq k + m$

         2.1.1. For each time unit $i = 1$ to m

         2.1.1.1. r informs tree vertex v_i.

         2.1.2. For each time unit $i = m + 1$ to $m + k$

         2.1.2.1. an informed root vertex informs another uninformed root vertex using any shortest path.

      2.2. If $k + m - 1 \geq \tau \geq k + 1$

         Let j be the largest index such that $\tau = t_j + j$

         2.2.1. For each time unit $i = 1$ to j

         2.2.1.1. r informs tree vertex v_i.

         2.2.2. At time unit $j + 1$, r informs another root vertex r_1.

         2.2.3. $\tau(r) = $ BROADCAST-MBG(G, r_1, $j + 1$).

         2.2.4. For each time unit $i = j + 2$ to $m + 1$

         2.2.4.1. r informs tree vertex v_{i-1}.

         2.2.5. If H_k is informed by time τ

            then OUTPUT: $b_A(r)$

            else FOLLOW steps 1.1 to 1.3

  3. TREE-BROADCAST(v_i, T_i) for $1 \leq i \leq m$.

Broadcast-mbg: Knodel graph $W_{k,2^k}$

INPUT: $G = (V, E)$, originator r_1, time at which r_1 is informed: t_{r_1}

OUTPUT: $\tau(r)$

BROADCAST-MBG(G, r_1, t_{r_1})

  1. Assume $r_1 = 2^k - 1$

  2. For each time unit $i = t_{r_1} + 1$ to $t_{r_1} + k - 1$

      2.1. For all $0, 1, ..., 2^k - 2$ do in parallel

         2.1.1. i sends to $2^l - 1 - i \bmod 2^k$, for $l = 1, 2, ..., k - 1$

  3. For all $0, 1, ..., 2k - 2$ except $2^k - 1$ do in parallel

      3.1. i sends the message to $2^k - 1 - i \bmod 2^k$

  4. Return $t_{r_1} + k$

Algorithm Complexity:

Broadcast-mbg takes $O(\log 2^k) = O(k)$ time to inform the root vertices.

Algorithm A: Step 0 takes m time to calculate τ. Steps 1.1 and 1.3 take constant time to run. Step 2.1.2 can be completed in $O(k)$ time. Also steps 2.1.1 and 2.2 run in constant time. Again, the tree broadcast algorithm in step 3 takes $O(|V| - 2^k) = O(|V_T|)$ time to run, where $|V_T|$ is the number of tree vertices in G. Thus, complexity of algorithm is $O(|V_T| + k) = O(|V|)$.

Theorem 1. *Algorithm A always generates the minimum broadcast time $b(r)$.*

Proof. The proof is easy for $\tau \le k$ and for $\tau \ge m + k$. When $\tau \le k$ then line 1.2 will require k more time after time 1 to broadcast within the mbg, and following line 1.3 it will take τ more time units after time 1, to complete broadcasting within the tree T. Since, $\tau \le k$ the algorithm will output $b_A(r) = k + 1$. From Property 1 it follows that $b(r, G) \ge k + 1$.

If $\tau \ge m + k$, by line 2.1. vertex r makes no delay in broadcasting within tree T, thus, all vertices of T will be informed by time τ. By line 2.1.2 of algorithm A all mbg vertices will be informed by time $m + k$. Since $\tau \ge m + k$ then the algorithm outputs $b_A(r) = \tau$. But $b(r) \ge \tau$ is an obvious lower bound on $b(r)$.

It remains to consider the case $k + 1 \le \tau \le m + k - 1$. Following the algorithm, the originator r informs the tree vertices at time units $1, ..., j, j+2, j+3, ..., m+1$ and informs a root vertex at time unit j. The algorithm completes broadcasting in mbg graph at time units $j+1, j+2, ..., j+k+1$. Then, vertex r completes broadcasting in tree T in $b_A(r, T) = \max\{t_1 + 1, ..., t_j + j, t_{j+1} + (j+2), ..., t_m + (m+1)\}$. Since j was the maximum value for which $t_j + j = \tau$, then $t_{j+1} + (j+2) \le \tau$, $t_{j+2} + (j+3) \le \tau$, $... t_m + (m+1) \le \tau\}$. Therefore, $b_A(r, T) = \max\{t_1 + 1, ..., t_j + j, t_{j+1} + (j+2), ..., t_m + (m+1) = \tau\}$. Thus, $b_A(r, G) = max\{j + k + 1, \tau\}$. If $\tau \ge j + k + 1$ then $b_A(r, G) = \tau$, which is an obvious lower bound on $b(r) = b(r, G)$. The last case, if $\tau \le j + k$ then the algorithm will generate broadcast time $b_A(r, G) = \tau + 1$. To prove the lower bound $b(r, G) \ge \tau + 1$ in this case, assume by contradiction that $b(r, G) = \tau$. Then, originator r had to inform its j neighbors in tree T, then inform all its neighbors in mbg to be able to complete broadcasting in G within $j + k$ time units. It follows then that r does not have more neighbors in T, and so $m = k$. Then, since $b(r, G) = \tau = j + k = m + k$. However, when $\tau = m + k$ the algorithm follows line 2.1 considered above. Thus, $b(r, G) = \tau$ is impossible in this case. Therefore, $b(r, G) \ge \tau + 1$, which is the output of algorithm A.   □

Broadcasting from a Root Vertex Other Than r: Let us assume that a root vertex u is at a distance d from vertex r, where $k \ge d \ge 1$. The algorithm D in G starts by informing along the path $\overline{ur}$ (the shortest among all paths between u and r). r receives the message at time d, and then it sends the message to the tree attached to it.

Broadcast Algorithm D:
INPUT: $G = (V, E)$, originator u, $b(r, T) = \tau$
OUTPUT: Broadcast time $b_D(u)$ and broadcast scheme for G
BROADCAST-SCHEME-$D(G, u, \tau)$

1. u informs along the path $\overline{ur}$ (the shortest among all paths between u and r) in the first time unit.
2. u continues to inform the other root vertices using any shortest path. r receives the message at time d.
3. TREE-BROADCAST(r, T).

Complexity Analysis:
Steps 1 and 2 can be completed in $O(k)$ time. The tree broadcast algorithm in step 3 takes $O(|V_T|)$time to run. Complexity of algorithm is $O(|V_T| + k)=$ $O(|V|)$.

Broadcasting from a Tree Vertex: Broadcast algorithm Q is similar to algorithm above, and the broadcast tree of G' originated at vertex r will be obtained from algorithm A.

Broadcast Algorithm Q:
INPUT: G', originator v in subtree T_i, r, τ_{m-1}, $m-1$, $t_1 \geq t_2 \geq ... \geq t_{m-1}$
OUTPUT: Broadcast time $b_Q(v)$ and broadcast scheme for G
BROADCAST-SCHEME-$Q(G', v, T_i, \tau_{m-1}, m-1, r, t_1 \geq t_2 \geq ... \geq t_{m-1})$
  1. $T_{G'}$ = BROADCAST-SCHEME-$A(G', r, \tau_{m-1}, m-1, t_1 \geq t_2 \geq ... \geq t_{m-1})$
  2. Attach $T_{G'}$ with T_i by the bridge (r, v_i) and let the resulting tree be labelled as T_v.
  3. TREE-BROADCAST(v, T_v).

Algorithm Complexity: Finding the broadcast time of a tree vertex in an arbitrary mbg of trees with one tree is equivalent to solving two problems: (1) Finding the broadcast time of a root vertex in an mbg of trees with one tree. As discussed before the complexity of this algorithm is linear. (2) Finding the broadcast time of a tree vertex in a tree which is also linear.

The correctness of the above two cases are similar to the proof of Theorem 1.

3   Linear Time 1.5-Approximation Algorithm for General Hypercube of Trees

Recall that [4] presents a $O(|V|)$ time 2-approximation algorithm for arbitrary originator vertex in any hypercube of trees $G = (V, E)$ containing up to 2^k trees rooted at the vertices of the k-dimensional hypercube. It is clear that to get a 2-approximation algorithm of the whole hypercube of trees G we have to run the $O(|V|)$ algorithm $|V|$ times from any originator, and then take the maximum of all broadcast times. The obtained algorithm will be 2 approximation and will run in $O(|V|^2)$ time. In this section we will present a $O(|V|)$ algorithm which gives 1.5-approximation for the broadcast time of any hypercube of trees G.

Our improvement in this section is twofold, we improve the complexity from $O(|V|^2)$ to $O(|V|)$ and we improve the approximation ratio from 2 to 1.5.

Assume graph G is an arbitrary hypercube of 2^k trees T_i rooted at r_i, $i = 1, 2, ..., 2^k$, where the base graph is forming a hypercube of dimension k, H_k. Denote by h_i the height of tree T_i rooted at r_i for all $i = 1, 2, ..., 2^k$. Denote the maximum value of the heights of all 2^k trees by h, and the maximum value of the trees rooted at the hypercube vertices by t. More formally, $h = \max\{h_i | 1 \leq i \leq 2^k\}$, and $t = \max\{b(r_i, T_i) | 1 \leq i \leq 2^k\}$. Our algorithm is very simple, for all $1 = 1, 2, ..., 2k$ we find both, its height h_i and the broadcast time $b(r_i, T_i)$ in $O(|V|)$ time using well-known)$(|V|)$ time algorithm for trees. And then in the same $O(|V|)$ time we will find the values of k, h and t defined above.

Theorem 2. *Broadcast time of graph G, $b(G) \leq h + k + t$, the above algorithm has 1.5-approximation ratio and runs in $(|V|)$ time.*

Proof. The fact that our algorithm runs in $O(|V|)$ is easy to see. Now, let's first prove that the broadcast time of any originator u in graph G is upper bounded by $h + k + t$, $b(u, G) \leq h + k + t$. Assume vertex u is in tree T_j, then broadcasting from u will inform r_j, the root of T_j by time unit h_j. Then within the next k time units r_j will inform all 2^k vertices of the hypercube H_k. Thus, after at most $h_j + k$ time units all hypercube vertices will be informed. Then, in the next $t = \max\{b(r_i, T_i) | 1 \leq i \leq 2^k\}$ time units the hypercube vertices will inform all vertices in their respective trees. Thus, $b(u, G) \leq h_j + k + t \leq h + k + t$. Note that, $0 \leq h_j \leq h$ and $0 \leq b(r_j, T_J) \leq t$, and the bound is correct for the cases $h_j = 0$ or $b(r_j, T_J) = 0$. This covers the case when the originator is r_j or a tree T_j is empty with $b(r_j, T_J) = 0$.

To get a lower bound on $b(u, G)$ note that if originator u belongs to a tree T_i with height h then any broadcast scheme from originator u will take at least $h + k$ time units, $b(u, G) \geq h + k$. Also, if the originator w is a root vertex r_l which has distance k from the root of tree T_p with $b(r_p, T_p) = t$, then clearly $b(w, G) \geq k + t$. Combining these two inequalities and applying $b(G) \geq b(u, G)$ and $b(G) \geq b(w, G)$ we get $2b(G) \geq h + k + k + t$. Thus, $b(G) \geq k + \frac{1}{2}(h + t)$.

If $h + t \leq 2k$, then $2k = h + t + 2x$ for some $x \geq 0$, and we get that $\frac{b_A(G)}{b(G)} \leq \frac{h+k+t}{k+\frac{1}{2}(h+t)} = \frac{k+2k-2x}{k+\frac{1}{2}(2k-2x)} = \frac{3k-2x}{k+k-x} \leq \frac{3k-1.5x}{2k-x} = \frac{3}{2}$.

If $h + t > 2k$, then consider broadcasting from originator u which is in a tree T_s and $dist(u, r_s) = h$. Then, $b(u, G) \geq h + t$ since broadcasting from vertex u has to go through vertex r_s and all vertices of a tree T_j with $b(r_j, T_j) = t$ must get informed by time $b(u, G)$. Thus, since $b(u, G) \leq b(G)$ for any originator u we will get $\frac{b_A(G)}{b(G)} \leq \frac{h+k+t}{h+t} \leq \frac{h+t+\frac{h+t}{2}}{h+t} = \frac{3}{2}$.   □

4   Linear Time Constant Approximation Algorithm in Graph of Trees with Known Broadcast Time of the Base Graph

Assume that we have an arbitrary graph H where its vertices are the roots of some trees. We call the resulting graph arbitrary graph of trees G (see Fig. 3).

Definition 1. *Consider an arbitrary graph $H = (V_H, E_H)$ where k of its vertices, denoted as V_r are the roots of the trees $T_i = (V_i, E_i)$ for $1 \leq i \leq k$ and $k \leq |V_H|$. We define the arbitrary graph of trees, $G = (V, E)$, to be a graph where $V = V_1 \cup V_2 \cup ... \cup V_k \cup (V_H - V_r)$ and $E = E_1 \cup E_2 \cup ... \cup E_k \cup E_H$. The vertices in H denoted as r_i will be called root vertices, where $1 \leq i \leq |V_H|$. The rest of the vertices will be called tree vertices.*

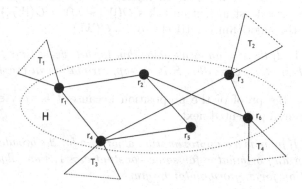

Fig. 3. Arbitrary graph of trees G

The first approximation algorithm is simple. When the originator is a root vertex r, then our algorithm S in G starts by informing all the vertices of H. When all the vertices in H are informed, each root vertex informs the tree attached to it.

When the originator vertex w belongs to some tree, then the algorithm S in G starts by informing along the path $\overline{wr}$. When r receives the message, the scheme informs all the vertices of H. When a tree vertex is informed then it follows the well known broadcast algorithm in trees [22], called A_T.

Tree Broadcast Algorithm A_T:
INPUT: originator r_i and tree rooted at r_i: T_i
OUTPUT: Broadcast time $b_{A_T}(r_i, T_i)$
TREE-BROADCAST(r_i, T_i)
  1. r_i informs a child vertex in T_i that has the maximum broadcast time in the subtree rooted at it.
  2. Let $\alpha_1, ..., \alpha_f$ be the broadcast times of the f subtrees rooted at r_i and $\alpha_1 \geq ... \geq \alpha_f$. Then, $b_{A_T}(r_i, T_i) = \max\{j + \alpha_j\}$ for $1 \leq j \leq f$.

Approximation Algorithm S:
INPUT: $G = (V, E)$ and any originator x
OUTPUT: Broadcast time $b_S(x)$ and broadcast scheme for G
BROADCAST-SCHEME-$S(G, x)$
  1. If $x = w$
    1.1. w broadcasts along the shortest path $\overline{wr}$ in time unit 1.

      r gets informed at time d.

      1.2. Starting at time $d+1$ onwards inform all the vertices in H.

  2. If $x = r$

      2.1. Inform all the vertices in H.

      3. TREE-BROADCAST(r_i, T_i) for $1 \le i \le m$. (m is the degree of r in tree T)

Algorithm Complexity:

Steps 1.2 and 2.1 take $O(m)$ time to inform the vertices in H. In steps 1.1 and 3, the tree broadcast algorithm takes $O(|V| - m) = O(|V_T|)$ time to run. Complexity of the algorithm is $O(|V_T| + m) = O(|V|)$.

Proposition 1. *If there is an exact algorithm for broadcast time from any originator of graph H, then algorithm S is a 2-approximation in graph G.*

Proof. We skip the proof of the proposition because it is very similar to the proof of Theorem 3 presented next.

Theorem 3. *If there is a c-approximation algorithm for the broadcast time problem in H from any originator for some constant $c > 1$, then algorithm S is a $2c$-approximation for any originator in graph G.*

Proof. **When the broadcast originator u is in base graph H, $u = r_0$:** Following our algorithm S originator u will follow the c-approximation algorithm, call it C, within the base graph H, then by time unit $b_C(u, H)$ all vertices of base graph H will be informed, and they will complete broadcasting within their respective trees by time unit $b_C(u, H) + \max\{b(r_i, T_i) | i = 1, 2, ..., |V_H|\}$. Let us assume that $\max\{b(r_i, T_i) = t$, then we have $b_S(u, G) \le b_C(u, H) + t$. On the other hand, it is obvious that $b_S(u, G) \ge b(H)$ and also $b_S(u, G) \ge t$ since all the vertices of graph H or the vertices of the tree, say T_j with $b(T_j) = t$, must be informed by the time unit $b_S(u, G)$. Thus, $b_S(u, G) \ge \frac{b(H)+t}{2}$, and we get, $\frac{b_S(u,G)}{b(u,G)} \le \frac{b_C(u,H)+t}{\frac{b(u,H)+t}{2}} = 2\frac{b_C(u,H)+t}{b(u,H)+t} \le 2\frac{cb(u,H)+ct}{b(u,H)+t} = 2c$ since $c > 1$ and also algorithm C was a c-approximation algorithm in the base graph H.

When the originator u is a vertex in tree T_i rooted at vertex $r_i \in V_H$: As above, following algorithm S the originator u from the tree T_i first will inform its root vertex r_i by direct path of length, say h_i, then will follow the c-approximation algorithm within the base graph H, and then starting at time unit $h_i + b_C(r_i, H)$ all vertices of the base graph H will start broadcasting within their respective trees. Thus, $b_C(u, G) \le h_i + b_C(r_i, H) + \max\{b(r_j, T_j) | j = 1, 2, ..., |V_H|\}$. Similarly, for the lower bounds, it is clear that $b_S(u, G) \ge h_i + b_C(r_i, H)$ and $b_S(u, G) \ge h_i + \max\{b(r_j, T_j) | j = 1, 2, ..., |V_H|\}$. Suppose that $\max\{b(r_j, T_j) | j = 1, 2, ..., |V_H|\} = t$.

    Thus, $b_S(u, G) \ge h_i + \frac{b(r_i,H)+t}{2}$, and finally we get that $\frac{b_S(u,G)}{b(u,G)} \le \frac{h_i + b_C(r_i,H)+t}{h_i + \frac{b(r_i,H)+t}{2}} \le 2\frac{ch_i + cb(r_i,H)+ct}{2h_i + b(r_i,H)+t} \le 2c$, since algorithm C is a c-approximation algorithm for base graph H, and $c > 1$. $\qquad\square$

Observation: Theorem 3 and Proposition 1 can be applied to many graphs known in the literature. In particular, for m-dimensional butterfly network BF_m [14], the m-dimensional shuffle-exchange graph SE_m [14] and the m-dimensional DeBruijn graph DB_m, constant approximation algorithms are known. Proposition 1 can be applied to hypercube [3], cube-connected cycle [14] and fully connected tree [12].

5   Conclusion and Future Work

The broadcast problem, more precisely, finding the broadcast time of any vertex in an arbitrary connected graph is very difficult. It remains NP-complete even for 3-regular planar graphs and for split graphs, graphs whose vertex set can be partitioned into a clique and an independent set. The broadcast problem is shown to be NP-hard to approximate within a factor $3 - \epsilon$. The best known approximation for broadcasting in general graphs is $O(\frac{\log(|V|)}{\log\log(|V|)})$. A long standing open problem is to present a constant approximation algorithm or to prove that it is NP-hard to approximate within a constant factor. Polynomial time algorithms for the broadcast problem are only known for some tree like graphs. In particular, there exist linear algorithms for trees, tree of cycles and necklace graphs, more generally in graphs where two cycles intersect in at most one vertex. Tree of cliques is the only graph where two cycles intersect in many vertices but there is a $O(n \log \log n)$ algorithm. However, it is a special case since in the clique all edges are available to be used for optimal broadcasting.

In this paper we consider graph of trees for which the exact or approximation broadcast algorithm for the base graph is known. First, we generalize the existing result on hypercube of trees for any minimum broadcast graph, and present an exact algorithm which runs in linear time. Next, we improve an earlier result on hypercube of trees by reducing both the algorithm complexity and the approximation ratio. The last result is a linear time 2-approximation algorithm to determine the broadcast time of any originator for the graph of trees when the broadcast algorithm for the base graph is known.

The immediate future work will be to generalize the result for mbgs on 2^k vertices to any broadcast graph (bg) on any number of vertices. Here the main difficulty is to prove a property similar to Property 1 from Sect. 2. Recall that bg on n vertices is graph with broadcast time $\lceil \log n \rceil$ from any originator.

References

1. Bar-Noy, A., Guha, S., Naor, J., Schieber, B.: Multicasting in heterogeneous networks. In: Proceedings of the Thirtieth Annual ACM Symposium on Theory of Computing, STOC'98, pp. 448–453. ACM (1998)
2. Beier, R., Sibeyn, J.F.: A powerful heuristic for telephone gossiping. In: Proceedings of the 7th International Colloquium on Structural Information Communication Complexity, SIROCCO'00, pp. 17–36 (2000)
3. Bhabak, P., Harutyunyan, H.A.: Broadcast problem in hypercube of trees. In: Chen, J., Hopcroft, J.E., Wang, J. (eds.) FAW 2014. LNCS, vol. 8497, pp. 1–12. Springer, Cham (2014). https://doi.org/10.1007/978-3-319-08016-1_1

4. Elkin, M., Kortsarz, G.: Combinatorial logarithmic approximation algorithm for directed telephone broadcast problem. In: Proceedings of the Thirty-Fourth Annual ACM Symposium on Theory of Computing, STOC'02, pp. 438–447. ACM (2002)
5. Elkin, M., Kortsarz, G.: Sublogarithmic approximation for telephone multicast: path out of jungle (extended abstract). In: Proceedings of the Fourteenth Annual ACM-SIAM Symposium on Discrete Algorithms, SODA'03, pp. 76–85. Society for Industrial and Applied Mathematics (2003)
6. Fraigniaud, P., Lazard, E.: Methods and problems of communication in usual networks. Discrete Appl. Math 53, 79–133 (1994)
7. Fraigniaud, P., Vial, S.: Approximation algorithms for broadcasting and gossiping. J. Parallel Distrib. Comput. 43(1), 47–55 (1997)
8. Fraigniaud, P., Vial, S.: Comparison of heuristics for one-to-all and all-to-all communication in partial meshes. Parallel Process. Lett. 9, 9–20 (1999)
9. Harutyunyan, H.A., Liestman, A.L., Peters, J.G., Richards, D.: Broadcasting and gossiping. In: Handbook of Graph Theory, 2nd edn., pp. 1477–1494. Chapman and Hall/CRC (2013)
10. Harutyunyan, H., Maraachlian, E.: Linear algorithm for broadcasting in unicyclic graphs. In: Lin, G. (ed.) COCOON 2007. LNCS, vol. 4598, pp. 372–382. Springer, Heidelberg (2007). https://doi.org/10.1007/978-3-540-73545-8_37
11. Harutyunyan, H.A., Maraachlian, E.: On broadcasting in unicyclic graphs. J. Comb. Optim. 16(3), 307–322 (2008)
12. Harutyunyan, H.A., Maraachlian, E.: Broadcasting in fully connected trees. In: Proceedings of the 2009 15th International Conference on Parallel and Distributed Systems, ICPADS'09, pp. 740–745. IEEE Computer Society (2009)
13. Hedetniemi, S.T., Hedetniemi, S.M., Liestman, A.L.: A survey of gossiping and broadcasting in communication networks. Networks 18(4), 319–349 (1988)
14. Hromkovic, J., Klasing, R., Monien, B., Peine, R.: Dissemination of information in interconnection networks. In: Du, D.Z., Hsu, D.F. (eds.) Combinatorial Network Theory, pp. 125–212 (1996). https://doi.org/10.1007/978-1-4757-2491-2_5
15. Hromkovic, J., Klasing, R., Pelc, A., Ruzicka, P., Unger, W.: Dissemination of Information in Communication Networks: Broadcasting, Gossiping, Leader Election, and Fault-Tolerance. Texts in Theoretical Computer Science, An EATCS Series. Springer, Heidelberg (2005). https://doi.org/10.1007/b137871
16. Jansen, K., Muller, H.: The minimum broadcast time problem for several processor networks. Theor. Comput. Sci. 147, 69–85 (1995)
17. Kortsarz, G., Peleg, D.: Approximation algorithms for minimum time broadcast. SIAM J. Discrete Math 8(3), 401–427 (1995)
18. Middendorf, M.: Minimum broadcast time is NP-complete for 3-regular planar graphs and deadline 2. Inf. Process. Lett. 46, 281–287 (1993)
19. Ravi, R.: Rapid rumor ramification: approximating the minimum broadcast time. In: Proceedings of the 35th Annual Symposium on Foundations of Computer Science, FOCS'94, pp. 202–213. IEEE Computer Society (1994)
20. Scheuermann, P., Wu, G.: Heuristic algorithms for broadcasting in point-to-point computer networks. IEEE Trans. Comput. 33(9), 804–811 (1984)
21. Schindelhauer, C.: On the inapproximability of broadcasting time. In: Jansen, K., Khuller, S. (eds.) APPROX 2000. LNCS, vol. 1913, pp. 226–237. Springer, Heidelberg (2000). https://doi.org/10.1007/3-540-44436-X_23
22. Slater, P.J., Cockayne, E.J., Hedetniemi, S.T.: Information dissemination in trees. SIAM J. Comput. 10(4), 692–701 (1981)

Author Index

Printed in the United States
by Baker & Taylor Publisher Services

Printed in the United States
by Baker & Taylor Publisher Services